Printing Paper and Ink

Printing Paper and Ink

Charles Finley, Ph.D.

Delmar Publishers

an International Thomson Publishing company I(T)P®

Albany • Bonn • Boston • Cincinnati • Detroit • London • Madrid
Melbourne • Mexico City • New York • Pacific Grove • Paris • San Francisco
Singapore • Tokyo • Toronto • Washington

Notice to the Reader

Publisher does not warrant or guarantee any of the products described herein or perform any independent analysis in connection with any of the product information contained herein. Publisher does not assume, and expressly disclaims, any obligation to obtain and include information other than that provided to it by the manufacturer.

The reader is expressly warned to consider and adopt all safety precautions that might be indicated by the activities herein and to avoid all potential hazards. By following the instructions contained herein, the reader willingly assumes all risks in connection with such instructions.

The publisher makes no representation or warranties of any kind, including but not limited to, the warranties of fitness for particular purpose or merchantability, nor are any such representations implied with respect to the material set forth herein, and the publisher takes no responsibility with respect to such material. The publisher shall not be liable for any special, consequential, or exemplary damages resulting, in whole or part, from the readers' use of, or reliance upon, this material.

Cover Design: Tim Kaage

Delmar Staff
Publisher: Robert Lynch
Acquisitions Editor: John Anderson
Senior Project Editor: Christopher Chien
Production Manager: Larry Main
Art and Design Coordinator: Nicole Reamer
Editorial Assistant: John Fisher

COPYRIGHT © 1997
By Delmar Publishers
a division of International Thomson Publishing Inc.

The ITP logo is a trademark under license.

Printed in the United States of America

For more information, contact:

Delmar Publishers
3 Columbia Circle, Box 15015
Albany, New York 12212-5015

International Thomson Publishing Europe
Berkshire House 168-173
High Holborn
London, WC1V 7AA
England

Thomas Nelson Australia
102 Dodds Street
South Melbourne, 3205
Victoria, Australia

Nelson Canada
1120 Birchmont Road
Scarborough, Ontario
Canada, M1K 5G4

International Thomson Editores
Campos Eliseos 385, Piso 7
Col Polanco
11560 Mexico D F Mexico

International Thomson Publishing GmbH
Konigswinterer Strasse 418
53227 Bonn
Germany

International Thomson Publishing Asia
221 Henderson Road
#05-10 Henderson Building
Singapore 0315

International Thomson Publishing—Japan
Hirakawacho Kyowa Building, 3F
2-2-1 Hirakawacho
Chiyoda-ku, Tokyo 102
Japan

1 2 3 4 5 6 7 8 9 10 XXX 02 01 00 99 98 97

Library of Congress Cataloging-in-Publication Data

Finley, Charles.
    Paper and ink for printing / Charles Finley.
        p.   cm.
    Includes bibliographical references (p.   ) and index.
    ISBN 0-8273-6441-5
    1. Paper—Printing properties. 2. Printing ink. I. Title.
Z247.F56 1996
676—dc20

95-17854
CIP

Dedication

An old saying holds that "if you see a turtle sitting on a fence post, you can be sure that he had some help." Accordingly, this book is dedicated to my parents and to my teachers—from grade school through dissertation—who chose to share their knowledge.

Contents

Preface

Although its copyright date is 1997, this textbook began to take shape in 1982 as a four-chapter supplement to a book that I was using in a course called Printing Papers. Like many graphic arts educators, I found no single text that included paper's history, manufacture, procurement, characteristics, and use in the pressroom. Each year, new chapters were written and eventually a stand-alone text for my course had taken shape. Nationally, however, graphic arts curricula tend to include paper and ink in a single course, often requiring that the students purchase more than one book. I use the term book instead of text because nearly all of the available materials about both paper and ink are written more for people working in the printing industry than for college students who are trying to prepare for work in the printing industry. *Printing Paper and Ink* is an attempt to provide a student-friendly textbook that deals with the complexity of both materials and how they interact to create the magazines, movie-star posters, and cereal boxes that contribute to what we call graphic communications.

The first nine chapters deal with paper. This examination begins with the nearly two-thousand-year evolution of the papermaking process with an emphasis on how the earliest stages of papermaking still exist, but as parts of a much more efficient, high-tech operation. A second chapter examines the sources of paper fiber for two reasons: it is an integral part of the papermaking process and considerable controversy exists as to the appropriate use of the nation's forest land.

The chapter on paper manufacture is supplemented with separate chapters dealing with coated and recycled papers because of the increased popularity of these products. The next four chapters focus on aspects of using paper after it has been made: physical characteristics, use on the press, classifications, and the calculations performed in ordering paper.

The next three chapters deal with the basics of ink—its ingredients, manufacture, and properties. These are followed by two chapters that have been included to provide background information for students who are not familiar with color theory and the printing processes. A chapter on estimating ink use for a specific job follows to complement the earlier chapter on estimating paper.

The final chapter examines issues that link the paper, ink, and printing industries with environmental concerns. This chapter is important for two reasons. First, environmental legislation definitely impacts those who manufacture, use, and order paper and ink. The cost of these materials, the availability of certain raw materials needed to produce them, and restrictions on their use and disposal are all influenced by these laws. Second, aside from their professional interest, people in print-related industries are also citizens and are thereby obliged to be informed about important governmental matters so that the opinions that they take into the voting booth are valid. A concerted effort has been made to present these highly controversial issues objectively.

In an effort to create a user-friendly text, the end of each chapter contains a list of **key words** and study questions that allow students to assess the breadth and depth of their understanding. A **glossary** and chapter-based **bibliography** for further reading can be found at the back of the book.

For students whose knowledge of science is spotty, appendices explain principles like relative humidity, specific gravity, and the difference between solutions and emulsions. This section of the book also contains a time line for the evolution of the papermaking process that can supplement Chapter 1.

One of the challenges of writing this book was deciding on the sequence of the chapters. For example, the chapter on ink ingredients necessarily makes reference to ink properties that are explained two chapters later. To alter the sequence would merely make matters worse because of the inextricable interrelationships that exist among the ingredients, manufacture, desired properties, and applications of both paper and ink. To minimize the reader's confusion, certain points are explained in more than one chapter. Often a brief explanation of a point is made, followed by a reference to a more thorough examination that is made in a later chapter.

The key words throughout the text are highlighted at two different levels, similar to outline form. Boldface terms are at the first level and italic terms are at a secondary or subordinate level. For example, the two main categories of printing processes—physical and chemical—are in boldface, while the members of these categories—engraving, letterpress, gravure, screen, and lithography—are in italic. The student can typographically distinguish the parts from the whole. When two terms are presented at the same time— such as **substance weight** (also called *basis weight*)—the word in boldface is the one that will be used subsequently in the book.

The second big challenge was to write about ink without using the chemical terminology that pervades other books on the subject. Because people who write about ink tend to be chemists, terms like ketone, urethane alkyd, toluidine red, and pentaerythritol are as meaningful to ink people as the term megabyte is to a computer buff. However, because printers and printing students are usually not chemists, these terms are merely impediments to their understanding of how inks work.

It is hoped that the reader has had a high-school chemistry course, but the book was written with the awareness that many readers bring with them a minimal background in chemistry. For this reason, the chemical jargon has been held to a minimum and emphasis has been placed on the broad concepts rather than the particular chemicals involved. As was mentioned, certain chemical principles are explained in the appendices and these points are referenced in the text.

Acknowledgments

Several individuals and organizations played crucial roles in the development of this text and their efforts should be acknowledged. First, Lisa Hahn of Flexo Tech, Phil Runge of Continental Inks, Byron Hahn of Braden-Sutphin Ink Company, Jenni Rogers of Potlatch Corporation, Tim Malone of Columbus Cello-Poly Corporation, Dave Brown of Sterling Paper, Jobe Morrison and Tony Rubio of Cross Pointe Paper Corporation, Brian Markam and Hart Swisher of Hammer Litho Corporation, the American Forest and Paper Association, Bruce Cuthbertson in the Office of Congressman John Kasich, Rob Nichols in the office of Congresswoman Deborah Pryce, John Jacob of the Sierra Club, Mark Dupre and Bill Birkett of the Rochester Institute of Technology, Bob Kitchens and Karl Perry of the U.S. Forest Service, Marla Mayerson, and John Debrosse, Paul Giusti, and Paul Wilcox of Mead Central Research were all helpful in providing technical expertise or other assistance. Joe Bond, Ed Bell, Susan Van Atta, and Candice Spangler played key roles in the preparation of the manuscript. Also, several Columbus State Community College students who used this book in its earliest forms offered valuable input.

Elena M. Mauceri's skillful copy editing and Cameron Poulter's book design are much appreciated. In addition, the manuscript benefited from the thoughtful review and recommendations of

Rich Bundsgaard, Arkansas State University
John Leninger, Clemson University
Dana Hayden, St. Augustine Technical Center
Jean Rosinski, Western Michigan University

History of Paper

AFTER STUDYING THIS CHAPTER, THE STUDENT SHOULD BE ABLE TO:

■ Identify stages in the modern papermaking process that occurred even when papermaking was in its infancy in China.

■ List pre-paper media for carrying writing.

■ Identify modern-day papermaking processes that were developed by the Japanese, Muslim, and European efforts.

■ Trace how the spread and development of the papermaking process was influenced by other events in history such as wars and plagues.

■ Describe the appearance and operation of the first paper machine.

■ Identify technological breakthroughs in papermaking that occurred after 1800.

If the question "What civilization invented paper?" were put to a group of people, the most popular answer would probably be that the ancient Egyptians invented paper. However, the **papyrus** that they developed was not a true paper at all. Instead, it was one of several pre-paper media used by early Man; among these were wood, metal, stone, leaves, ceramics, bark, papyrus, parchment, and cloth. What prevents all of these materials from qualifying as paper is that they were not formed from a liquid suspension of individual fibers into a mat or sheet.

Developed before 2,200 B.C., papyrus was made by slicing the plant's stalk from end-to-end into very thin blades and then laminating them into something of a weave. The end result appears to be more of a cloth than a solid sheet (see Figure 1–1). But even though this ancient medium was not structured like the writing surface that we are so familiar with today, it did provide some terms that have endured. The word *paper* comes from the Latin term papyrus. Also, the Latin word for the message written on a sheet of papyrus was *biblia* (derived from the word for the inner fiber of the papyrus plant), which is the source of the word Bible.

It was, in fact, the Chinese who—over 2,000 years later—first created a true paper by actually reducing a raw material into individual fibers and then forming them into a mat or sheet. Amazingly, the general process that they used over 18

1

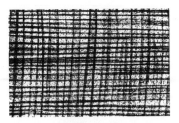

Figure 1–1 This close-up of a sheet of papyrus reveals its layers of thin strips laid at right angles to one another. *Courtesy of Dr. Jules Heller.*

centuries ago remains fundamentally the same, although it has been made more efficient.

The Invention of Paper

Credit for the invention of paper is given to Ts'ai Lun (pronounced Sy Lun), a private counsel to the Chinese Emperor Ho Ti, for developing the process of papermaking around 105 A.D.

The method that Ts'ai Lun employed involved eight steps (see Figure 1–2). First, branches of the mulberry bush were cut with a sickle-like knife and soaked in a tub to soften the outer bark for removal. After this bark was peeled off, the inner portion was taken to a stream for washing. Next, this inner portion was placed into a container of water and cooked—thus using the heat and steam pressure that built up inside the container to soften the limbs, creating a mushy substance. After a second washing, this pulplike material was beaten with paddles to further soften and break the pulp into individual fibers and to flatten the fibers as well. During this stage, the mulberry bark mixture was often augmented with other materials such as old fishnets, hemp, rags, and other barks.

When the fibers had been sufficiently broken down, they were mixed with water to form a slurry and poured into a tub. A cloth screen, stretched tightly across a wooden frame (similar to what is used today for screen printing), was then dipped into the tub and filled with a water and fiber mixture. As the screen was lifted up from the slurry, the water poured down through the openings in the fabric, leaving the fibers to mat together on top of the screen (imagine fishermen lifting a net full of fish from the water). After draining for several seconds, this collection of matted fibers was solid enough to be lifted from the screen and laid in the sun for drying. As the sheets dried, the sun's rays also acted as a bleaching agent, lightening the paper to a moderate degree.

Before the development of this papermaking process, the Chinese wrote on strips of bamboo (difficult to write on and awkward to store) and on silk (too expensive to encourage widespread use). Ts'ai Lun's new process was obviously much more efficient and less expensive than either alternative. Thirty-seven years after its introduction, paper became the most widely used medium, having replaced both wood and bamboo.

Ts'ai Lun's subsequent history has an interesting and ironic twist. Rewarded by the Emperor for his invention, he was given a series of promotions within the Emperor's court.

Figure 1–2 The first papermakers cut mulberry plants and cooked these branches to loosen and remove the bark. Then, the inner bark was further cooked to a pulp, washed, and beaten. After the sheets were formed on a screen, they were dried and trimmed.

3

He was later asked by the Empress to launch a "smear" campaign against another member of the Imperial family, presumably using his new invention to comply with the Empress's request.

When a new Emperor came to the throne, Ts'ai Lun's smear campaign proved such a personal embarrassment that he committed suicide by drinking poison. He remains revered by Chinese students today, however, more for the paper that he developed than for the libelous statements it may have soon carried.

It should be noted that while paper was becoming universally used within the Chinese Empire, it was still unheard of outside those boundaries. For example, a letter from St. Augustine dated 390 A.D. contains his apologies for having to write on papyrus instead of parchment, which was considered a more formal medium. Paper was still unknown in Europe and it would be another seven centuries before paper would be manufactured there.

Papermaking Moves outside China

In the seventh century the secret of papermaking moved outside of China, but it went east instead of west as Buddhist monks carried manuscript books made from mulberry paper into Japan. By 610 A.D.—some 500 years after its invention—the Japanese began to make paper and before long the craft was practiced throughout the island nation. Records indicate that the Japanese were the first to repulp old paper—a practice that is today called recycling.

For six centuries paper remained a mysterious substance everywhere outside of China and Japan. One of the cultures that was fascinated by imported paper but unable to figure out its creation was the Persian Empire. But in 750 A.D. a battle was fought between these Muslims and the Chinese at Samarkand in Turkestan (now part of Uzbekistan and formerly part of the Soviet Union). The Muslims won the battle and captured numerous prisoners; among them were several Chinese papermakers who revealed their secrets. Samarkand turned out to be a natural center for the manufacture of paper; not only did the Tharaz River provide an ample water supply, but hemp and flax (plants that provide linen fiber) flourished in that area.

Contributions of the Muslims

Not content merely to duplicate the process that they had acquired from the Chinese, the Muslims made three signifi-

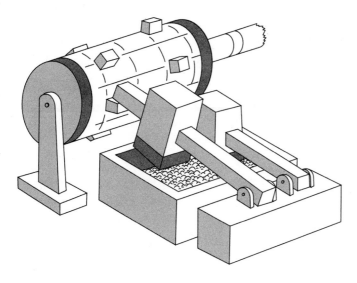

Figure 1–3 The early stamping mills relied on water power to turn a spoked wheel that lifted the stamper arms before releasing them. The stampers' falling against the pulp in the granite tub flattened the fibers.

cant contributions to the industry. First of all, unaware of the Japanese process of **repulping** old paper, they reinvented the concept that is today called *recycling*. Second, they were the first to manufacture paper from **linen fiber,** a substance still recognized as an excellent paper fiber. Third, they developed a more efficient method of beating the fibers by harnessing water power to replace what had been a manual operation. This device, the **stamping mill** or **hammer mill,** used falling water to turn a large wheel that would raise and then drop a large stone upon the pulp, which was in a granite tub (see Figure 1–3).

By this point, several diverse uses had been found for this curious invention. Although paper had been originally designed as a writing surface, the Chinese soon saw other applications; by 875 A.D. there was toilet paper and, by 969 A.D., playing cards and paper money. In the eleventh century Muslim street merchants were providing buyers of vegetables and other products with paper wrappings. The significance of this phenomenon is that paper had become so plentiful and cheap that it could be used in this manner by street vendors. Two factors brought this about: first, the practice of recycling allowed paper fibers to be used repeatedly; second, an enterprising group of nomads was raiding ancient Egyptian burial tombs and removing the linen cloth bands from mummies as a source of fiber. Because each was wrapped in up to 300 yards of linen, the supply was abundant.

Figure 1–4 The secret of papermaking moved from China to Japan, then across the Muslim world before becoming part of Europe.

As papermaking become a permanent industry of the vast Persian Empire, the process spread westward to other Muslim cities such as Baghdad and Damascus and then into Africa at Egypt and across the top of that continent all the way west to Fez in Morocco. This journey across the Muslim world took five centuries, but it was necessary because the Straits of Gibraltar were the entry points for the papermaking process to enter Europe in the twelfth century (see Figure 1–4). The link to Europe were the Moors of southern Spain, who were Muslim.

The Arrival of Paper in Europe

It should be pointed out that the first paper made in Europe was not made by people of European background. The Moors held little in common with the rest of the population of Spain, and exhibited a fiercely independent attitude that remains today. First of all, they were Muslims living in a nation—indeed a continent—that was predominantly Christian, and Europeans were very slow to adopt any new product associated with the Muslim cultures, for which they held much suspicion. Bear in mind that this was the time of the seven major Crusades (1096–1270), during which thousands of European Christians died in an effort to capture the Holy Land from the Muslims. Although this period of European history was not conducive to the acceptance of anything from the Middle East, those leaders and knights who did return (for example England's Richard the Lion-Hearted and France's King Philip) brought awarenesses of Eastern technology that foreshadowed the end of feudal Europe and the Dark Ages.

At this time Europe's documents were still being recorded on **parchment**—a surface made from the skin of sheep (a college degree is still commonly referred to as a "sheepskin"). To make parchment, the skin is split into two layers; the outside is tanned into leather and the inside portion is washed, treated with lime, rubbed, scraped, washed again, and stretched tightly across a frame for a final scraping and rubbing. A second medium of that period was **vellum.** Similar to parchment, vellum was a product of calfskin; it underwent roughly the same processing, but the skin was never split and the marks of the hair follicles usually produced an irregular surface and appearance. From this description, it is apparent that Europe had long needed a better method of producing a surface for writing (printing had not yet been developed). Imagine rows of monks or clerks bent over tables, laboriously copying texts, and you have the general picture of graphic communications in twelfth-century Europe.

During the time of the Crusades military campaigns were launched to push the Muslims out of Spain and to "free Spain from the infidel." Finally, by the twelfth century, the Moors had been pushed back from some provinces and within the conquered territory the Europeans discovered paper mills, so the secret of papermaking was finally theirs.

Now part of the Christian world, the process of papermaking rapidly moved to the seat of that world—Italy, home of the Vatican. Italy became the primary producer of paper in Europe, but at first the new product was not considered worthy of carrying important communications. In some regions it was illegal for paper to be used for public documents, but these decrees did not stifle its growing acceptance. During this time of Italian dominance of European paper production, two developments took place. One was the creation of **guilds,** forerunners of today's unions, which closely guarded their members' craft and were very competitive with guilds of other regions. This competition gave birth to the second development—the **watermark.** This decorative feature of paper started out as simple designs—predictably, small crosses to please Vatican officials. Later, the watermarks began to be initials that served as a trademark for the particular guild that produced the sheet, similar to the cattle brand of the American frontier. Figure 1–5 shows some early watermarks.

Two types of screens were developed, **laid** and **wove.** The difference between them lies with the pattern of the wires

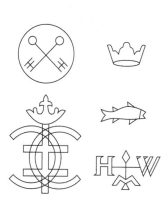

Figure 1–5 Early watermarks commonly used symbols of Christianity and the monarchy.

7

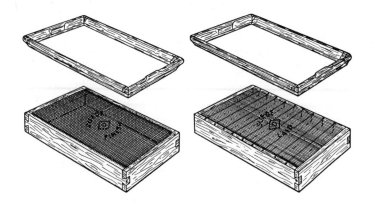

Figure 1–6 Two wires commonly used in wire patterns were the wove (a) and laid (b) screens. The removable frame shown above each screen was called the deckle. Today, bond paper can be ordered in wove or laid finishes and text paper can be ordered with a deckle edge.

and this distinction is transferred to the finish of the sheet (see Figure 1–6).

Technical Developments

From Italy, the process of papermaking moved to France, where the king encouraged the industry's development. Soon France became the source of the finest quality paper in the world and the major exporter. The demand for paper continued to grow steadily, but in 1454 an event triggered an enormous growth in the need for paper: Johann Gutenberg produced the Bible with movable type. This development made printing a much more practical process and, once printing became more affordable, French and Italian paper mills began an even greater volume of business (see Figure 1–7). This prosperity continued for around two centuries before the internal strife that preceded the French Revolution caused many of that nation's businessmen (paper manufacturers included) to feel insecure about their economic future.

Fearing that they would have a lot to lose in a civil war, several papermakers emigrated from France and established their operations in Germany and Holland. Both of these countries had an abundance of water, but the German mills had a major advantage over those in the flat country of Holland. Because they were located in rougher terrain, the German mills could harness the streams that came cascading down the mountains to drive their stamping mills. The Dutch, however, overcame this disadvantage in 1680 by developing an oval-shaped wooden tub that macerated rags into individual fibers by circulating them in water around the tub and under the blades of a large paddle wheel. This beater required so little energy that one windmill could easily power five of them. In fact, it was so efficient that it soon began to replace the stamping mills in Germany that had forced its invention. Understandably, the new beater proved to be a winner and, though improved through the years, remains much like the original; even its name—the **Hollander beater**—is unchanged (see Figure 1–8).

This major development in the industry made it possible to prepare more pulp for papermaking, but at the same time it created two new problems. First, because the new beaters could process rags faster, a *shortage of rags* began to develop. In 1665 England was hit with the Great Plague, which killed some 70,000 Londoners in that year alone. Searching for a cause, officials suspected that rags imported by papermakers had brought the disease. It was decided that rags should be replaced as the source of fiber by virgin cotton and linen and, to save these materials for papermaking, a decree was issued in the following year requiring only wool to be used in burying the dead; this made available an additional two tons of cotton and linen each year. Other alternatives were also sought. Jute was introduced as a supplemental source of fiber. In 1798 experiments began using wood, straw, and old paper, but no breakthroughs were made; therefore, maintaining an adequate supply of fiber remained a chronic

Figure 1–7 This early French cut shows the vat man and coucher at work along with a worker stacking dried sheets. A series of stampers can be seen in the background.

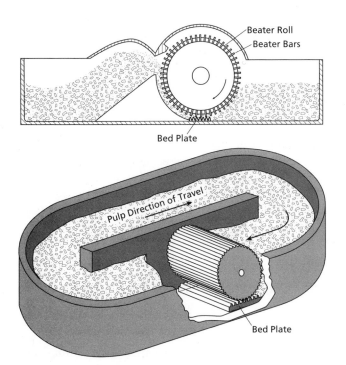

Beater Roll
Beater Bars
Bed Plate

Pulp Direction of Travel
Bed Plate

Figure 1–8 These two views of a Hollander beater reveal how the pulp was flattened between the beater bars and the bed plate.

9

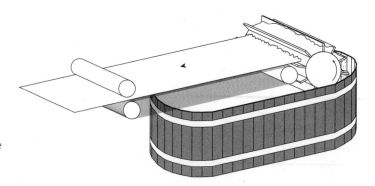

Figure 1–9 The first Fourdrinier machine used a paddlewheel to move pulp onto a moving screen. The device was driven by a hand crank.

problem as the demand for paper grew. By 1800, Germany's 500 paper mills were turning out 1,200 tons annually. Spain was operating 200 mills; even Russia had 26 mills that required 800 tons of rags a year. In America, the state of Connecticut alone annually consumed over 320 tons of rags in its 16 mills.

At the other end of the beater was the bottleneck at the vat. The second problem created by the Hollander beater was the gap in technology related to the process of *forming the sheets*. As late as 1800 sheet formation was still being done exclusively with screen molds that were dipped into vats of pulp, not unlike the original practice begun in 105 A.D. This slowness in forming the sheet created the second problem of trying to keep up with the new beater's ability to prepare pulp. Fortunately, in 1800 a Frenchman named Nicholas-Louis Robert patented an invention that would revolutionize the paper industry—the **paper machine.** Consisting of a continuous wire screen that operated something like a conveyor belt, this new device poured pulp onto the screen at one end and, as the slurry moved along on top of the screen, the water poured through it, forming a progressively more solid sheet that soon passed under a felt-covered roller (Figure 1–9). This roller pressed still more water from the sheet and, by the time it reached the other end of the conveyor belt, it was solid enough to be carried away for drying. The remarkable feature of this breakthrough was that a continuous stream of paper could be turned out instead of individual sheets.

The enormous potential was obvious, but the device was not without its operational problems and a couple of years later, it was sold to two London stationers, the Fourdrinier brothers, who hired an engineer to construct a large machine based on Robert's plans. After several years and $300,000, they created a machine that was commercially practical. In 1812 the first "Fourdrinier" paper machine was put into operation in a mill at Two Waters, England.

It might be noted here that neither the inventor, Robert, nor the developers, the Fourdrinier brothers, ever profited from the device. Amid the early success, the volatile Robert became obsessed with the potential wealth from his invention. He argued constantly with his partner and invested so many of his resources into the device that he found himself so far in debt that he had to sell out for practically nothing. The Fourdriniers earned a place in history because the machine still carries their name, but nothing else came their way because of a loophole in their patent which allowed others to construct near replicas without having to pay royalties.

Another interesting note is that Robert's motivation for inventing the machine was not to ease the world's demand for paper, but rather to ease his constant bickering with the members of the handmade papermakers' guild in the French mill where he was a manager. This sort of labor-management strife did not end with the new technology. Years later and across the channel, it is said that, in anticipation of their jobs being threatened, local handmade papermakers rioted outside the Two Waters mill mentioned earlier. The windows are said to have been boarded up to protect the device from the mob.

The concern of the mill's workers may have had some validity. In the early 1800s, it was observed that a paper machine with a 30-inch wire could match the production of four vats and that a 54-inch machine could replace 12 vats. An on-going effort to increase production spurred the development of wider and faster machines. By 1897, the largest paper machine had a wire that was 162 inches wide and 60 feet long and the fastest machine ran at 500 feet per minute.

But what may have been seen as a threat to a few was clearly good news for the rest of civilization and it came just in time because, in 1804, a Bavarian playwright named Alois Senefelder invented a printing process called lithography. Now that printing could be done from stones rather than expensive metal plates, the new process made printing more affordable and further accelerated the world's demand for paper.

The **cylinder paper machine** was developed during the same time period, but with a very different mechanism. In 1809, Englishman John Dickinson created a device that consisted of a screen-covered drum that revolved inside a pulp-filled vat. As the screen moved through the pulp, fibers

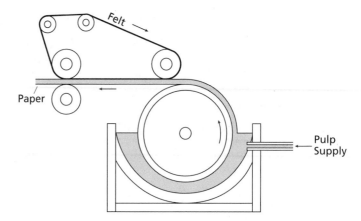

Figure 1–10 The cylinder machine allows fibers to form on a moving screen that lifts them up to the underside of a moving felt and is still used to form paperboard.

accumulated on it to form a sheet that was rotated from the slurry and then transferred to the underside of a felt conveyor belt (see Figure 1–10). The cylinder machine was successful and in 1817 was the first paper machine to be used in America, predating the first American-operated Fourdrinier machine by ten years. In 1824, a cylinder machine was developed to produce cardboard. High-speed cylinder machines are still used to form paperboard and other heavyweight grades.

The development of the **drum drier** was the next major technological breakthrough in papermaking. Understand that even though the paper machine produced a continuous strip of paper, as this strip came off the machine it was cut into sheets and literally hung up to dry. The natural solution to this would appear to be a drying system of one or more heated drums against which the stream of paper could be run until dry enough to be wound into a roll. The drums usually contained a wood or charcoal fire as a heat source, but eventually fires were replaced by steam. Although this concept for drying a web of paper may seem obvious now, it took 11 years after the invention of the Fourdrinier machine before such an invention was made and another 18 years before it was perfected in 1839 by an Englishman named Robert Ranson.

Because cellulose fibers are very absorbent by nature, even the earliest papermakers sought methods to make their sheets more resistant to ink and thereby limit the feathering-out of the printed image. This process is called **sizing** and it was first practiced in eighth-century China where sheets were coated with rice flour. By the fourteenth century, Europeans were dipping their handmade sheets into a sticky liquid boiled from animal hides, a practice that continued for nearly 500 years. However, hand dipping sheets into a tub of sizing was made obsolete by the Fourdrinier machine and drum drier, so the concept of internal sizing was developed in the early 1800s. Instead of treating only the surface of a finished sheet of paper, **internal sizing** occurs much earlier, during beating, when *rosin*, a sticky substance derived from softwood trees, is added to the pulp. With the aid of a precipitant, the rosin coats individual fibers and makes them less likely to absorb moisture after they are formed into a sheet.

The internal sizing of paper became even more important with the growth of offset lithography because unsized fibers can absorb so much fountain solution that the fiber-to-fiber bond weakens, especially during subsequent passes through the press.

A second type of sizing is **surface sizing,** a process that is performed within the paper machine between the two sections of drum driers. Surface sizing is merely the application of a coating of liquid starch to both sides of the stream of paper after it has been through the first set of driers. However, the act of sizing the sheet gets it wet again, so the paper then goes through the second set of drum driers before it is finally rolled up.

With the Fourdrinier machine and drum drier working as a single unit, nowhere was this new technology more readily accepted than in America. By 1845, 22 states were producing paper from Maine to Georgia and as far west as Illinois and Missouri. Massachusetts had 89 mills producing 600,000 reams a year. And, of all the mills in America, only two were still forming paper by hand. All the other mills were using paper machines. By contrast, Germany was still employing over 1,000 vats for the hand dipping of sheets.

Wood Becomes a Fiber Source

With paper machines turning out increasingly wider rolls at faster speeds, the supply of fiber again became a concern. Up to this time rags were still the primary source, but were unable to satisfy the growing need. The concept of making paper out of trees, therefore, began in earnest. As early as 1719, it was observed that wasps make their paper-like nest material by chewing wood fiber, but no way of reducing whole trees into individual fibers was known until around 1840 when a German and a Canadian, acting independently, developed **wood grinding machines.** Similar to the groundwood process used today, the logs were forced sideways against large rotating stones that simply tore away the fibers to fall into a tank of water. Although this invention would lead to a giant industry, groundwood paper was slow to gain acceptance in some regions. For one reason, it had the dirty grey appearance much like the groundwood paper used for today's newspapers so some publishers of the day rejected it as being inferior to the whiter rag stock. However, groundwood paper had two major selling points: (1) it printed well; and (2) it was much less expensive than rag paper. Needless to say, eventually newsprint became the medium for dailies all over the world.

For publications requiring greater permanence, newsprint was inadequate because it reduced practically the entire log into pulp—not just the fiber. This efficiency made groundwood paper inexpensive, but because certain chemicals that bond the tree's fibers together were not removed, the sheets would discolor and turn brittle as newspapers do today. Much effort was expended by several groups in an effort to remove these chemicals, leaving only the fibers. Unfortunately, the pioneers in this venture made the early mistakes that other people would later learn from on the way to success and financial rewards.

Around the time of the Civil War, a plant near Philadelphia was successfully boiling wood shavings in a large pressure cooker with water and caustic alkali. After this boiling, the shavings and solution were reduced to a pulpy slur which was then washed, drained, and bleached. This procedure was called the **soda process** and it worked satisfactorily for woods with low resin content such as hemlock and poplar. Various improvements in the chemistry were developed in succeeding years so that pine, fir, and other woods could also be cooked. These breakthroughs resulted in the **sulfite** and **sulfate processes** of chemical pulping which later became the more popular processes.

Groundwood and chemical pulping of wood did not come a day too soon. Prior to their development, people were experimenting with materials that now seem to be a laughable—if not ludicrous—source of cellulose fiber. For example, straw was often used, but its fibers produced a weak sheet that also passed light too easily. Seaweed, beet root, cactus, grass, and cornstalks were also subjects of experimentation. In fact, an 1879 issue of *Scientific American* argued that paper be made from cow dung.

The lengths to which paper makers have gone to maintain an adequate supply of cellulose fiber is evidenced by an adventure involving a Maine papermaker named I. Augustus

Date	Number of Vats	Annual Tons per Vat	Number of Machines	Annual Tons per Machine
1805	760	21.70	6	92.8
1835	430	26.08	82	298.5
1865	109	30.40	390	103,700.0
1900	104	37.40	428	647,764.0

Table1–1 The increase in both the number of paper machines in America and their productivity during the nineteenth century is revealed. *Source:* Papermaking.

Stanwood. During the Civil War, Stanwood, like most papermakers, experienced a shortage of fiber for making his brown wrapping paper. Not one to lack ingenuity, he took a lesson from history and had brought across the ocean several shiploads of ancient Egyptian mummies. He had no use for the bodies, but the linen wrappings that encased them were unwound, beaten into pulp, and turned into the brown wrapping paper that he sold to butchers and grocers in New England. At this time, the supply of these mummies was considered endless and the only other interested party was the Egyptian railroad which used them as the primary fuel for its locomotives. Unfortunately, the cloths from these ancient tombs caused a cholera epidemic among the mill's employees, which ended the enterprise. As was stated earlier, the pulping of wood came not a day too soon.

Today, 98 percent of the fiber for the world's paper comes from wood. Annually America's forests alone produce four billion cubic feet of wood to supply the modern paper machines that can produce over 3,000 feet a minute. Overall, America's mills turn out over 200,000 tons of paper and paperboard every day for the estimated 14,000 different products made from paper (see Table 1–1). As is reported by the American Forest and Paper Association, paper is used to publish more than two billion books, 350 million magazines, and 24 billion newspapers in the United States each year.

Summary

In summary, it is apparent that the spread of the papermaking process through the world and its technological advances were neither steady nor the result of cooperation among men or nations. Instead, a combination of border wars, religious strife, jealousy, political upheaval, and greed were the catalysts for the evolution of Ts'ai Lun's creation into the modern technology of today.

Questions for Study

1. Why does papyrus not qualify as paper?
2. Outline the basic steps involved in the ancient Chinese culture's production of paper.
3. What contributions were made by the Japanese and Muslim cultures to the papermaking process?
4. Why was so much of the paper produced in early Muslim cities made of linen fibers?
5. How did the stamping mill work?
6. How did the papermaking process enter Europe and how did it become part of the Christian world?
7. What brought France into prominence as a manufacturer of paper? What caused much of the industry to later move to Holland and Germany?
8. What motivated Holland to develop the tub beater known today as the Hollander beater?
9. Identify the main function of the first watermarks.
10. Trace the development of the paper machine.
11. List the various experimental sources of fiber for papermaking.
12. Why was the drum drier such a significant development in the evolution of papermaking?
13. How quickly did the paper machine find acceptance in early America? Why do you suppose this was the case?
14. Why did it take seven centuries for Europe to learn the secret of papermaking?
15. What was the first method of reducing a mature tree to paper pulp?

Key Words

papyrus	wove	surface sizing
parchment	drum drier	wood grinding machine
laid	sizing	repulping
Fourdrinier machine	stamping mill	linen fiber
internal sizing	watermark	guilds
vellum	Hollander beater	

Sources of Fiber for Papermaking

2

AFTER STUDYING THIS CHAPTER, THE STUDENT SHOULD BE ABLE TO:

■ Explain why two of the three categories of natural materials cannot generate paper fiber.

■ List the sources of cellulose fiber and evaluate them for their desirability in papermaking.

■ Trace the per capita consumption of paper in America since 1920 and project it into the future.

■ Explain why America witnesses a decrease in total acreage of forest land each year, while also realizing gains in total timber.

■ Define silviculture and explain its components.

■ Compare and contrast the major tree harvesting methods.

■ Distinguish between even and uneven forest stands.

The first chapter revealed that humans have experimented for centuries with scores of fiber sources in an effort to produce both inexpensive and functional papers. Geographic location and economic factors have greatly influenced this search, as nations have tried to utilize local resources as often as possible. However, as broad as this range of materials has been, 95 percent of all paper today starts with the felling of a tree and this fact is a source of concern to many persons who wonder if the demand will deplete the earth's forests or play havoc with its ecology. This concern is understandable and poses the question: Are there other sources of paper fiber that we should be using instead of trees? To address this concern, a systematic examination of the potential fiber sources is in order.

Categories of Fiber

As most schoolchildren are aware, the earth's natural materials are divided into three broad categories—animal, vegetable, and mineral. The first category, animal fiber (or hair), is of no value to papermakers. This is because hair is a protein material, and therefore, has two properties that prevent its use: (1) it repels water; and (2) when matted together with other hair, does not hold the fiber-to-fiber bond after

17

drying. The second major category to be examined, mineral, is not very helpful either because, although there are mineral fibers such as asbestos, they are very expensive to produce and are mainly used when fireproofing is the primary need. By the process of elimination, we find ourselves limited to vegetable fiber, also known as cellulose fiber.

Appropriateness of and Sources for Cellulose Fiber

Cellulose fiber lends itself very well to papermaking. The properties that make it so suitable are that these fibers (a) readily accept water and (b) dry to form a strong bond with one another. Shaped like miniature soda straws, cellulose fibers come in various lengths, from 1/16-inch to over an inch, depending upon the species of the plant. As far as papermaking is concerned there are four basic groups of plants that produce suitable cellulose fiber: bast, seed hair, grass, and wood.

Bast Fiber

One of the earliest types of cellulose fiber ever used to make paper is **bast fiber,** which is gathered from the flax plant and is more commonly known as *linen*. Linen fibers are comparatively expensive because they are found only within the stem of the plant where they form something of an "inner skin" and only 5 percent of the stem is comprised of these linen fibers.

Humans have been using linen for centuries; the early Egyptians, for example, wrapped their mummies with linen cloth. The reason for linen's popularity is the length of the individual fibers. Averaging just under 1 inch, linen fibers offer excellent strength and durability. Because of high cost, papermakers have historically relied on old rags as their main source of linen; today, textile cuttings provide most of the linen fiber used in producing paper. This high cost also dictates that only a few kinds of specialty paper contain any linen and, even then, it is used sparingly in the mixture. Cigarette and other lightweight papers commonly have some linen fiber that provides the strength needed in such a thin sheet. Bank notes and high quality bonds can also contain some linen fiber to impart permanence. Other examples of bast fiber are jute and hemp fibers which are obtained from burlap mill trimmings and old cordage, respectively. Like linen, these fibers have good length and are usually added to lightweight paper, such as bible and cigarette paper, to provide added strength.

Seed Hair Fiber

Seed hair fibers are the light, hairlike extensions from some plants' seeds that help the wind carry these seeds through the air. **Cotton** is the representative of this group that is used in papermaking. Because its fiber is even longer than that of linen (over an inch), cotton fiber produces a very strong and durable sheet. Once a primary source of paper fiber, cotton was obtained through the collection of rags that were then sold to the paper mills. Generally, this practice has long since been abandoned and today's papermakers depend upon the scraps from textile mills as their chief source for cotton fiber. The resultant reduction in the supply makes papers with cotton in them comparatively expensive and usually limited to high grade bond papers, bank notes, maps, and other applications where permanence and strength are important. In the case of bond paper, the cotton content is watermarked into the sheet, which identifies it as being 25, 50, 75, or 100 percent cotton fiber.

Grass Fiber

Various **grasses** are used throughout the world as sources of cellulose fiber. One that is well-known is *straw.* Because of the shortness of the fiber ($\frac{1}{16}$-inch), paper made from straw tears easily and is fairly easy to see through. For these reasons, straw fiber is not a first choice in papermaking, but in some countries is used along with longer fibers where wood is in short supply. England, for example, augments its wood fiber with straw and, during World War II, relied heavily upon the latter because it was about the only source indigenous to the British Isles.

In addition to short fibers, straw presents still other problems. Although papermaking is, of course, a year-round industry, straw is an annual crop and requires large volume storage for up to a year. During this time, it would have to be kept dry enough to prevent decay, yet not so dry as to allow spontaneous combustion.

A second member of the grass family that is used in British and other Eastern hemisphere mills is **esparto** (ehs-PAR-toe)—a plant found in eastern Spain and parts of West Africa. Like straw, esparto has a short fiber ($\frac{1}{16}$-inch) and produces weak paper, but unlike straw, it is not so easily seen through. Because it grows only on the other side of the Atlantic, America's mills do not use esparto, although some paper made in England can be comprised of over one-third esparto fiber.

A third type of grass fiber is **bagasse** (buh-GAS), obtained from the discarded stalk of sugar cane. Although sometimes used in papermaking, it is not an economically practical source because the sugar mills burn it to power their equipment, thus making it fairly expensive for paper mills.

Research is underway to develop another grass as a practical source of paper fiber. In the Imperial Valley of California, other parts of the Southwest, and the Southeast, plots of **kenaf** (kuh-NAF) are being grown and studied to ascertain its potential in papermaking. Kenaf is a hibiscus plant that grows 12 to 15 feet tall in a growing season and has an excellent fiber yield. Studies indicate that it can produce over twice as much pulp per acre as southern pine trees and is ready for harvest in four or five months instead of 20 years.

Because kenaf requires less chemistry and energy to be converted to pulp than wood, it is arguably more environmentally friendly. Also, it produces long bast fibers as well as short woody fibers. Because kenaf is not a wood, it is more easily pulped and bleached than woods. Test results indicate that paper made from kenaf fiber can have higher brightness, opacity, tear strength and tensile strength than paper made from southern pine.

But kenaf is not without its drawbacks. First, like straw, it is an annual crop and presents the problem of storage. Also, it can be susceptible to seed rot. Most important, kenaf fiber can be twice as expensive as wood fiber because of the long distances between the kenaf fields and the nearest pulp mills. Kenaf can only be grown domestically in the Southern states and the cost also reflects the economic influence of a very limited supply. For these reasons, the future of kenaf in America is uncertain.

Nations without extensive forests or effective forest management view kenaf as an economic blessing that will reduce their need for imported pulp material. Brazil, India, Pakistan, Thailand, Italy, France, Australia, and the United States are performing laboratory and field studies. In 1979, Thailand went so far as to construct an $80 million paper mill that relies entirely on kenaf for its pulp.

Wood Fiber

We have seen that none of the fiber sources examined is commercially practical; each either produces paper with undesirable properties or is too expensive. So, by the process of elimination, the paper industry must look to **wood** as the primary source of paper fiber.

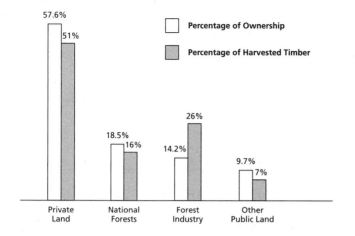

Figure 2–1 Seventy-two percent of America's timberland is held by private ownership. The "other government" category includes Bureau of Land Management, Native American, and state and local government lands. *Source: U.S. Forest Service.*

The use of mature trees as a fiber source was late in coming. Papermaking had been going on for well over 1,600 years before the first mature tree was used to provide the fiber in 1844. This was largely because, until the introduction of the groundwood process, no means of reducing a whole tree down to individual fibers had existed. However, as the technology developed, the use of wood quickly grew and today it provides 95 percent of the world's paper.

Forest Supply and Demand

One of the reasons for wood's popularity is the abundance of forests in the world. In the United States alone there are over 731,000,000 acres of forest land, approximately one-third of the total land area. Of this land, nearly 60 percent is owned by private individuals, while 15 percent is owned by the *forest industry,* commercial operations that use wood in the production of lumber, furniture, paper, cardboard, paneling, plywood, firewood, etc. Because it owns less than one-sixth of the timber, the forest industry must rely heavily on *privately owned* timber as well as some *public lands* for adequate wood supplies; in fact, more than half of America's timber comes from private land.

Figure 2–1 illustrates how the distribution of wooded land ownership compares with the actual supply of harvested timber. While the forest industry owns only 15 percent of the trees, it produces 26 percent of the timber, and accounts for 42 percent of reforestation. Note that nearly one-fourth of the 12.8 billion cubic feet of timber harvested in America during 1987 came from public lands.

Sources of Fiber for Papermaking   21

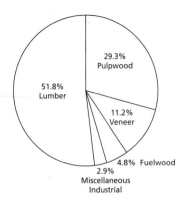

Figure 2–2 Paper and paperboard products comprise nearly a third of the uses for America's harvested timber.

Figure 2–2 reveals that the primary users of timber are the lumber and pulp industries, together consuming 84 percent of the harvested trees. The pulp industry alone uses nearly one-third of America's timber and, therefore, maintains a keen interest in the availability of this national resource, not only for now, but well into the future.

America's need for paper has skyrocketed during the last 60 years. In 1920 (when figures were first compiled), the nation required 6.1 million cords of pulpwood, while by 1990, the quantity had risen to 99.1 million cords—over 16 times as much (see Figure 2–3). This is, of course, attributed to increased use of paper products by individual Americans. Since 1920, the annual per capita consumption of paper and paperboard products has jumped from 145 pounds per person to 699 pounds in 1990. By the year 2030, it is projected that each person in America will use an average of 1,296 pounds of paper products annually, and this is the domestic market alone. During the relatively short time period of 1977 to 1988, the nation's exports of pulp materials increased by 88 percent. When this increasing demand for forest products is viewed together with the steady decline in America's forest lands (see Table 2–1), the challenge of maintaining an adequate supply of wood becomes apparent.

While it is true that America's forest land is shrinking every year, it is also true that America has over 23 percent more cubic feet of growing timber than it had in 1952; in short, the reduction in acres has been greatly offset by much

Figure 2–3 America's consumption of pulpwood (in millions of cords) increased 34 percent between 1977 and 1990. Despite a drop of less than 1 percent in 1991, the projection is for a continued climb in the usage rate. *Source: U.S. Forest Service.*

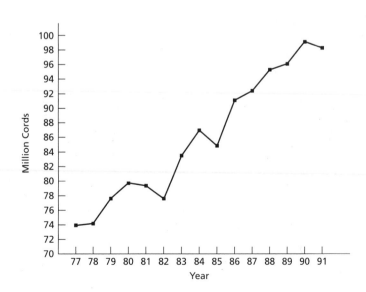

	1952	1962	1970	1977	1987	Percent Change
North	154	156	154	152	153	−0.6
South	204	212	204	200	195	−4.4
West	150	150	147	139	133	−11.3
Total	508	518	505	491	481	−5.3

Table 2–1 Timberland (in million acres) in the United States shows a pattern of decreasing in most regions. Observe that the South has the greatest number of acres of forest. The percent change is for the 40-year period. *Source: U.S. Forest Service.*

	1952	1962	1970	1977	1987	Percent Change
North	24.1	28.2	32.0	35.3	35.6	+11.5
South	32.8	38.2	48.6	56.7	53.6	+63.4
West	23.4	32.2	39.2	44.2	46.5	+69.7
Natl. Average	27.4	32.2	39.2	44.2	46.5	+69.7

Table 2–2 This table contains the annual net volume of America's growing stock (in billion cubic feet) by region between 1952 and 1987. Observe that forest yield has increased in all regions and more than doubled in the West. The percent change covers the 35-year period. *Source: U.S. Forest Service.*

larger increases in yield per acre. An examination of Table 2–2 reveals that each region of the nation has witnessed increases in timber productivity. The bottom line here is that America is not running out of timber. This increase in yield can be attributed to a combination of advances in science, improved forest management, and governmental involvement.

The Role of Silviculture

The figures presented in the previous five paragraphs were compiled by the U.S. Department of Agriculture's Forest Service, a governmental agency that constantly monitors the planting, growth, and harvesting of the nation's timber. Fortunately, the Forest Service and the forest products industry have been cooperating since the 1920s to ensure adequate supplies of mature timber to meet these projected needs. Their efforts have revolved around improvements in three areas—(1) forest management; (2) harvesting procedures; and (3) utilization of waste materials. The portion of forest management that deals with the first two of these areas is called **silviculture,** a component of forestry that is key to providing adequate supplies of timber because its goal is *improved timber yields.* Silviculture is the science of growing, harvesting, and regenerating forests and it begins in the tree nursery where superior trees are developed—trees that will grow faster, taller, and straighter; have a high resistance to disease and insects; and yield more usable wood.

Developing Superior Trees

The first method for superior tree development is **grafting,** a procedure that begins in the forest where trees with desired characteristics are located. Cuttings from the tops of these trees are grafted onto the root systems of two-year old trees at a nursery and, as they mature, pollen from male cones is used to pollinate female cones to produce seeds with the desired hereditary traits. Grafting allows the pollen to be gathered from 15-foot trees in a nursery instead of from 30 or 40-foot trees scattered throughout a forest.

Federal, state, and private organizations are actively involved in silviculture with researchers working to develop trees that will thrive in each part of the country. In the Northeast, tree research is underway on the Eastern Scotch, Norway spruce, sugar maple, poplar hybrid, and a number of hardwoods. In the South and Southwest, work is going on with the slash pine, loblolly pine, longleaf pine, shortleaf pine, gums, tulip poplar, cottonwood, and several oaks. Species undergoing research in the Great Lakes region include pines, spruces, poplars, and numerous hardwoods. In the Central states much attention is going to ash, cottonwood, maples, oaks, sycamore, walnut, yellow poplar, and several pine varieties. Emphasis in the Western region is given to the Western white pine, ponderosa pine, and Douglas fir (see Figure 2–4).

A second method for obtaining superior trees is **hybridization,** the artificial development of species that did not previously exist. By crossing species, scientists are able to produce trees with genetic combinations that do not occur in nature. Clearly, the goal is to produce a tree that has the best qualities of two or more species. Often this procedure also can result in **hybrid vigor,** the condition of the hybrid offspring obtaining greater growth or other desirable characteristic than that of either parent.

Mutagenesis is the creation of genetic diversity through the artificial acceleration of the natural mutation rate by subjecting seeds, pollen and tissue to ultraviolet light, x-rays, high energy radiation, or chemicals. Because the pro-

Figure 2–4 Three of the most highly valued trees that grow in the Pacific Northwest are (from left) the Douglas fir, ponderosa pine, and grand fir. All three species are excellent for conversion to lumber. *Courtesy of Will Nelson.*

cess is very time consuming and researchers cannot predict which tree characteristics will be affected or the direction the effect will take, mutagenesis has yet to play a major role in silviculture.

More success has occurred with **polyploidy,** the production of trees with three or more complete sets of chromosomes. By treating tree tissue with chemicals, the chromosome numbers can be doubled, resulting in a tree that is polyploid, meaning that it has more than the standard single pair of chromosomes. One of the most notable successes in this area is the development of giant poplars with increased fiber length and growth rate. Foresters are hopeful that even more developments will result from further work with polyploidy.

Seeds that result from all of these methods are planted in large quantities at tree **nurseries** (see Figure 2–5). After a few years, the seedlings that spring up are then transplanted to areas previously harvested. More than 1.5 billion seedlings were planted in the U.S. in 1991 alone—43 percent by the forest industry, 39 percent by private landowners, and 18 percent by governmental agencies. In this way the felled trees are replaced by a stand of an improved species.

Figure 2–5 An automatic irrigation machine moves over thousands of pine seedlings in a U.S. Forest Service nursery. *Courtesy of U.S. Forest Service.*

Once the young trees begin to grow out on the planting site, the forester will monitor the stand, watching for overcrowding, disease, and insect infestation while ensuring proper fertilization, and appropriate harvesting techniques. All of these actions are aimed toward increasing the area's yield of timber.

Studies conducted in various regions with a wide range of species indicate that five factors dictate the handling of a particular species. They are:

1) degree of tolerance to shade;
2) windfirmness of the root system;
3) ability to grow in pure or mixed stands;
4) growth rate in even-aged and uneven-aged stands; and
5) ability to easily reproduce.

Not only do these factors influence the decision of which species will be planted and how it will be managed while growing, but they also will influence which method will be used to harvest the trees upon maturation.

Harvesting Methods

Not all forests lend themselves to being harvested; if a delicate environment exists with respect to the topography, climate, or soil, then logging would be harmful. These areas are left to remain in a wilderness condition and attention is turned to other forests. Therefore, part of the forester's skill is determining how best to manage the country's forests so that they can provide a wildlife habitat, offer wooded recreation areas, and supply needed commercial timber.

Silviculture plays a big role in this overall program of management because when more timber can be produced from a given acreage of forest, less total acreage is required. In other words, when a paper company can get more pulp wood from its own forest holdings it does not have to look so much to private and government lands.

Choosing the best method for **harvesting** trees is a very important way to improve the productivity or yield of a forest. Several ways of cutting a stand of timber exist and each one has its advantages. The primary factors that determine which cut will be used are (a) the *species* of the trees involved and (b) to what degree the owner is concerned about *yield* versus preserving the *appearance of the forest*.

Intermediate Cutting

One way that a timber stand can be improved is through **intermediate cutting,** which consists of removing certain trees to encourage the growth of others. There are two ways of accomplishing this. The first method is called **thinning.** Similar to how a gardener removes some plants that are growing too close to one another and competing for nutrients, the forester can increase the overall growth of a stand by removing a few trees from a dense stand (see Figure 2–6).

A second type of intermediate cutting is the **improvement cut,** which involves the removal of trees considered undesirable by species, form, or condition. Examples of an improvement cut would include the removal of diseased, dead, dying, or wind damaged trees. The owner could also "weed-out" trees that are of a lower commercial value.

A type of intermediate cut is the **selection cut** (or selective cut) in which individual trees are evaluated against several criteria including (a) proximity to other trees, (b) desirability of the species, (c) health, (d) formation, and (e) maturity (see Figure 2–7). No single factor (such as diameter) determines a given tree's fate. For example, a mature tree with little growth potential may not be harvested if the sudden increase in exposure that would result from its removal might threaten younger trees nearby. Also, the selection cutting of a mixed stand of trees would not allow the removal of all trees of a superior species because such a harvest would leave behind only inferior species and thereby reduce the future value of the forest.

Regeneration Cutting

Although intermediate cuts enhance forest productivity, an even greater increase can result from a **regeneration cut.**

Figure 2–6 Two foresters inspect a plantation of loblolly pine after thinning. The timber stand will now be more productive than if the cut had not been made. *Courtesy of U.S. Forest Service.*

Figure 2–7 The large stump surrounded by mature trees indicates that this forest received an intermediate cut, specifically a selective cut. *Courtesy of U.S. Forest Service.*

There are three main methods of performing a regeneration cut, but they share the same primary goal: to create the best conditions for a *particular species* to renew itself. With most regeneration cuts, the forester allows nature to do most of the work of forest renewal.

The decision of which regeneration cut to use involves several factors: aesthetic concerns, economic factors, and the ecology must all be considered, along with the characteristics of the tree species involved.

One way to perform a regeneration cut is to use the **shelterwood cut** method. For species whose seedlings require protection from frost, wind, and intense sun, this method leaves enough mature trees standing to ensure a successful regeneration (see Figure 2–8). Typically, 12 or 13 trees per acre are left standing while the remaining trees are removed. After the new growth has become established and the shelter is no longer needed, the mature trees are removed so that the young stand will have maximum access to light and nutrients.

The second approach to regeneration is the **seed tree cut** method which leaves some mature trees behind, but only half as many as the shelterwood cut method (see Figures 2–9 and 2–10). The seed tree method is limited to species that do not require as much protection, and only a few mature trees are left standing to provide seeds for natural regeneration. The seed tree cut is most effective with conifer species that are firmly rooted and grow best in the absence of shade. It is also used when an adequate supply of seeds or seedlings does not exist on the floor prior to cutting.

There are instances in which neither shelter nor seeds need to be supplied by remaining mature trees. Often the forest floor contains a rich supply of viable seeds that merely need the catalyst of sun and rain to begin a new forest. In this instance, the **clearcut** method is the most efficient. As the name implies, a clearcut removes all trees except saplings from a defined area (see Figure 2–11). Afterward, the area regenerates itself from (a) seeds already in the ground, (b) seeds blown in from adjacent areas, or (c) sprouts from roots or stumps. Studies by the Forest Service have found that, where appropriate, clearcutting provides the highest yield of all methods. One reason for this is that the logging operation and equipment roughens the forest floor just enough to expose mineral soil and actually improve seed germination and promote rapid early growth. Second, the even-age stand that results will not be bothered by subsequent trips into the forest with heavy equipment to remove

Figure 2–8 A thin canopy of hardwood trees remains after a shelterwood cut; these trees will be left in place until the new growth no longer needs their protection. *Courtesy of U.S. Forest Service.*

mature trees left behind by the earlier cutting. Because subsequent cutting operations necessarily damage some of the young growing trees, clearcutting is the most efficient method of timber harvesting.

Clearcutting must be performed in a manner that will protect the soil and ensure regeneration of the forest. In the past, clearcutting has been performed irresponsibly where the tree species or soil type was inappropriate. For example, in an effort to quickly rebuild Chicago after the great fire of 1871, a several square mile white pine forest growing in the sandy soil of Michigan's west coast was clearcut and left to

Figure 2–9 Only a handful of trees are left standing after a seed tree cut of this Michigan hardwood forest. These remaining trees will be cleared out after their released seeds become established.

Figure 2–10 A seed tree cut was considered appropriate for this soft-wood forest in the Southern Appalachia region. *Courtesy of U.S. Forest Service.*

the elements. After the roots deteriorated, winds eventually stripped the soil cover to create expansive sand dunes (see Figure 2–12).

Since that time, much has been learned and applied to the practice of clearcutting. The Douglas fir, for example, never should be clearcut in areas larger than just a few acres and the bare clearings that result must be surrounded by uncut trees to allow for seeds from surrounding trees to blow into the barren area and generate new growth. When cut in patches or strips, however, the Douglas fir is an ideal species for clearcutting because its young growth needs no shelter-

Figure 2–11 Three-year-old loblolly pine trees grow in an area that earlier had been clearcut in North Carolina's Sumter National Forest. *Courtesy of U.S. Forest Service.*

Figure 2–12 Although they are picturesque, the sand dunes at the Silver Lake State Park near Hart, Michigan also reveal what can happen when a forest is carelessly clearcut.

ing and flourishes in open areas. After the new trees have become established, the mature growth in the surrounding areas are no longer needed and can then be removed.

Increasingly, clearcuts are being made in irregularly-shaped patches that relate to the contour of the terrain

Figure 2–13 Shown here are two approaches to clearcutting in patches. In the second drawing, the clearcuts are blended into the contour of the land, an approach that increasingly is used when the area is visible from a road. *Courtesy of U.S. Forest Service.*

instead of the more traditional rectangles. These more natural-looking treatments are more aesthetic (see Figure 2–13).

A second advantage of clearcutting in strips is that wildlife is never completely without the forest that provides its cover and food supply. U.S. Forest Service studies have also shown that, if planned judiciously, even the visual problems with clearcutting can be minimized.

The efficiency and timber yield that clearcutting brings makes it a favorite method of commercial loggers where practical. Research on Ohio's Vinton Furnace Experimental Forest by the U.S. Forest Service has shown clearcutting to be the best method for oaks, poplar, ash, and walnut trees—all species indigenous to that part of the country. An experiment that began in 1953 divided an area into six five-acre plots that were cut with six different harvesting methods. Yearly, researchers studied each section for growth rate and amount of timber being produced. Of the six harvesting methods, the complete clearcut—removal of everything over 4-½ feet tall—clearly showed the greatest productivity. Also, despite the hilly terrain of the test site, no soil erosion had occurred following the clearcutting because the roots left behind easily held the soil while new growth was emerging.

Figure 2–14 The even-age stand pictured here allows each tree adequate sunlight and soil nutrients. Contrast this area with the one pictured in Figure 2–15.

Even and Uneven-Age Stands

Any first regeneration cut results in a new forest that likely will be more productive than what preceded it. The increased productivity is due to the new trees' being the same age and roughly the same height. Such a forest is called an **even-age stand** (see Figure 2–14). No longer will older and taller trees form a canopy that robs the majority of the forest of the light and water needed for rapid growth. Even-age stands commonly consist of a single species that is considered most approriate to that particular region or commercially desirable. A natural forest is an example of an **uneven-age stand** (see Figure 2–15). Some environmental groups have opposed all regeneration cuts, arguing that replacing a mixed-stand forest with a single-species forest can greatly alter the *ecology* of the area. Their concern is that eliminating all but one species of trees impacts on the presence of other forest vegetation, which can make the regenerated forest inappropriate for certain animal species that

Figure 2–15 Although the floor area of this uneven-age stand is covered with young trees, their growth is stifled by the dense ceiling created by the mature trees. As a result, the productivity of the entire forest is reduced.

had been living there. Chapter 16 looks at environmental groups' concerns more thoroughly.

Reducing Wood Waste

The third area that foresters are concerned with involves efficiency in utilizing nearly all of a felled tree. In America's earlier years, the forests seemed so expansive that exhausting the supply seemed impossible. During that time loggers would carry off only the most valuable part of trees—usually just the main trunk—and leave the limbs and branches behind to rot on the ground. Also, waste timber was discarded at the saw mill (see Figure 2–16). Today, a

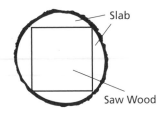

Figure 2–16 Above is a cross section view of a saw log that is about to be sawn into lumber. The first cuts remove the slab, which is usually chipped for pulping.

Figure 2–17 In less than a minute, the blades of a whole-tree chipper reduce a mature tree to chips that are blown into a truck for transport to a pulp mill.

Figure 2–18 A feller-buncher machine cuts a tree near the ground, carries it to a clear area, and lets it fall where it waits to be dragged to the whole-tree chipper.

concerted effort is underway to use as much of the tree as is possible. For example, nearly 20 times more wood residue produced at sawmills was purchased by paper mills in 1970 than only 20 years before. Also, instead of processing only 5-foot logs hauled out of the forest to the mill, papermakers are processing the entire tree right on the logging site, a technique known as **whole-tree chipping** (see Figure 2–17). This innovation allows access to the cellulose fiber from the entire tree rather than the trunk alone—an increase in usage efficiency of from 25 to 200 percent, depending on the size and species of tree. Also, instead of using chain saws to cut the tree 2–3 feet above the ground, tractors using

Product	Quantity
Personal checks	460,000
Sheets of letterhead bond	89,000
No. 10 envelopes	61,370
One-gallon milk cartons	4,000
Issues of *National Geographic*	1,200
Copies of *Sunday New York Times*	250

Table 2–3 This table reveals the number of various products that can be produced from a single cord (128 sq. ft.) of lumber. A cord can generate over two tons of paper. *Source:* Statistical Record of the Environment.

hydraulic scissorlike blades can sever the tree at ground level—further minimizing waste (see Figure 2–18).

Summary

In summary, potential sources of fiber for making paper were examined and, by the process of elimination, trees were found to be the most feasible source when considering quality of fiber and economic concerns. A look at the current and future needs for paper—and therefore trees—showed that a sophisticated program of silviculture is needed to ensure an adequate fiber supply 20 to 50 years from now. Table 2–3 offers a glimpse at the diverse printed products derived from wood and underscores the need to increase the volume of timber that is generated from a diminishing amount of forest land. Fortunately, the forest industry and governmental agencies have made great strides in improving forest productivity and their research and cooperation is continuing.

Questions for Study

1. What are the three general categories of the Earth's materials and why are the members of two of them inappropriate for use as paper fibers?
2. How are seed hair fibers different from bast fibers relative to where they are found on a plant?
3. What characteristics make cellulose fibers well-suited to making paper?
4. Why do trees constitute 95 percent of the world's source of paper fiber? Why is wood used over straw, bagasse, esparto, and kenaf?
5. Define silviculture and explain how it works. How do grafting, hybridization, mutagenesis, and polyploidy fit into the process?

6. List the rationale behind these tree harvesting methods: improvement cut, shelterwood cut, seed-tree cut, and commercial clearcut. What five factors dictate how a forest should be harvested?
7. Which method is most efficient as far as timber yield is concerned?
8. What considerations must be made when using the clearcutting method of tree harvesting?
9. What is the projection for America's timber needs in the next century? What is being done to ensure an adequate supply?
10. Who owns most of the forest land in America: the timber industry, the government, or private individuals?
11. In what areas is America working to reduce timber waste?

Key Words

bast fiber	esparto	grafting
grasses	silviculture	mutagenesis
kenaf	hybrid vigor	intermediate cutting
hybridization	nurseries	selection cut
polyploidy	improvement cut	seed tree cut
thinning	shelterwood cut	whole-tree chipping
regeneration cut	even-age stand	harvesting
clearcut	cotton	uneven-age stand
seed hair fibers	bagasse	

The Manufacture of Paper

3

AFTER STUDYING THIS CHAPTER, THE STUDENT SHOULD BE ABLE TO:

■ List and explain the various techniques of pulping wood.

■ Explain the rationale behind multistage bleaching.

■ Explain why concern over dioxin is causing many pulp mills to modify their bleaching process.

■ Describe the beating/refining process and explain its function.

■ Describe the components of a modern paper machine from headbox to the finished log and their functions.

■ Define the finishing operations that paper can receive after it is formed on the paper machine.

■ Describe the ways in which paper is packaged for shipment.

An understanding of the earliest methods of papermaking is also a fairly good understanding of the steps that paper mills go through today. This is because technological advances have resulted in *improvements in,* rather than a *substitution for,* the basic approach invented in second century China. Today's mills are geared up to produce very high volumes of consistently high-quality paper, but the basic processes of cooking, beating, forming, and drying are still followed.

Pulp, Paper, and Integrated Paper Mills

Most paper and paperboard mills do not perform the pulping operation. Instead, they purchase pulp from **pulp mills** that usually chip the pulpwood that arrives, cook the chips into pulp, bleach it, and form thick, very rough sheets that are sold to paper mills. **Paper mills** that purchase pulp are generally not very large in volume. These mills repulp the sheets into a light fiber-water mixture in large machines that resemble a kitchen blender. Most of the desired physical characteristics that the mill desires for the finished sheet are then imparted through additives and refining the pulp before the fibers are matted together on the paper machine. A paper mill that contains both pulping and papermaking operations is called an **integrated mill;** generally speaking, the large volume papermaking operations are found within integrated mills.

The Six Major Components of Modern Papermaking

The six major components of an integrated mill can be seen in Figure 3–1. The next two paragraphs offer a very brief overview of them.

The wood lot converts the wood to chips and stores them. Chips are then cooked in the digester and the resulting pulp is washed to remove noncellulose materials. Multistage bleaching whitens the pulp before it is refined (beaten) and sent on to the paper machine. After the sheet is formed, it is dried, sized, dried again, and calendered before being rolled into a log that could be 25 feet long. The next stop in the process involves the finishing operations that begin with the log being split into smaller rolls. At this point the roll may be coated, polished to a gloss, embossed, or receive a combination of these treatments. Paper to be printed on a web press can then be wrapped and shipped. Paper to be printed as sheets must then be cut into sheets, trimmed to the desired size, then packaged on skids or in cartons, and shipped.

The sixth major component of an integrated mill is the recovery plant that converts nearly all of the waste materials from the pulping and bleaching operation into either energy. Bark chips that are either too large or too small, and the chips' noncellulose fiber material are burned to generate steam used to heat the drier's drums. Also, the spent cooking chemicals from the pulping operation are converted into new cooking chemistry. Because of the work done at the recovery plant, very little waste material is generated in the production of paper.

With the exception of the recovery plant, the basic operations of the modern integrated mill perform the same basic functions as their counterparts in ancient China. The processes are automated, the equipment is huge, and the production volumes are amazing, but paper is still made by chopping, cooking, forming, and drying. The remainder of the chapter provides a detailed examination of each operation.

The Debarking Process

When pulpwood logs arrive at the wood lot, the bark must be removed because it contains no cellulose fiber. This process of **debarking** can occur in different ways. In the Pacific Northwest and Canada, water jets under great pressure blast the bark off, but the more widely-used method uses the *debarking drum*. Logs enter the 40-foot horizontal revolving drum at one end and, as they make their way to the other end, lose their bark by rolling and banging against each other and the steel bars of the drum's walls (see Figure 3–2). As the bark comes loose, it falls through the openings between the bars and onto a conveyor belt that collects it for fuel. In the 10 to 15 minutes that it takes for the logs to make their way through the drum, they become completely debarked and ready for the chipper.

Chipping

Reducing debarked logs into bite-size chips is the job of the **chipper,** a device that functions something like a food processor and can dice up a log in a matter of seconds. Except for paper that is made with the groundwood pulping method (which will be explained later), chipping is absolutely necessary because the chips will later be cooked and their small size allows the liquid to penetrate through and break down the wood into individual cellulose fibers (see Figure 3–3).

It was mentioned in Chapter 2 that, in an effort to get as much fiber as possible from each tree, the process of **whole tree chipping** is becoming increasingly popular. Instead of cutting the tree into 5-foot lengths and leaving behind the smaller limbs and branches, the tree is severed at ground level and dragged to the chipper where the entire tree is reduced to chips which are blown into a tractor trailer for efficient hauling. The load of chips that arrives at the mill then must be screened to separate out pieces of bark, leaves, and dirt.

Wood arrives at paper mills by various means; both individual farmers and commercial loggers can haul 5-foot lengths of pulpwood that will be unloaded and later chipped. Also, the already-mentioned tractor trailer rigs can roll in with their loads of chips loaded at the logging site. These rigs are usually unloaded by backing them onto a platform that is tilted, causing the chips to fall onto a conveyance mechanism that carries them to a storage pile (see Figure 3–4). Obviously, mills that rely solely on getting their wood in the form of chips are able to skip the debarking process altogether and go directly to the pulping stage. Mills often maintain a 30-day supply of chips (see Figure 3–5).

Pulping of Wood

The actual reduction of the wood—whether in the form of logs or chips—into individual fibers is known as pulping: There are three general means of pulping: they are (1) mechanical, (2) chemical, or (3) a combination of mechanical and chemical. Because mechanical is the oldest form, it will be examined first.

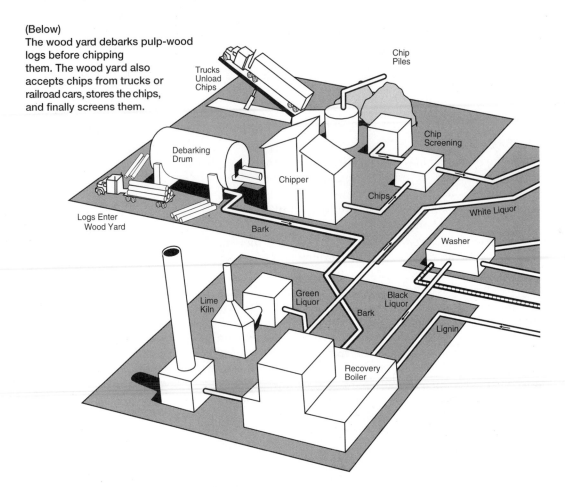

(Below)
The wood yard debarks pulp-wood logs before chipping them. The wood yard also accepts chips from trucks or railroad cars, stores the chips, and finally screens them.

(Above)
The recovery boiler accepts lignin and spent chemicals (black liquor) from the brown stock washers. It also accepts bark and rejected chips from the wood yard. The lignin, bark, and chips are burned to generate steam and the black liquor is converted to green liquor, which is then converted to a new supply of white liquor for the digester. Excess steam, along with carbon dioxide, is released from the smoke stacks.

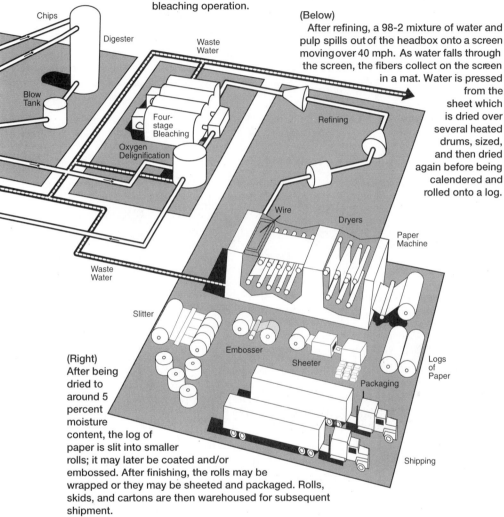

(Below)
Chips, water, and white liquor are loaded in the digester for cooking. The separated fibers and spent chemicals are then blasted into the blow tank before being pumped to the washers where the spent chemicals are separated from the fibers and sent to the recovery boiler

(Below)
Oxygen is often used to remove most of the lignin prior to multistage bleaching. Oxygen delignification allows the elimination of elemental chlorine from the bleaching operation.

(Below)
After refining, a 98-2 mixture of water and pulp spills out of the headbox onto a screen moving over 40 mph. As water falls through the screen, the fibers collect on the screen in a mat. Water is pressed from the sheet which is dried over several heated drums, sized, and then dried again before being calendered and rolled onto a log.

Chips

Digester

Waste Water

Blow Tank

Four-stage Bleaching

Oxygen Delignification

Refining

Wire

Dryers

Paper Machine

Waste Water

Slitter

Embosser

Sheeter

Packaging

Logs of Paper

Shipping

(Right)
After being dried to around 5 percent moisture content, the log of paper is slit into smaller rolls; it may later be coated and/or embossed. After finishing, the rolls may be wrapped or they may be sheeted and packaged. Rolls, skids, and cartons are then warehoused for subsequent shipment.

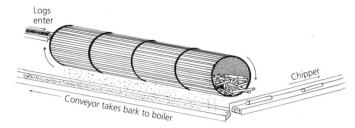

Logs enter

Chipper

Conveyor takes bark to boiler

Figure 3–2 The debarking drum is simply a large rotating cylinder of steel bars that allows pulp logs to tumble and rub against one another, causing the bark to fall away in chunks. Note how the debarked logs are conveyed to the chipper and that a second conveyor belt carries pieces of bark to the boiler to be used as fuel.

Figure 3–3 Optimum chip size allows the cooking liquor to penetrate all of the wood. Note that some chips are viewed from the side. The shape on the ends reflects the 45-degree angle of the chipper's knife.

Mechanical Pulping

Paper can be mechanically pulped in several ways. The earliest method of **mechanical pulping** was devised in 1840 by Frederick Keller, a German. His process involved the tearing away of fibers from the face of a debarked log by a large grindstone. His invention proved successful in both Europe and North America and, today, paper pulped with this technique is called *stone groundwood* (**SGW**). Figure 3–6 illustrates one of the variety of ways that logs can be pressed against the revolving grindstone. The water is constantly sprayed onto the stone to cool it. Water in the form of hydroelectric power is also the usual source of the tremendous force needed to turn these grindstones. Therefore, groundwood pulping is most popular in Canada and the Pacific Northwest with its large rivers. The advantage of the groundwood pulping is its high productivity. Because this

Figure 3–4 A chip truck is tipped into the air to cause the chips to slide out and onto a large conveyor belt. After being unloaded, chips are blown onto a storage pile. *Courtesy of* Cadillac Evening News.

Figure 3–5 A bulldozer moves chips around in the chip pile to prevent spontaneous combustion. In the background, new chips are blown onto the doughnut-shaped pile. *Courtesy of Hammermill Papers—Division of International Papers.*

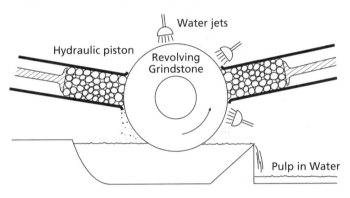

Water jets

Hydraulic piston

Revolving Grindstone

Pulp in Water

Figure 3–6 A pocket groundwood pulper uses hydraulic pistons to hold debarked logs against a large revolving grindstone. Jets of water keep the friction from producing too much heat.

process does not remove the noncellulose materials from the pulp, 90–95 percent of the wood becomes usable fiber. This ratio compares very favorably to the 45–55 percent efficiency of chemical pulping and is a chief reason why groundwood pulping is popular in Canada and the northern United States where timber's value is enhanced by shorter growing seasons.

There are, however, some disadvantages to the groundwood pulping process. First, the grinding, tearing action of the stone severely damages the fibers, making them weaker and shorter than do other means of pulping. Second, groundwood pulping does not attempt to remove noncellulose matter from the pulp. Although the presence of other materials improves the amount of pulp produced from a given log, these unwanted chemicals greatly impair the quality of the finished sheet. The chief culprit of these chemicals is **lignin**—nature's resin that bonds individual fibers together within the wood. Lignin that is left in paper combines with oxygen to form an acid, so papers made from groundwood pulp, such as newsprint, are quick to discolor and deteriorate. This effect is lessened significantly, though, when groundwood paper is coated on both sides, as is often the case with magazine stock.

In an effort to produce stronger paper, two other methods of mechanical pulping have been developed. *Refiner mechanical pulp* (**RMP**), first produced in the 1950s, begins with reducing the log to chips which are then mixed with water in a 30–70 ratio and pumped into a disk refiner. Inside, the chips are forced between rotating disks and the friction-produced heat softens the lignin, causing the fibers to lose their bond and separate with less damage than with SGW pulp; the result is a stronger sheet. A process introduced in 1968 is *thermomechanical pulping*—more commonly known as **TMP**. Like the refiner groundwood process, TMP pumps a mixture of water and chips between rotating plates but not before it is subjected to very *high temperature* and pressure which loosens the fiber-to-fiber bond. The end result is even less damage to the fibers and a stronger finished sheet.

Another new development in mechanical pulping is *pressurized groundwood* (**PGW**), developed in Finland in 1977. Basically, PGW is a sophisticated form of stone groundwood pulping in which the grinding occurs under intense heat and air pressure because the cold water spray of SGW pulping is replaced by very hot water. The intense heat and pressure softens the lignin and produces pulp with less damage to the fibers as compared with SGW pulp. PGW pulp mills produce newsprint and coated stocks.

A new pulping process from Sweden augments the mechanical pulping process with chemistry. *Bleached hemi-thermomechanical pulping* (**BCTMP**), is similar to the TMP process except that the chips are also treated with a mild chemical prior to refinement. The chemical additive softens the chips more than hot water alone can, resulting in less fiber damage and stronger paper than TMP or any other purely mechanical process can produce.

Chemical Pulping

In sharp contrast to mechanical pulping methods, **chemical pulping** is able to produce paper that is substantially free of lignin and the other nonfibrous wood components. Today, over 70 percent of wood pulp is produced by first converting logs to chips and then using chemicals, heat, and pressure to dissolve the lignin and allow the fibers to fall away from one another. Because chemicals do the pulping instead of a grindstone or other physical means, the fibers fall away from one another intact and this lack of tearing and abrasion results in longer and stronger fibers. Although chemical pulping was developed to produce stronger paper, the fact that it also separates the cellulose fibers from most of the non-fibrous matter is another major benefit because lignin-free paper has much greater permanence.

The actual chemistry used in the cooking can vary from one mill to another. The oldest method of chemical pulping is the *soda process*, which was developed by an Englishman in 1851 but not performed commercially until a mill was set up in Pennsylvania in 1855. The process gets its name from its primary ingredient, caustic soda or lye, which are common terms for sodium hydroxide (NaOH). Because it works best with low-resin woods, the soda process was used primarily to pulp hardwoods. The high pressure and temperature of the cooking process forces this strong base (or alkaline) inside the chips where it dissolves the lignin. The early success of the soda process was in part due to its ability to produce paper for half the cost of making it from old rags. However, maintaining the cooking agent at the desired strength throughout the cooking time was not possible, so some of the chips were not completely digested, while some of the fibers were actually dissolved in the process.

The *sulfate process*, begun in Germany in 1884, is a refinement of the soda process. The process begins with sodium sulfate (hence, the name sulfate process) being reduced to sodium sulfide, Na_2S, which is included in the cooking liquor with the same sodium hydroxide used in the soda process. The genius of including sodium sulfide is that it acts as a replenishing agent. During the cooking process, the original supply of sodium hydroxide would gradually become depleted if the sodium sulfide were not concurrently reacting with the water to form more sodium hydroxide. Expressed as an equation, this replenishment reaction is:

$$Na_2S + H_2O \rightarrow NaHS + NaOH$$

This ability to create new amounts of the chief cooking ingredient as needed allows the strength of the sodium hydroxide to remain at a controlled level, which results in stronger pulp than soda pulping can produce because too much NaOH can weaken the fiber itself. In fact, another term for paper made by the sulfate method is *kraft*, the German and Swedish word for "strength." Originally, the sulfate method produced paper that was used in wrapping and packaging—applications where strength is critical. Sulfate pulp could not be whitened, however, until the development of multistage bleaching in the 1930s. Today it is the leading method of pulping for nearly all types of paper.

The term *sulfite bond* is commonly found in paper merchants' catalogs to denote bond papers that have been made through the chemical pulping of wood chips. The primary significance of the term is to distinguish these bonds from *rag bonds*—bonds made from cotton. The term sulfite bond is derived from the *sulfite process* of cooking chips with an acidic agent in contrast to the alkaline agents used in the soda or sulfate processes.

Although the sulfite process was a popular technique for chemical pulping between the 1880s and the 1930s, the process was found to have several drawbacks. First, the acid attacked the fibers and produced a weaker pulp. Second, the digesters required an acid-resistant brick lining that increased the cost. Third, the process did not work effectively with Southern softwoods with high resin content. Last, stream pollution was a difficult problem. For these reasons, the sulfite process has largely been replaced by the sulfate (or kraft) process; however, the term sulfite bond is still used when referring to all bond papers made from the chemical pulping of wood.

Combinations of Pulp

Another blending of fibers occurs when softwood chemical pulp and hardwood chemical pulp are combined to produce a better sheet than either type could by itself. Softwood or coniferous trees (such as pine, fir, or spruce) have longer and thicker fibers than do hardwood or deciduous trees (such as maple, oak, or beech) (see Figures 3–7 and 3–8). Therefore, paper made from softwood fibers will be stronger than hardwood paper; on the other hand, the smaller hardwood fibers are less bulky and will produce a smoother sheet. When combined, the smaller hardwood fibers act to fill in the low places between the larger softwood fibers and the finished sheet benefits from each. The ratio between softwood and hardwood fibers varies with the type of paper being made.

The Digester

As stated earlier, the process of chemical pulping uses the combination of chemicals, heat, and great pressure to dissolve the lignin—inside the **digester,** a device that functions like a 50-foot tall pressure cooker (see Figure 3–9). Once inside the digester, the chips and cooking chemistry (**white**

Figure 3–7 These hardwood fibers have been chemically pulped. Compared with softwood fibers, they are short and thin (100x enlargement). *Courtesy of Champion International.*

Figure 3–8 Even with less magnification, these chemically pulped softwood fibers are longer and thicker than the hardwood fibers of Figure 3–4. 150x. *Courtesy of Champion International.*

liquor) are locked in and heat is applied to create pressure that forces the cooking liquor into the pieces of wood. After the batch of chips has been cooked for the proper amount of time, the pulp expelled from the bottom of the digester and another batch of chips and cooking liquor is pumped in and processed. As might be expected, the term for this process is the *batch* method. A newer development in this process is the *continuous digester* which works in basically the same way except that the chips are digested continuously as they make their way through a series of large tubes.

The Role of the Blow Tank

In either case, after the digesting is complete, the cooked chips and clumps of fibers are pumped onto the next stop, the **blow tank,** for further reduction. The pulp enters the blow tank under the tremendous pressure built up inside the digester and when it smashes against the far wall of the tank, the impact combines with the sudden drop in pressure to cause the softened chips to "explode" into mostly individual fibers.

The Process of Washing the Stock

From here the pulp is pumped to **brown stock washers** where nearly all of the noncellulose material is removed (see Figure 3–10). The salvageable cooking chemicals are reclaimed and used again while the noncellulose wood components (lignin, resins, etc.) are burned in the mill's boilers. After this cleaning process, the fibers that result constitute only about half of the original log that entered the mill. Although the percentages vary with the tree species and cooking method, a basic formula for the makeup of wood is 50 percent fiber, 30 percent lignin, 16 percent carbohydrates, and 4 percent proteins, resins, fats, and other impurities. Therefore, a clear trade-off must be made between quantity and quality when comparing groundwood and chemical pulping.

Chemical Recovery After Sulfate Pulping

After the chips have been pulped in the digester, the noncellulose materials are washed from the fibers and carried away from the brown stock washer. Because of its dark color, this collection of spent cooking chemicals as well as lignin, turpentine, and other organic materials from the wood is referred to as **black liquor.** To dispose of these chemicals would be economically wasteful as well as environmentally insensitive, so pulp mills have installed a very complex, but

Figure 3–9 The digester (medium gray tower) converts chips into individual fibers by heating them with water and a cooking liquor. *Courtesy of Hammermill Papers—Division of International Papers.*

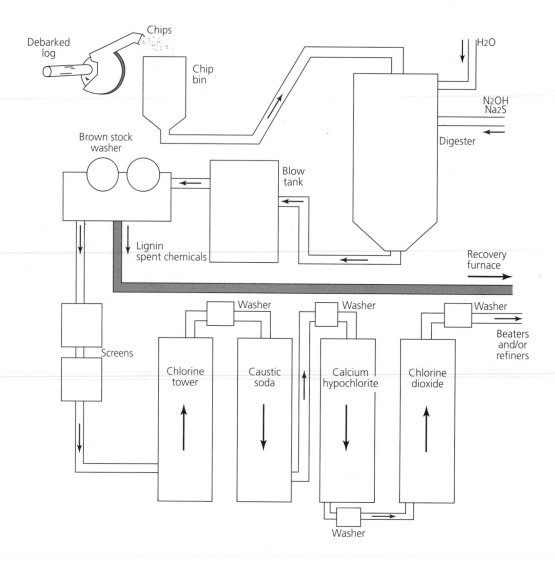

Figure 3–10 This flow chart traces the early stages of chemical pulping. Note how the brown stock washers separate the cellulose fibers from other materials. Also observe that the pulp is washed between each bleaching stage. From here, the pulp goes on to the beating/refining stage.

equally amazing, recovery system that recycles much of the cooking chemicals and burns the organic materials to produce energy (see Figure 3–11).

The first step is to thicken the black liquor in an **evaporator** by extracting enough water to increase its solids content from 16 to 65 percent so that it is ready to burn in the **recovery furnace.** In a smelting operation within the furnace, the lignin and other organic materials burn while the cooking chemicals accumulate at the bottom of the furnace. After the organic matter has burned away, the cooking chemicals— Na_2CO_3 (sodium carbonate), Na_2SO_4 (sodium sulfate), and

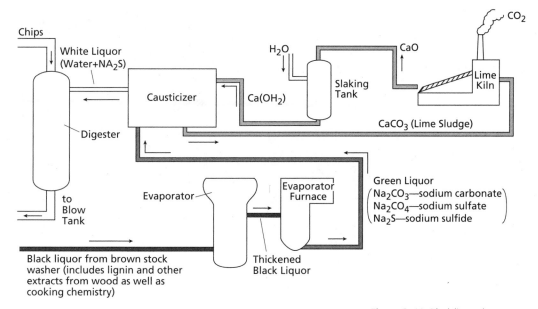

Chips

White Liquor
(Water+NA₂S)

Causticizer

Ca(OH₂)

H₂O

Slaking
Tank

CaO

Lime
Kiln

CO₂

Digester

CaCO₃ (Lime Sludge)

to
Blow
Tank

Evaporator

Evaporator
Furnace

Green Liquor
(Na₂CO₃—sodium carbonate)
Na₂CO₄—sodium sulfate
(Na₂S—sodium sulfide)

Black liquor from brown stock
washer (includes lignin and other
extracts from wood as well as
cooking chemistry)

Thickened
Black Liquor

Figure 3–11 Black liquor (spent cooking chemistry) that was separated from the pulp at the brown stock washer is carried through an elaborate process that recovers key chemistry to generate a new supply of white liquor.

Na_2S (sodium sulfide)—make a lighter slurry called **green liquor** that is sent to the **causticizer** where the Na_2CO_3 (sodium carbonate) is combined with $Ca(OH)_2$ (calcium hydroxide) to form a new supply of NaOH (sodium hydroxide or caustic soda) which is the primary cooking agent for the chips. Therefore, by adding calcium hydroxide to the spent cooking liquor, fresh cooking liquor is generated as demonstrated in the formula:

$$Ca(OH)_2 + Na_2CO_2 \rightarrow 2NaOH + CaCO_3$$

Observe that, in addition to the sodium hydroxide, the reaction also creates $CaCO_3$ (calcium carbonate or *lime sludge*) which precipitates out of the causticizer and is carried onto the **lime kiln** and burned to recover the calcium in the form of CaO (calcium oxide) and the by-product CO_2 (carbon dioxide) as shown in the formula:

$$CaCO_3 \rightarrow CaO + CO_2$$

The calcium oxide then moves to the **slaking tank** where it combines with water to form a new supply of $Ca(OH)_2$ (calcium hydroxide) as shown:

$$CaO + H_2O \rightarrow Ca(OH)_2$$

The calcium hydroxide, you will recall, is then used to react with the sodium carbonate in the green liquor to form a new supply of sodium hydroxide (*white liquor*) for the digester.

Figure 3–12 Giant tanks contain the proper concentrations of the four-stage bleaching chemistry used to whiten either virgin or recycled pulp. The actual chemistry and sequence vary from mill to mill. *Courtesy of Cross Pointe Corp.*

The bottom line of this intricate system is that the only wasted component of the black liquor is the carbon dioxide gas emitted from the lime kiln and some mills capture and use even that in their process.

Pulp Bleaching

Unless the paper being manufactured is designed for packaging (i.e., paper bags), the next step after the brown stock washer is **bleaching.** The brownish stain that remains on the fibers is evidence of lignin traces that must be removed before white paper can result.

Although papermaking has included a bleaching process of one means or another since ancient times, it was not until *chlorine* was discovered in the eighteenth century that this step became effective. Chlorine was first used in a single application during which the pulp was soaked in a chlorine-water solution. Even though this soaking adequately whitened the sheet, the oxidizing of the chlorine also severely weakened the fibers, thereby negating to some degree the main reason for chemical pulping in the first place.

In an effort to whiten the pulp without weakening it, the **multistage bleaching** process was developed in the 1930s (see Figure 3–12). More than one sequence is used today, but the most common procedure works as follows. The pulp/water mixture is pumped into a tank with *chlorine gas* which bonds with the traces of lignin to produce chlorolignin and an acid which is promptly washed away before it can harm the fibers. During the second stage, this chlorolignin is dissolved and washed away with *caustic soda* or more accurately sodium hydroxide, NaOH. The third stage sees the pulp brought into contact with *calcium hypochlorite*, $Ca(OCl)_2$, which combines with the sodium hydroxide to oxidize any remaining colored pigments. Finally, the nearly-white pulp is treated with chlorine dioxide to enhance brightness. Each bleaching stage is always immediately followed by a washing process to carry away noncellulose material and prevent excessive exposure of the pulp to the chemistry.

An exception to the practice of multistage bleaching is the handling of groundwood pulp. To reduce cost, a single application of **hydrogen peroxide** is made and then neutralized with no further bleaching. Although the hydrogen peroxide does an adequate bleaching job without weakening the fiber, groundwood paper (newsprint, for example) never attains the degree of whiteness as does, for example, a bond or offset sheet produced with multistage bleaching.

At this point in the papermaking process the bleached pulp may be dried for shipment elsewhere. This operation occurs in facilities called *pulp mills* which specialize in producing market pulp to be sold to paper mills that do not perform their own pulping. After bleaching, the pulp is dried to a moisture content of 15 percent or less and baled for transit. With *integrated mills,* where the pulping and papermaking are linked, the pulp is simply pumped onto the next department, the beating/refining stage.

Chlorine-Free Paper

After testing the pulp and wastewater at over 100 paper mills, the EPA announced in 1989 that chlorine bleaching of wood pulp generates 200 different chlorine compounds, one of which is 2,3,7,8 tetrachlorodibenzo-p-dioxin (TCDD) or, more commonly, **dioxin.** Sometimes called the Darth Vader of chemicals, dioxin was, in the early 1980s, considered to be the most powerful carcinogen ever tested and, because of its link to cancer in lab animals, even the minute presence of dioxin is still cause for concern, partly because little is known about how the compound works and what exposure levels are hazardous. However, while scientists continue to research the issues and debate their findings, the paper industry began a move to virtually eliminate the slight traces of dioxin presently in its paper products and wastewater (usually measured in parts per quadrillion). For some environmentalists, even the smallest of traces are unacceptable as reflected by three Greenpeace activists who, in July of 1994, scaled halfway up the 47-story Time-Life tower to hang a banner protesting *Time* magazine's use of chlorine-bleached paper.

The chlorine compounds that include dioxin are **absorbable organic halides** (AOX) which result from the combination of *chlorine* and *lignin*. Therefore, dioxin's presence in paper is attacked with a two-part approach—reduce both the lignin content of the pulp entering the bleaching stage and the amount of elemental chlorine used in bleaching. To reduce the amount of lignin leaving the digester, some mills have successfully modified the chemistry of their cooking liquor. After cooking, the pulp then goes through **oxygen delignification,** a process that can further reduce lignin content by 50 percent. However, the major effort in reducing AOX compounds is the replacement of the chlorine stages of the bleaching process with alternative chemistry. Chlorine can exist alone (Cl_2, known as *elemental chlorine*) or in compounds such as chlorine dioxide (ClO_2), but it is much more likely to produce AOX compounds in its elemental form. Mills that have replaced elemental chlorine with chlorine dioxide in the first bleaching stage report an 80 percent reduction in the AOX level of their pulp. International Paper reported in 1993 that its 11 U.S.-based bleached paper mills have reduced dioxin production by 95 percent over the past several years and estimates its annual collective dioxin discharge to be less than one ounce. This success has prompted the prediction that, by the year 2000, 60 percent of America's chemical pulp will be bleached without elemental chorine.

A few mills have gone so far as to eliminate chlorine entirely by replacing their chlorine dioxide and calcium hypochlorite stages with treatments of **ozone** or hydrogen peroxide, both of which are expected to gain wider acceptance within the industry. In fact, the 1989 volume of elemental chlorine used in pulp bleaching was reduced by over 37 percent in 1994. Major changes in the bleaching process will be forced upon pulp mills that do not virtually eliminate dioxin and other chlorolignin compounds from their wastewater when new Environmental Protection Agency standards are scheduled to go into effect in 1996. Chapter 16 looks at this controversy in more detail.

Paper that is produced without the use of chlorine in any form is referred to as **totally chlorine-free** or TCF. In contrast, paper bleached with chlorine dioxide or another chlorine compound instead of elemental chlorine is referred to as **elemental chlorine-free** or ECF. As of mid-1994, no more than three American mills were producing TCF paper, while ECF paper was much more common.

The reduction of lignin in prebleached pulp and the replacement of elemental chlorine will eliminate the dioxin issue from papermaking, but not without an impact on printers. First, these process changes cost between $20 and $50 million per pulp mill and are not offset by increased production or reduced operating costs. Therefore, TCF paper usually costs more than chlorine-bleached paper. Second, smaller mills may not be able to comply with the potentially stringent dioxin standards that may be imposed by the EPA, possibly eliminating certain sheets from the marketplace. Third, brightness levels generally decrease slightly when chlorine is eliminated from the bleaching process. For communication-grade papers, it is predicted that brightness levels may drop by 3–4 percentage points by the year 2000. In short, paying higher prices and choosing from a smaller selection of less bright papers is the worst case scenario that printers and their customers may face in order to be more environmentally responsible.

The Beating and Refining Stages

Chapter 1 explained that the ancient Chinese would beat their pulp with sticks in an effort to flatten the fibers and break up any remaining clumps. Today's **beating** operation performs these two functions and also serves to "fine-tune" the fibers so that they will produce paper with the desired physical characteristics. A key to this is the degree to which the fibers are roughed up and frayed, resulting in threadlike extensions called **fibrils** (see Figures 3–13, 3–14, and 3–15). The degree of fibrillation will influence the strength of the paper because the fibrils of different fibers interlock with one another, increasing the bond. Other characteristics that are affected by the amount of beating are bulk, opacity, and uniformity. Clearly, this stage of the operation plays a key role in determining many of the end product's characteristics. For example, blotter paper receives practically no beating, while heavy beating produces the strong and nearly transparent glassine. Chapter 6 will discuss this relationship further.

The appearance and basic function of today's beaters are similar to those developed in Holland during the seventeenth century; in fact, they are still referred to as *Hollander beaters*. Shaped like an oval tub, the beater circulates a pulp-in-water suspension between a paddlewheel-like device

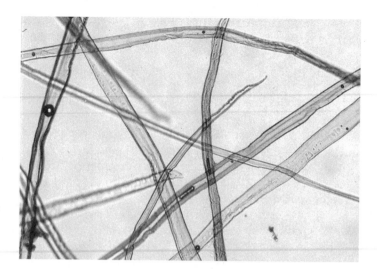

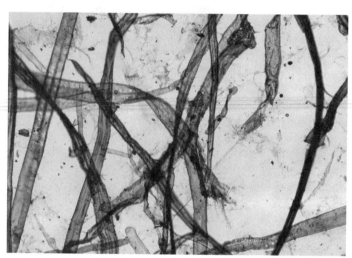

Figure 3–13 A microphotograph of unbeaten fibers reveals their pristine condition after chemical pulping (100x enlargement).

Figure 3–14 These fibers have received moderate beating to flatten them and create a mild degree of fibrillation (100x enlargement).

called a beater roll and raised steel knives. The proximity of the revolving beater roll to these knives, the time the pulp spends in the beater, and the speed of the beater roll can be altered to produce a paper with certain characteristics (see Figure 3–16).

Refiners, first developed in 1860, serve the same basic function as do beaters, but they work on a different principle. Instead of looking like a giant whirlpool tub, most refiners consist of a cone-shaped metal casing into which pulp is pumped (see Figure 3–17). The casing remains stationary while the pulp is caught between the casing and a revolving element called a *plug*. Fitting together something like one Dixie cup inside of another, the refiner beats the fiber with

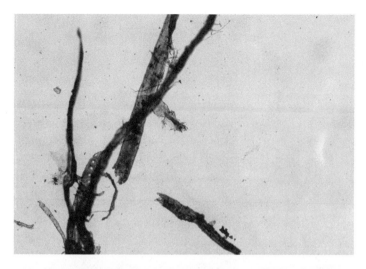

Figure 3–15 The extreme fibrillation and pronounced shortening of these heavily beaten fibers is evident (100x enlargement).

Figure 3–16 Recently cooked and bleached pulp floating in a Hollander beater resembles cottage cheese. The pulp will continue to circulate through the beater and between the beater bars until it receives the correct degree of refining.

the friction caused by the horizontal bars on both the casing and the rotating plug. Refining can be controlled by adjusting how far the plug penetrates the casing as well as the placement of the bars on the plug. **Jordan** and **Claflin** refiners are examples of this type. A second type is the **disk** refiner in which two disks rotate in opposite directions, with the grinding action of the bars performing the beating function.

Figure 3–17 · The Jordan refiner is conical in shape. The pulp is pumped in through pipe A and leaves through pipe B.

Whether conical or disk, refiners allow the papermaker to pump a steady stream of pulp in at one end and out at the other, thereby creating a continuous operation rather than the batch system required of the beater. Because of this increased efficiency, refiners are the favored method for "fine-tuning" the pulp.

Additives

There is at least one advantage to having a Hollander beater, however, and that concerns the introduction of **additives.** Because the beater is an open tub that holds a known amount of pulp, it serves as an easy and efficient place to measure out a certain amount of dye, adhesive, filler particles, or sizing, and also ensures that the additive will be evenly distributed throughout the pulp. These additives, like the beating and refining processes, play critical roles in producing paper with specific physical characteristics. *Starch* or other adhesives improve the fibers' bonding; *fillers* enhance the sheet's opacity; dyes provide the color; and *rosin* or other agents impart internal sizing to the fibers in an effort to improve the finished sheet's resistance to liquids.

Internal sizing coats individual fibers to reduce their natural absorbency. Paper that has not received internal sizing may pick up so much dampening solution from lithographic printing that the sheet's fiber-to-fiber bonds are impaired. Traditional internal sizing consists of coating the fibers with *rosin;* this technique requires the addition of aluminum sulfate particles to precipitate the rosin onto the fibers. The biggest drawback to rosin sizing is that it produces an acidic sheet that turns brittle over a few decades. Many mills have converted to **alkaline sizing,** which coats the fibers with a synthetic polymer that creates a sheet with more permanence. Chapter 6 discusses the roles of pulp additives in more detail. In all cases, these must be added before the pulp is formed into a sheet on the paper machine.

Final Cleaning

After the "recipe" for a particular grade of paper has been followed, the pulp goes through a final **cleaning** stage just before its formation into a sheet. At this point *centrifugal cleaners* and/or *screens* are used to isolate any dirt particles or clumps of fibers that may have eluded previous operations. Purity is critical now because the next step is the actual formation of the sheet.

The Paper Machine

Chapter 1 described the papermaking machine that was first used by Robert in 1800. Although today's paper machines are fascinating marvels of modern technology, their basic function has not changed. In fact, the most common machine operating today is still called a **Fourdrinier.** The pulp is held in reserve and distributed onto the wire by the headbox. Much technical development has gone into the **headbox** so that it can keep the fibers in a uniform suspension and maintain a constant supply of stock onto the wire; this latter task is made more difficult with high-speed machines that have a 24-foot wide wire traveling over 3,000 feet per minute. From the headbox, a jet of stock shoots out onto the full width of the screen through the *slice*—a control device aimed at ensuring good formation in the finished sheet (see Figure 3–18).

As the stock comes onto the screen, it is approximately 99 percent water, but the actual percentage varies with the desired weight of the paper being made. This ratio of water-to-

Figure 3–18 A sheet of paper is beginning to take form as it flows from the headbox on the right and moves on the wire to the left under the dandy roll of this small Fourdrinier machine.

Figure 3–19 The dandy roll contains a series of raised wire designs that impart the watermark to certain business papers.

fiber is instantly reduced, however, as the water pours down through the wire, leaving the fibers to mat together as they travel atop the screen. The rapid removal of water is further hastened by the rollers that support the wire. When these **table rollers** come in contact with the water falling through, they act to accelerate its removal. About two-thirds of the way down the length of the wire, the pulp passes over **suction boxes** that remove still more water so that within the two seconds it can take to traverse the length of the wire, the pulp's water content has been reduced from 99 to 75 percent. During this time, it has also passed under the **dandy roll**, which breaks up any air bubbles that could create a weak spot on the paper and also can serve to carry the design found on watermarked papers. These air bubbles can result from the spattering of pulp as it hits the wire. Watermarking is accomplished by embossing an image into the pulp while it is still mostly water, thereby forcing some fibers out of the way. Before the image area fully regains these displaced fibers, the sheet begins to solidify, creating the translucent quality of the watermark (see Figure 3–19).

Although still 75 percent water, the pulp is solid enough to leave the wire at the **couch** (pronounced KOOCH) **roll** and move on to the wet press stage for further water removal (see Figure 3–20). Even though it is now off the wire, the finished sheet will maintain some evidence of the two seconds spent moving across it. This structural difference is expressed in the terms **wire side** and **felt side** of a sheet, referring to the bottom and top sides, respectively. Because the

Figure 3–20 The large roller on the right is the couch roll, where the wire turns down and and heads back toward the headbox while the web of paper is lifted off and onto the felt that will support it through the wet press stage. *Courtesy of Cross Pointe Paper Corp.*

water drainage is always down, the wire side of the sheet loses significantly more fibers and filler particles than does the top side. Second, the subtle impression of the wire remains on one side of the sheet and can be discerned by the trained eye—especially on uncoated stock.

A different kind of effect—fiber alignment—is also imparted to the finished sheet during its time on the wire. Through a combination of the rapid forward motion and a sideways shake of the wire, the fibers tend to line up parallel to the direction of the wire's travel, thus producing the characteristic of **grain direction,** the predominant alignment of fibers.

Twin-Wire Machines

Today, most paper is formed on the Fourdrinier machines that have just been described. However, one limitation of these devices concerns how fast the wire can travel before problems develop with fiber formation. As the machine runs faster, supplying a uniform distribution of fibers onto the wire becomes difficult, resulting in clumps of unevenly dispersed fibers and poor formation. A device that overcomes this problem is the **twin-wire machine** (or *gap former*), which shoots the jet of pulp up between two wires that form the sheet. The use of two wires allows more control over the

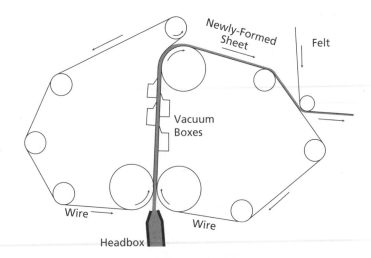

Newly-Formed Sheet

Felt

Vacuum Boxes

Wire

Wire

Headbox

Figure 3–21 The twin-wire machine shoots the stream of pulp vertically between two moving screens before transferring the sheet onto the felt. Because water is removed in two directions, the machine can run at a faster speed than a Fourdrinier machine.

Figure 3–22 This twin-wire paper machine is producing newsprint. At left are the hood-covered drier sections that are part of all paper machines. *Courtesy of Valmet Corporation and Steve Rabey.*

pulp and avoids the spattering effect associated with a Fourdrinier at very high speeds; also, water is vacuumed off from both sides (see Figures 3–21 and 3–22). These factors make twin-wire machines much faster than Fourdriniers and also produce paper without the "two-sided effect" because both sides of the sheet were formed against a wire. Although introduced only 30 years ago, twin-wire machines are commonly replacing Fourdriniers in paper mills around the country. A third type of paper machine is the **hybrid machine** which has a Fourdrinier section followed by a twin-wire section. Current technology allows a single high-speed machine to produce paper at the rate of 4,000 feet per minute and 1,000 tons a day.

Figure 3–23 Moving felts come together, sandwich the wet paper web between them, and squeeze out approximately 10 percent of its moisture. *Courtesy of Cross Pointe Paper Corp.*

Using Pressure to Remove Water

From the wire, the paper moves into the **wet press** section of the paper machine where the sheet is supported by felts and carried between rollers designed to exert pressure and force out more water (see Figure 3–23). A suction press is used to accelerate the water's removal. Two other functions of the wet press are (1) a smoothing effect because fibers riding high are eased down into the valleys and (2) a way to control the caliper or bulk of the finished sheet by adjusting the pressure exerted.

The Drier Section

Now down to 65 percent water, the paper web enters the **drier** section—the longest part of the paper machine. Here, the paper web weaves its way over and under two or three dozen steam-heated rollers that are around 5 feet in diameter. These rollers or "cans" bring the paper down to a 5 to 10 percent moisture level. Exacting control is crucial here because improper drying can create a host of irregularities within the finished sheet. To increase control, giant hoods are placed over the drying section, thereby reducing drafts and heat loss.

Not all paper is dried with a series of drier drums. Certain wrapping papers are given a glossy surface on one side by being dried against a single highly polished 12-foot drum called a **Yankee drier.** As the paper dries against the smooth surface of the drum, the smoothness is imparted to that side of the sheet. Tissue and creped paper achieve their texture with a Yankee drier used in conjunction with a doctor blade

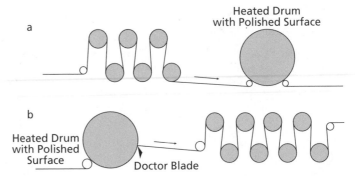

Figure 3–24 (a) Yankee drier is used with drum driers to impart a glossy surface to the sheet; (b) an incompletely dried paper being "creped" as it is scraped from a Yankee drier and dried further by drum driers.

Figure 3–25 An application of surface sizing is rolled onto paper between drier sections of the paper machine. *Courtesy of Cross Pointe Paper Corp.*

that scrapes the paper from the drum which is moving a little faster than the paper is being pulled by the subsequent drier drums (see Figure 3–24). This difference in speed determines the degree of crepe.

Surface Sizing the Paper

Paper intended for writing or printing usually receives a refinement called **surface sizing,** the application of a film to give the sheet a harder surface. The most common sizing agent is *starch,* which is applied after the first section of driers has brought the paper down to around a 10 percent moisture level. As it is pulled through the size press, the sheet receives a thin application of sizing on both sides which is made more uniform by rubber covered rollers (see Figure 3–25). After this step, the paper will stand up better to handling such as erasing, be less likely to have loose surface fibers, and—most important—be able to accept writing or

Figure 3-26 One of many scanning sensors found on a paper machine monitors the sheet as it moves through the various sections. *Courtesy of Valmet Corporation and Steve Rabey.*

printing ink without the image "feathering out" due to excessive absorbency.

A second set of driers withdraws the water that the sheet regained during sizing and brings it down to the exact moisture level prescribed for that run of paper. This range can vary from 1 percent to as high as 10 percent, depending upon the type of paper being produced and the subsequent operations that may follow.

Quality Control Scanners

During its trip through the paper machine, the paper passes through several sensing devices that scan and measure one or more of its physical properties and send these data to the computer console of the machine operators (see Figure 3–26). Moisture content and caliper are examples of properties that can be measured during the sheet's manufacture, while tear strength, opacity, and absorbency are properties that must be tested after the formed and dried paper has been removed from the machine. A giant roll of paper that is tested and found to be outside of mill specifications can be repulped.

Calendering

Before the sheet leaves the machine, it moves through **calender stacks** which compress it between cast iron rollers and smooth out the sheet to produce fewer lumps and reduce the variance between the *wire side* and *felt side* surfaces. From the calender stack, the sheet is wound into a roll or *machine log* that is the width of the Fourdrinier screen. As this reel fills, it will be taken off-line and a new reel started with no disruption of the machine's production.

Figure 3–27 A just-completed short log of paper is prepared for further finishing. *Courtesy of Cross Pointe Paper Corp.*

Finishing Operations

After a log of paper has been produced (see Figure 3–27), it is taken to a **rewinder** where it can be unwound and slit into several shorter rolls of various desired lengths. These shorter rolls may be ready at this point for shipment to web printers or these rolls may be ready for further finishing. One such process is **supercalendering.** Similar to the calendering operation that the paper received at the tail (or dry) end of the paper machine, supercalendering sends paper that has been coated between rollers that alternate between steel and a more resilient material—usually cotton (see Figure 3–28). The resilient cotton rolls are compressed by the steel and then expand back to their original diameter as a given area rolls beyond the contact point. The bump or nip in the resilient roll at the contact point results in a slight variance in the speed of the rolls, with the steel rollers going slower. Because the cotton rollers are not as slippery as the steel rollers, the paper moves at the same speed as the cotton rollers and, therefore, a little faster than the steel, thereby polishing the coating that is on the side of the sheet that is against them.

If both sides of the sheet require a gloss, the paper must be turned over and sent through again for the other side unless a double finishing supercalender machine is used. This device simply changes the ordering of the steel and resilient rollers at the midpoint of the stack so that both sides become glossy. It is this polishing of the sheet's coating that produces what is termed enamel or glossy paper.

Another finishing operation that some paper receives is **embossing.** Not unlike the process of embossing performed by printers using a die and counter, paper is given a textured

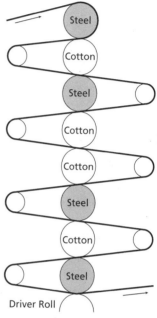

Figure 3–28 A double-finish super-calender machine (above) will polish both sides of the sheet because the two adjacent cotton rollers in the middle of the stack switch the side of the paper that is against the steel rollers; (left) a supercalender machine in the process of giving paper its gloss. *Courtesy of Valmet Corporation.*

surface on both sides as it passes between matching male and female rotary embossing rollers. When a less sharply-defined texture is desired, only one embossing roller is used in conjunction with a smooth-surfaced pressure roller. Embossing produces the interesting textures such as linen, stipple, coral, and emboweave.

To enhance brightness, gloss, and ink receptivity, paper is often coated with a thin layer of clay or another material. Several methods of producing coated paper exist and, for this reason, Chapter 4 examines the subject in depth.

Cockling is a finish given to certain bond papers so that they will resemble the handmade sheets of earlier times. Cockling is a texturing of the surface that results from the uneven drying that occurred when sheets were once hanged in a loft. Today cockling is created by running paper through a tub of starch or some other liquid, then removing only a

Figure 3–29 Carefully spaced slitter knives transform a log of paper into several smaller rolls that, after finishing, will either be wrapped as rolls for web presses or cut into individual sheets for sheet-fed presses. *Courtesy of Cross Pointe Paper Corp.*

Figure 3–30 Rolls of paper wait to be wrapped and shipped to printers with web-fed presses. *Courtesy of Cross Pointe Paper Corp.*

portion of the moisture with drum driers. Cockling is done "off-line" from the paper machine; it is an additional step and, like all finishing operations, adds to the cost of the paper.

Paper that will be used on a web press is shipped in rolls of various widths. Paper that will be run through sheet-fed presses, however, must next be **slit** (see Figure 3–29) into a certain dimension and then **sheeted** to the other dimension. This latter operation has traditionally been performed by a machine known as a *cutter*. As the web moves through the

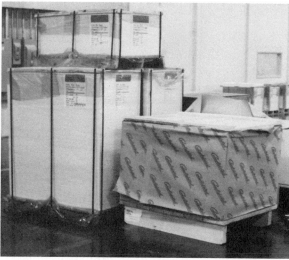

cutter, a revolving knife cuts the paper into sheets during each revolution on all four sides by a guillotine trimmer to ensure uniformity of size within the pile. A newer method of converting rolls into sheets is the rotary sheeter, which can produce individual sheets with such a degree of dimensional uniformity that no subsequent trimming is necessary.

Packaging Paper

Whether shipped out of the mill as rolls or sheets (see Figures 3–30 and 3–31) all paper must be wrapped to minimize damage from handling and variation in moisture content. The sides of rolls are protected by heavy moisture-proof paper **headers** that seal the edges and then similar material is wrapped around the circumference. Sheet-fed paper is often *ream* (500 sheets) wrapped in specially treated paper and then placed inside either a *junior carton,* which contains eight to ten reams or a *carton,* which is packed so that its weight usually will be between 120 and 150 pounds. If a large amount of paper is needed for a sheet-fed press, money can be saved by ordering it by the *skid,* a pile of paper loaded onto a pallet and wrapped as a unit, thereby avoiding the cost of ream and carton wrapping. It is becoming common practice for printers to order a smaller skid of the same size paper so that it can be loaded directly into the feeder system of the press.

The crucial role that moisture control plays in successful printing is examined in detail in Chapter 7, but let it suffice here to say that the precision that went into controlling the moisture content of the sheet throughout its formation, drying, and finishing can be negated if the paper is not protected during its transit, storage, and time spent in the pressroom.

The Productivity Challenge

At the turn of this century America was producing just over two million tons of paper and paperboard annually. By 1992, this figure had soared to 82 million tons distributed among numerous categories and grades (see Table 3–1). However, the growth in production has not equalled the rise in domestic consumption, thereby requiring that America import more paper than it exports. As Figure 3–32 reveals, the pattern of paper consumption is clearly one of continued growth. To keep pace with future demand, the industry is faced with the challenge of improving productivity while maintaining quality and conforming to increasingly stringent environmental regulation.

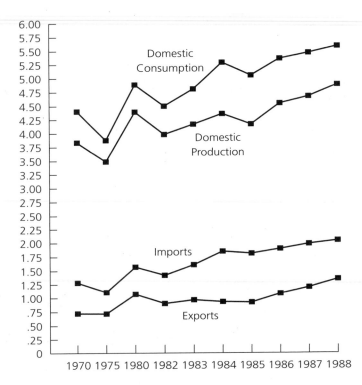

Figure 3–32 This chart tracks America's domestic production, consumption, tion, and foreign trade of pulp products (in million cubic feet) from 1970 to 1988. *Source:* Statistical Abstract of the United States 1992.

72

	Production in Tons	Percent of Total
Paper		
Newsprint	6,988,000	7.8
Uncoated groundwood	1,974,000	2.2
Coated	8,723,000	9.8
Uncoated free sheet	13,000,000	14.6
Other printing/writing	1,668,000	1.9
Packaging and other papers	4,725,000	5.3
Tissue	6,098,000	6.9
Subtotal	43,174,000	48.5
Paperboard		
Unbleached kraft board	22,515,000	25.3
Semichemical medium	5,963,000	6.7
Bleached paperboard	4,985,000	5.6
Recycled paperboard	12,302,000	13.8
Subtotal	45,765,000	51.4
TOTAL	88,939,000	99.9

Table 3–1 Over 88 million short tons (1 short ton = 2,000 pounds) of paper and paperboard were produced by American mills in 1994, a 4.7 percent increase over the previous year. *Source:* Pulp and Paper.

Questions for Study

1. Why must bark be removed from wood before pulping begins?
2. What are the three main means of pulping and how do they work? What are the advantages and/or disadvantages of each?
3. What is the main chemical used in the soda pulping process?
4. What is the role of sodium hydroxide in the sulfate method of pulping?
5. Why has the sulfite method of pulping largely been replaced by the sulfate method?
6. Explain the need for combinations of softwood and hardwood fiber in papermaking.
7. Trace the flow of pulp from the digester to the bleaching stage. Describe the need for each process.
8. Why does most paper receive multiple stages of bleaching?
9. List the reasons for beating pulp.
10. How do refiners perform their function?

11. Identify some common additives that go into paper.
12. Describe how paper is formed on a Fourdrinier machine; include terms such as headbox, slice, wire, suction boxes, dandy roll, and couch roll.
13. How is the function of the wet press section different from that of the dryer section?
14. At what point in the papermaking process is the paper surface sized? Calendered?
15. Explain why twin-wire machines are more productive than Fourdrinier machines.
16. How is supercalendering different from calendering?
17. Paper shipped in cartons or on skids is made for what type of printing press? What about paper shipped in rolls?

Key Words

debarking
paper mill
mechanical pulping
RMP
sulfate process
blow tank
recovery furnace
black liquor
multistage bleaching
AOX
TCF
fibrils
Claflin refiner
alkaline sizing
table rollers
couch roll
grain direction
Yankee drier
supercalendering
hybrid machine
chipper
integrated mill
chemical pulping

TMP
BCTMP
sulfite process
brown stock washers
green liquor
lime kiln
hydrogen peroxide
oxygen delignification
ECF
beating
disk refiner
headbox
suction boxes
wire side
twin-wire machine
surface sizing
embossing
SGW
pulp mill
lignin
PGW
digester
evaporator

white liquor
slaking tank
dioxin
ozone
refiners
Jordan refiners
additives
Fourdrinier
dandy roll
felt side
wet press
calender stacks
headers
causticizer
bleaching
internal sizing
cleaning
drier
rewinder
cockling
slit
sheeted

Coated Papers

AFTER STUDYING THIS CHAPTER, THE STUDENT SHOULD BE ABLE TO:

■ Explain the four primary reasons for coating paper.

■ Analyze the components of the mixture used to coat paper.

■ Explain the blade and air knife coating techniques and the properties that each imparts to the finished sheet.

■ Explain the cast coating process and how the process influences production costs.

■ Distinguish pressure, wash, and film coated sheets from sheets that qualify as coated paper.

■ Explain how the coating weight of a sheet affects various physical characteristics of that sheet, such as folding strength.

One legacy of the 1980s that should continue into the next century is an increased demand for printers to work with coated papers, a pattern reflected in the decade's paper production figures. From 1980 to 1990, America's overall paper and paperboard production increased 20 percent (68 million to 82 million tons). For the same period, the production of coated paper grew 43 percent, more than twice as fast. This trend increases the need for printers to understand that various coated sheets may be profoundly different in appearance, runnability, printability, strength, and cost. Printers also must be able to advise their customers on the implications of stock selection relative to what each customer is expecting from the printed sheet.

Several factors are involved in determining how a coated sheet will perform: (a) how the coating was applied to the sheet; (b) the size of the coating particles; (c) the amount of polishing; and (d) the amount and type of adhesive used. All of these factors vary among the various coated sheets available. Therefore, the ability of the printer to ask the right questions can save press time, reduce waste, and keep clients happy.

The development of paper coatings makes possible a number of printing applications that would not be possible otherwise, the use of fine-screen halftones being the most prominent.

The first evidence of coated paper appears in second century China. At this time the sheets had an application of a coating mixture of clay and water that was brushed on by hand. The coating of paper appears to have begun in Europe after an English patent was granted for the process in 1764. The coating mixture of plaster of Paris, stone lime, white lead, and water was applied with a brush—a method that continued until around 1850 when the demand for coated paper justified the development of machines to speed the process. The arrival of the *halftone plate* into the letterpress printing industry heightened the call for a smoother sheet and served, therefore, as the chief proponent for coating.

Why Paper Is Coated

Today, there are four primary reasons for coating paper. The first reason is to improve the sheet's **printability,** its ability to receive ink and create a faithful reproduction of the image. Coating the fibers with a layer of pigment provides a better surface on which to apply ink than exists with fibers alone. The ink will not spread out from the point where it is placed nearly so much if the paper has first been coated, providing, for example, a cleaner halftone dot. At the same time, ink will not soak into the sheet so much, but instead will stay up on the surface. This improved *ink holdout* provides a more vivid ink film and a printed job with more snap. As ink holdout improves, less ink is required because a smaller percentage of the applied ink soaks into the sheet and, in this sense, less is wasted.

The second reason for coating is to improve the sheet's **smoothness.** The application of the coating mixture, followed by a polishing action, reduces irregularities in the surface of the sheet as the peaks are polished down and valleys are filled in (see Figure 4–1). A more uniform or "level" surface is very desirable in the letterpress process which transfers its image by pressing the raised part of its plate against the paper. Although offset lithography, with its compressible blanket, does not require as much smoothness as does letterpress, its image transfer still benefits from a smoother surface.

Rough, uncoated paper will reflect light's rays but at various angles, thereby producing what is called **diffuse reflection.** As the surface is smoothed by applying coating and then polishing the sheet, the angles of reflection become more consistent and **specular reflection** is the result. The waxing of an automobile to get a shine is a similar process.

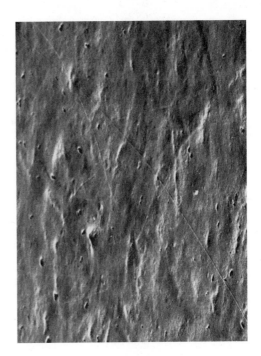

Figure 4–1 At left is a sheet of uncoated paper and at right is the same type of paper after being coated and supercalendered. Both photographs are 200-power magnifications. *Courtesy of Mead Central Research.*

This smoothness also results in the third reason for coating paper: it enhances the **gloss** level of a sheet. Gloss is a measurement of the amount of light that is reflected from a surface to a particular point (see Figure 4–2). The smoother the sheet is, the more consistently the surface will reflect light that strikes it from a point light source.

An analogy can also be drawn here to a baseball shortstop playing on different infields. As the ground balls coming toward him on a poorly kept field strike rocks and other irregularities, they ricochet in unpredictable directions, taking the proverbial "bad hop." In contrast is a well-maintained, and therefore smooth, field where the bounce is consistent. Just as a rocky, unkept field will send ground balls careening in a number of directions, so will the bumpy surface of paper fibers scatter light's rays in several directions. This dispersing of light keeps objects from being glossy, and at the same time it is the *consistent reflection* of light from a smooth surface that produces gloss.

A fourth reason for coating paper concerns the **brightness** of the sheet. Paper made from groundwood or other lightly bleached chemical pulp can be made much brighter by the application of a coating layer of white pigments. This procedure is more cost effective than heavy bleaching of the fiber and is commonly found in magazines.

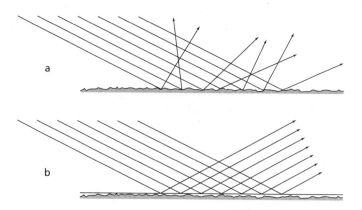

Figure 4–2 Illustration (a) is a representation of light striking a surface with poor microsmoothness. Notice that all of the reflected light takes the form of diffuse reflection. Compare this with the highly polished surface of illustration (b). Although no more light is being reflected, the smooth surface creates specular reflection and this consistency of reflection angle produces gloss.

The Components of Coating

An analysis of the coating found on this type of paper would reveal two main components—**pigments** and **adhesives.** Of these, the pigment material is the primary ingredient because it supplies the *color, smoothness,* and *ink holdout* to the paper.

The most common raw materials for coating pigment are *clays* and *calcium carbonate* (limestone). Both materials are readily available in nature and, when ground to very tiny particle size and bleached, they provide an inexpensive, but effective, smoothing and whitening agent.

The most popular clay for coating paper is *kaolin* clay which, in North America, is found in large amounts in a belt between Macon and Augusta, Georgia. When kaoline clay is heated properly, it takes the form of an ashy dust known as *calcined kaoline.* Because calcined kaoline's microscopic particles are platelike in shape, they are better than mere ground clay at reducing the passage of light through the sheet, thereby improving opacity.

Calcium carbonate is a plentiful natural resource and is also available as a by-product of certain methods of chemical pulping. Used much more sparingly than either clay or calcium carbonate is *titanium dioxide* which, when added to clay, increases the brightness and opacity of the coating; however, its high cost prevents the use of this synthetic pigment in anything but very small percentages.

No less important and, in fact, more expensive than the pigments are the coating's adhesives because they provide the pigment-to-paper bond. There are three main categories of sources for adhesives—*starches, proteins,* and *synthetics.* With corn as its chief source, starch is a carbohydrate and has been one of the most common bonding agents in the paper industry. Starch is very effective in letterpress printing; however, because of the role that moisture plays in lithography, a water-soluble adhesive such as starch is inappropriate.

The advent of the offset printing industry created a demand for coating adhesives that were not water-soluble; fitting this bill were the proteins. Animal glue was a very popular source in the 1800s but around the turn of the century it was used less and less and is now mostly limited to playing cards and wallpaper. Primarily derived from skim milk and soy beans, today's protein adhesives are strong, compatible with offset presses, and do not depend on animals as a source.

Synthetic adhesives were largely developed after World War II and include latex, vinyls, acrylics, and man-made rubber. Their strong points are good gloss and ease of distribution throughout the mixture. These synthetics can lend themselves to emulsion-type coatings which, due to their high solids content, have less water and require less drying time as part of their application. A high solid content also means a greater application of coating on the finished sheet.

In addition to pigment and adhesives, several other **additives** are included in the coating mixture. These additives improve the uniform distribution of components, reduce foaming, and further increase the coating's resistance to the moisture of an offset press.

The Effects of the Adhesive Level

The proportion of adhesive in the coating mixture is determined in part by the printing process for which the sheet is being produced. Of the primary printing processes, lithography requires the greatest surface strength because of the high-tack ink that is involved. Letterpress is a distant second and gravure is the least demanding. Paper can be produced with an adequate adhesive level to withstand the pull of lithography's high-tack (sticky) inks, but, as the surface strength is improved by increasing the amount of adhesive, there are necessarily several side effects.

One of these side effects concerns the *gloss* of the sheet. An increase in the amount of adhesive naturally means a proportional decrease in pigment and it is the pigment particles that the supercalendering polishes into a gloss. Neither proteins nor starches polish as well as the pigments that they replace, so the gloss level is reduced. An exception to this can be certain latex adhesives that soften due to the friction in the supercalender stacks and polish along with the pigments.

The amount of adhesive also affects the sheet's opacity and brightness. The adhesive fills in the air pockets between the particles of pigment, so there are fewer interfaces to scatter the light, resulting in greater transmission of light and reduced opacity. Because less light is reflected, brightness suffers as well. More adhesive also increases ink holdout and increases drying time. Unless special synthetic adhesives are used, this lowered porosity retards the escape of steam during the use of heat-set inks, with blistering being a danger.

Another by-product of increasing the coating adhesive is that the brittle nature of most adhesives will be imparted to the entire sheet, resulting in a tendency to crack at the fold. This hardness will also increase the sheet's stiffness. The exception here is the group of soft latex adhesives mentioned in the discussion on gloss that actually soften the sheet, improve its strength at the fold, and reduce the sheet's stiffness.

To summarize, increasing the percentage of adhesive in the coating will produce a sheet with (a) greater surface strength, (b) lower gloss, (c) less opacity and brightness, (d) increased brittleness, (e) lower fold strength, and (f) possibly impaired printability. All of these result because an increase in adhesive necessarily comes at the expense of pigment. Clearly, the ratio between pigment and adhesive is a delicate matter because it impacts upon the finished sheet's appearance and performance.

Various Methods of Applying Coating

Not only does the "recipe" of the coating have a multifaceted effect on the sheet, but so does its method of application. Because paper was first coated in second century China, a host of procedures have been employed to improve the speed of the application and quality of the product. Since coating was originally brushed onto the paper by hand, increasing the speed of application was not difficult. The arrival of the letterpress halftone plate spurred demand for coated paper, so machines that applied a coating onto a continuously moving roll of paper and then used brushes to spread it out were quick to become popular. Beginning around the 1920s, a number of other methods for automated coating applications were developed, each having advantages in producing a certain type of sheet. These developments have continued, resulting in several methods that are currently in use for the

Figure 4–3 A trailing blade coater uses a steel blade to smooth the sheet's coating layer and, along with supercalendering, create high gloss. *Courtesy of Valmet Corporation.*

production of coated papers, and each method produces a somewhat different sheet.

Buyers of coated paper today find themselves inundated with information promoting a given sheet, data on its creation, and claims for its performance. Terms such as pressure coated, blade coated, brilliance, and ink holdout are commonly encountered. The uninformed buyer, therefore, cannot know which sheet is best for a given application or what kind of performance can be expected.

Two methods that are popular today are the *blade* and the *air knife* coaters. Each produces a unique product and the end use of the sheet determines which is preferable. Developed in 1950, the **blade coater** resembles a rotogravure printing press because after the coating is applied to the sheet, a flexible steel blade scrapes off the excess (see Figure 4–3). The scraping action has a tendency to produce a very smooth sheet because coating is left in the valleys of the sheet's surface (see Figure 4–4a). It should be noted, however, that comparatively little coating is left on the peaks and this lack of uniformity can create an inconsistent ink film during printing.

Developed in 1930, the **air knife coater** rolls the coating onto the sheet and then directs a stream of air against it to remove the excess and smooth out the coating. The trade-off of this process is that, by being uniform in thickness, the

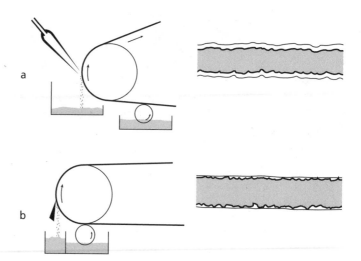

Figure 4–4 Illustration (a) shows how the air knife coater rolls the coating mixture onto the paper before the excess is blown off by the air blast. Also shown is a representation of the dried coating layer on the sheet. Observe the uniform thickness of the coating. Illustration (b) reveals the trailing blade coater's operation and the type of coating layer that results. Note the leveling effect as well as the inconsistency of the coating thickness.

coating conforms to the original sheet's surface contour and does not fill in the valleys; smoothness is thereby sacrificed (refer back to Figure 4–4b). This uniformity of the coating layer is a prime concern to the printer in a couple of instances. One of these is a printing job requiring large areas of solid ink coverage. Because the coating thickness is more uniform across the sheet, the ink holdout is also more uniform. A second application of the air knife coater is the carbonless forms industry (NCR paper) in which the coating contains the dyes that produce the copy's image. Here, a variation in the thickness of the coating would result in poor copies.

As the reader may surmise by now, each method produces a desirable effect and, when used in *combination*, the advantages of both are realized. For example, when paper is first blade coated to fill in the valleys of the surface and then air knife coated to produce a uniform layer, a sheet that has excellent smoothness plus good printability can be the result (see Figure 4–5). Of course, the extra production time increases the sheet's cost.

An even smoother, and therefore glossier, sheet can be produced by a slow, expensive process known as **cast coating.** Developed in 1937, this method produces such popular brands as Kromekote and Mark I. Cast coating bears little resemblance to the other coating methods and works more like a photographer's drum dryer for producing glossy prints. After the coating is applied, the paper moves in contact with a highly polished chrome-plated drum that revolves slowly. Because the drum is hot, the coating is cast

Figure 4–5 A close examination of this cross section photograph of a double-coated sheet of paper reveals the two layers of clay. *Courtesy of Mead Central Research.*

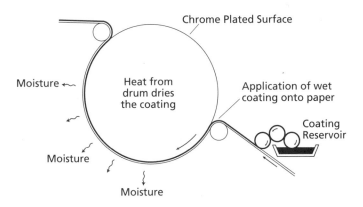

Figure 4–6 The cast coater is a unique coating process. The coating is rolled onto the top surface of the paper and the web moves onto a slowly revolving, heated chrome-plated drum. A casting of the drum's surface is made onto the coating's surface so it becomes as smooth as the chrome surface itself.

against its polished surface; when the sheet is taken off the drum, one is as smooth as the other (see Figure 4–6). The slow production speed makes cast coated paper expensive; however, the unmatched printing quality makes it popular in annual reports and other publications in which appearance is a priority.

In addition to these three common methods of applying coating, there are others less frequently used. Developed in the 1930s and at one time very popular, the **roll coater** is used primarily today to lay down a base layer for another coating method to coat over. This process rolls the coating onto both sides of the sheet at the same time and the splitting action of the coating layer as the sheet moves away from contact with the roller creates a lumpy pile pattern upon drying (see Figure 4–7). Although polishing the sheet helps reduce the pile, a subsequent pass through a blade coater is the best remedy. Also, rarely used today are the **brush** and **spread shaft** methods. In the first, brushes are used to spread out the coating, not unlike the way a fine

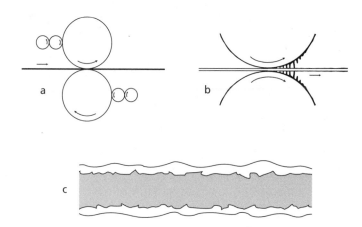

Figure 4–7 The roll coater (a) applies a coating layer to both sides of the sheet simultaneously. Illustration (b) shows how the splitting of the coating mixture creates the thick layer and heavy piling that is seen in illustration (c). This pile pattern is the main reason that the roll coater today is mostly used in laying down a base coating for a second method (usually trailing blade) to cover later.

paintbrush uniformly spreads its medium. The spread shaft coater applies coating to the sheet and then sends it against steel shafts that are revolving in the opposite direction of the web's travel to distribute the mixture. Although the product quality of this technique is excellent, both methods have been replaced by faster processes.

We have taken a look at the most common ways to apply coating to a basestock and noted the characteristics of their products. To summarize, the air knife coater's application of a uniform coating layer make it well-suited for printing applications involving large areas of solid ink due to the uniformity of the sheet's ink holdout. Although blade coaters put down a layer of uneven thickness, their filling-in of the surface's low places produces a sheet of excellent smoothness. The roll coaster can apply a large amount of pigment onto the basesheet, but it leaves a bumpy texture upon drying.

By combining the blade method with one of the other two, the qualities of each can be achieved. Some mills also get excellent results from multiple applications with the same coating method. For example, a mill in Scotland produces a premium sheet that has been blade coated three times. Because of the number of coating operations involved, paper with multiple coating layers can be expected to cost more than those with a single layer. An exception to the rule is cast coated paper, a very desirable sheet because of its high gloss and printability. Its high cost is due to the slow speed of its manufacturing process.

It should be noted that some papers are described as being *wash coated, film coated,* or *pressure coated.* These sheets have merely been sized with a mixture that contains a small amount of coating pigment and cannot be considered true coated sheets. For a sheet to qualify as coated, the weight of the coating layer needs to be at least 50 percent pigment material.

The Effects of Coating Weight

There are several other considerations when buying coated paper. One of these is the displacement by weight of paper fiber by the coating itself. This means that if a 70-pound (Sub. 70) sheet of offset paper carries 10 pounds of coating on each side, then only 50 pounds of the sheet is actually cellulose fiber, the material that provides the sheet's strength. Hence, the more coating that is applied to improve gloss, ink holdout, and smoothness, the weaker the sheet will necessarily become, resulting in possible web breaks, poor folding strength, or low burst strength. Also, because coating pigment is heavier than the same volume of fiber, a coated sheet has less caliper than an uncoated sheet of the same basis weight.

Summary

To review the makeup of the coating mixture itself, more adhesive (at the expense of pigment) will improve the coating-to-base-sheet bond, but will sacrifice some brilliance, opacity, folding quality, and smoothness. Conversely, increasing pigment and reducing adhesive will produce a sheet with a greater tendency toward surface pick due to the pull of lithographic ink.

The best approach to this series of trade-offs is to be aware of the nature of the printing job and the papers under consideration. The old adage that asserts "there is no such thing as either bad paper or bad ink, but only poor combinations of the two" appears to have merit where coated papers are involved.

Questions for Study

1. What technological development in printing increased the need for coated paper?
2. How does coating improve a sheet's printability?
3. What is the difference between diffuse and specular reflection?
4. Gloss is the result of what condition?
5. List materials that are commonly used as coating pigments.
6. What are the advantages and/or disadvantages of each type of coating adhesive?

7. How does the proportion of coating adhesive affect a sheet's gloss? Its surface strength? Its capacity?
8. For what type of paper is the blade coater well suited? What about the air knife coater?
9. How is cast coating different from the blade and air knife coating processes? Why is it more expensive?
10. Why is roll coating seldom used as the only method of coating a sheet of paper today?
11. Why is it that sheets identified as being wash coated, film coated, or pressure coated do not qualify as true coated papers?
12. How would a 50-pound (Sub. 50) coated sheet be different from a 50-pound uncoated sheet relative to strength, all other factors being equal?

Key Words

diffuse reflection	cast coating	printability
adhesive	brush coating	smoothness
air knife coater	pigment	gloss
roll coater	blade coater	brightness
specular reflection	spread shaft coater	additives

Recycled Paper

5

AFTER STUDYING THIS CHAPTER, THE STUDENT SHOULD BE ABLE TO:

■ List and evaluate the various rationales for paper recycling.

■ Cite examples of governmental encouragement of recycling at national, state, and local levels.

■ Explain the four criteria for assessing the practicality of recycling a given material.

■ List and define the categories of secondary fiber.

■ Distinguish between preconsumer and postconsumer waste.

■ Explain the economic factors that influence the market value of waste paper.

■ List and explain the stages involved in producing recycled paper.

The arrival of the Industrial Revolution changed not only the way societies *produced* goods, but it also *created new viewpoints and priorities* for various sectors of society. For the manufacturer, the efficient production of marketable goods required the creation of a demand of these goods. Generally, consumers eagerly embraced products that brought increased convenience to their lives. This partnership between producers and consumers has produced many desirable outcomes, such as growing economies for growing populations and an improved standard of living for each succeeding generation. However, at its worst, a modern society's narrowed focus on inexpensive and disposable goods causes it to lose sight of potential long-term effects. This myopia usually lasts until that society comes face-to-face with a side effect so distasteful that convenience and productivity suddenly seem less necessary.

In the case of the paper industry, the consumers' pursuit of convenience has resulted in a steady rise in the demand of pulpwood products. From 1920 to 1980, the per capita consumption of paper and paperboard grew from 145 to 650 pounds, an increase of 348 percent. This growth in consumption was naturally accompanied by huge growths at both ends of the producer/consumer chain: the volume of needed *raw material* and the volume of *postconsumer waste*. To supply this raw material, vast and ever-increasing amounts

Method of Solid Waste Disposal	Million Tons	Percent
Waste that is recycled	17	11
Waste incinerated to generate energy	10	6
Waste simply incinerated	5	3
Waste buried	126	80
Total	158	100

Table 5–1 In 1986, 7 times as much of America's solid waste stream was buried in landfills than was recycled. *Source: U.S. Environmental Protection Agency.*

of timber land have been required to feed the demands of American business for paper goods. For example, the annual need for paper products of only one fast-food chain has been equated to 315 square miles of forestland; a single Sunday edition of a nationally read newspaper could require up to 175 acres of trees.

At the consumer end, America's population currently generates over 500,000 tons of trash every day, creating an increasingly burdensome problem of what to do with it all (see Table 5–1). Until the last two decades, only a small percentage of Americans gave any thought to either problem. Nonetheless, awareness gradually grew during the 1970s and 1980s. As a result, most Americans have heard the ongoing discussions about timber harvesting, landfilling, and the need to recycle paper products. In fact, from changes in how Americans go about the ritual of disposing of their trash, both at home and at the office, to the increased governmental and consumer-driven mandates for paper products that include reclaimed fiber, the recycling movement has become firmly entrenched in our society.

A Historical Perspective

Although recycling seems like a new and very contemporary activity, the reclamation of discarded materials as a source of paper fiber has been practiced for centuries. The first chapter of this book revealed that the Japanese were the first to recycle paper when they augmented their supply of wood pulp with old papers, fishnets, and hemp as far back as 1035 A.D. Centuries later the Muslims reinvented the procedure. Later still, the practice of reclaiming discarded sources of cellulose fiber was employed throughout Europe and America. In fact, old rags were once the primary source of fiber for papermaking.

However, the emergence of the groundwood process in 1841 was the technological breakthrough that made the pulping of mature trees both possible and economically desirable. The subsequent arrival of chemical pulping further accelerated the emergence of forests as the primary source of cellulose fiber. In America and Canada 50 years ago, forestland and vacant acreage for dumping appeared to be inexhaustible, so concerns over fiber conservation and solid waste reduction were secondary to the demands of a growing nation's economy of consumption. The only exception to this pattern occurred during World War II when domestic paper recycling reached its all-time high of 35.3 percent reclamation, but with the end of the war Americans were quick to shed this level of austerity.

The economic boom of the postwar era once again pushed aside concern for conservation in favor of plentiful and inexpensive paper products. It was not until the 1970s that industries that were dependent upon abundant supplies of paper and paper products began to feel the pinch of curtailed production from America's paper mills. For many Americans, the paper shortages of the 1970s were a "wake-up call" that an endless supply of paper could no longer be taken for granted.

The Rationale for Recycling

The *recycling logo* that has become so common in modern society was introduced in 1971. During the decades of the 1970s and 1980s, conservationists, governmental agencies, and other groups concerned with the increased recycling of waste materials promoted the reclamation of paper as a means of *saving trees*. In fact, it was said that "recycling 120 pounds of paper can save a 500 pound tree." The truth is, however, that the depletion of America's forestland was never a serious threat. Because of the nation's prudent management of its timber lands—both public and private-growth has actually exceeded harvest by 37 percent. In 1990 alone, for example, over 1.9 billion tree seedlings were planted—nearly half of these by the forest industry. With this rate of planting coupled with the advances in silviculture discussed in Chapter 2, America's supply of pulpwood has never been in jeopardy.

The link between recycling and tree conservation was made in the 1970s and 1980s because such a case could be understood by most Americans. Actually, the need to recycle had been created by a much less romantic issue—the disposal of the 500,000 tons of trash generated in America ev-

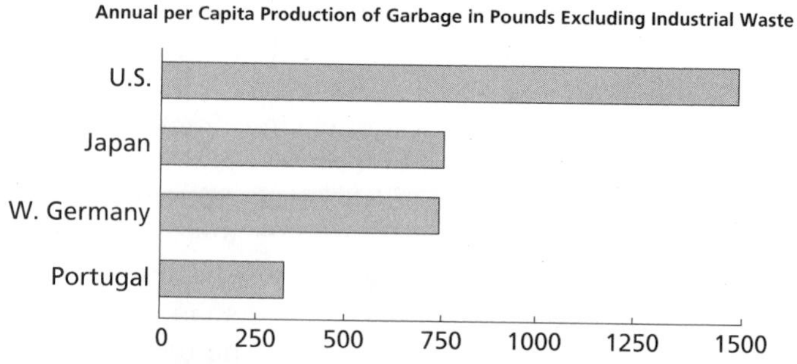

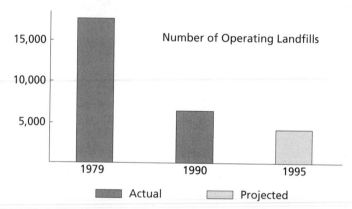

Annual per Capita Production of Garbage in Pounds Excluding Industrial Waste

Figure 5–1 The average American generates over twice as much garbage as the per capita average of Japan and West Germany, both highly industrialized nations. *Source: National Solid Wastes Management Association.*

Number of Operating Landfills

Figure 5–2 The number of this nation's operating landfills is dwindling at an alarming rate. This pattern is expected to continue, leaving local governments with fewer options for waste disposal.

ery day. Although the creation of waste is a global problem, the United States generates one-half of the world's solid and industrial waste and 95 percent of its hazardous and special wastes. In fact, the average American generates two to three times more waste than people in other industrialized nations with a comparable standard of living; ours is indeed the "throw away society" (see Figure 5–1). America's approach to processing its waste is also atypical when compared with other industrialized countries. For example, Japan recycles 40 percent of its solid waste, as compared with America's 10 percent; Japan buries only 27 percent of its solid waste, while America buries 80 percent.

Even if the practice of burying 80 percent of the nation's solid waste were the best option, the practice cannot continue for two simple reasons: (1) America's landfills are quickly reaching capacity; and (2) concerns about water table pollution are causing communities to prevent the opening of new landfills anywhere nearby (see Figure 5–2). Also, with landfill space becoming increasingly scarce, tipping fees for municipalities and private enterprises to dump their wastes are rising and straining budgets. From 1988 to 1991, the national average tipping fee increased 17 percent to

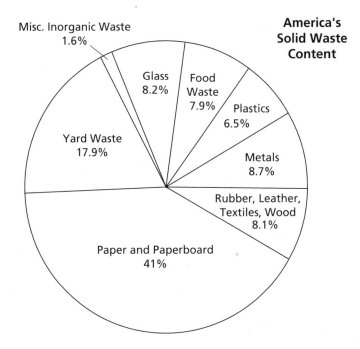

America's Solid Waste Content

Misc. Inorganic Waste 1.6%

Glass 8.2%

Food Waste 7.9%

Plastics 6.5%

Yard Waste 17.9%

Metals 8.7%

Rubber, Leather, Textiles, Wood 8.1%

Paper and Paperboard 41%

Figure 5–3 Examination of this pie chart reveals the potential impact that paper recycling can have on the reduction of the solid waste stream. *Source: U.S. Environmental Protection Agency.*

$26.50 per ton. Because the Center for Biology of Natural Systems estimates that 84 percent of municipal trash that is currently being hauled to landfills could be recycled, it is obvious that a key to solving the landfill problem is to transform as much trash as is possible into a useful raw material (see Figure 5–3 for the composition of America's solid waste stream).

In addition to forest conservation and solid waste reduction, a third motive for recycling is to reduce America's importation of paper and pulp, which in 1991 amounted to 17,297,702 tons. Regrettably, much of this volume comes from developing nations that are less committed to forest preservation than is the United States. For example, in 1989 the United States imported 400,000 tons of paper from Brazil, a nation that during the late 1980s was destroying its rain forests at an estimated annual rate of 23,000 square miles.

Three more motives for increasing paper recycling are to conserve energy, reduce pollution, and save jobs. An examination of Table 5–2 reveals how producing a metric ton of newsprint from old newspapers (ONP) instead of from virgin fiber requires less than half the expenditure for power, thereby conserving energy.

From the standpoint of pollution reduction, the primary alternative to recycling or landfilling solid waste is incinera-

Cost Per Short Ton of Newsprint		
	Virgin Pulp	ONP
Raw material	$ 60	$ 83
Power	$143	$ 60
Other process costs	$149	$144
Total cost	$352	$287

Table 5–2 A comparison of operating costs between producing a short ton of newsprint from virgin fiber and from old newspapers. The above data do not include investment-related costs such as constructing the facilities. Observe that virgin fiber, although less expensive than de-inked recycled fiber, is more expensive to process and ultimately more costly. A short ton is 2,000 pounds and should be distinguished from a metric ton, which is 2,204.6 pounds. *Source: Andover International Associates.*

tion. Japan, for example, burns 22 percent more trash than it buries. Several American municipalities and private firms have followed Japan's lead and built waste-to-energy incinerators. However, in 1987 the Environmental Protection Agency reported that incineration plants can produce hazardous air emissions, so the comparative value of this means of dealing with solid waste is uncertain. The final rationale for recycling argues that reclaiming solid waste creates approximately 36 jobs for every 10,000 tons of material recycled, compared to the six jobs required to dispose of it.

Governmental Involvement

A desire to encourage paper recycling has prompted governmental action at the federal, state, and local levels. In 1976, Congress passed the Resource Conservation and Recovery Act to encourage the use of recycled paper. In 1991, the EPA established guidelines for the recycling of wastepaper and standards for minimum content of reclaimed fiber in paper grades. The American Forest and Paper Association (AFPA) also established and reached a 1995 target date for the reclamation of 40 percent of America's wastepaper products. Had this goal not been reached, the law would have mandated individual targets for specific categories of paper. At present there are no federal requirements, only guidelines for the percentage of recycled fiber (see Figure 5–4).

Legislation has also been enacted at the state level to increase both the supply of and demand for secondary fiber. Oregon was first to enact legislation, with a 1983 law that requires cities with populations over 4,000 to provide curbside collection of separated materials. By 1992, all 50 states

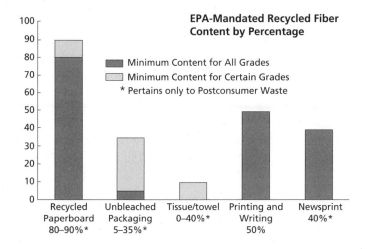

Figure 5–4 Guidelines for minimum recycled fiber content for paper purchased by governmental agencies have been established by the Environmental Protection Agency. Where a range (e.g. 5–33 percent) exists, different grades within that classification have different requirements.

had followed suit and established some form of regulation to encourage paper recycling. To increase the demand, some state governments have mandated that their agencies show preference to paper stocks made with recycled fiber. Connecticut has gone so far as to require that, by 1998, all newspapers sold in that state contain 90 percent recycled fiber—even newspapers published elsewhere, such as the *New York Times*.

At the local level, cities around the nation have sought to slash the volume of solid waste that they must pay to bury. Already mentioned, dozens of cities have established waste-to-energy incinerators, but hundreds more have opted for the less expensive strategy of the curbside collection of separated-by-type waste materials such as paper, glass, and plastic. In 1989, Los Angeles began to collect paper from its city hall offices and three private office buildings for recycling. At the time, Los Angeles was generating nearly 4,000 tons of office wastepaper daily.

Criteria for Acceptable Secondary Fiber

As desirable as is the reclamation of used paper in the effort to reduce the solid waste stream, not all materials, not even all types of paper, are appropriate for recycling. In a free market economy, the practicality of using secondary fiber as a raw material in the production of recycled paper is determined by four basic factors: *mass, degree of contamination, homogeneity,* and *location.*

For a material to have **mass,** it simply cannot be scattered across a wide area. Therefore, old newspapers do not begin to take on a significant value until they have been collected from hundreds of thousands of homes and accumulated into bales. For this reason, successful waste paper programs require that either volunteers bring their supplies to a central location or that a curbside collection service be initiated.

A second key factor is **degree of contamination:** staples, heavy clay content, varnishes, adhesives, ultraviolet inks, and the plastics found in envelope windows head a long list of substances that can make a load of waste paper more difficult, and therefore impractical, to process. When a shipment of waste paper arrives at a mill, it is inspected for contaminants and their presence can cause some or all of the shipment to be rejected.

Homogeneity measures the degree to which a volume of wastepaper is uniform in type. Although mixed paper lots are not as highly valued as a homogeneous shipment, some mills use them in the production of corrugated paperboard, tissue, and other products that emphasize bulk over appearance. Until very recently, wastepaper carrying the ink/toner of photocopying machines was impractical to recycle because the toner particles that are actually fused to the paper's fibers could not be separated. However, technological breakthroughs are making this common type of office-produced wastepaper a more viable source of fiber.

The final factor, **location,** measures the distance between the raw material and the processing plant. For logical reasons, pulping mills that depend on virgin fiber are usually located near forests, while mills that rely on secondary fiber are generally built near large population centers.

When these four criteria are considered, not all components of the nation's volume of solid waste are good candidates for recycling. Still, there are several paper products in the waste stream that are profitably reclaimed, and that list continually grows longer as ongoing research succeeds in overcoming problems in collection, transportation, cleaning, de-inking, and finding an appropriate end product.

Sources of Secondary Fiber

Several paper and paperboard products at various stages of production and utilization are used as sources of secondary fiber (see Figure 5–5). One of the earliest sources has been paper that never made it out of the mill. Referred to as **pulp substitute,** this paper may be stock that is below grade specifications, was damaged at some point, represents old inventory, or is trim waste. Still another source of pulp substitute is *broke*, the incompletely dried stock that results from a web break on the paper machine. Other sources of pulp substi-

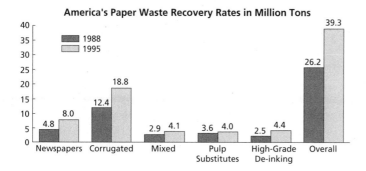

America's Paper Waste Recovery Rates in Million Tons

Legend: 1988, 1995

Category	1988	1995
Newspapers	4.8	8.0
Corrugated	12.4	18.8
Mixed	2.9	4.1
Pulp Substitutes	3.6	4.0
High-Grade De-inking	2.5	4.4
Overall	26.2	39.3

Figure 5–5 This bar chart shows paper recovered from the waste stream in 1988 for five categories. Also shown is a projection of recovery rates for the year 1995. Corrugated products have a high recovery rate because they accumulate at retail firms and warehouses and, therefore, are easier to collect than other paper products.

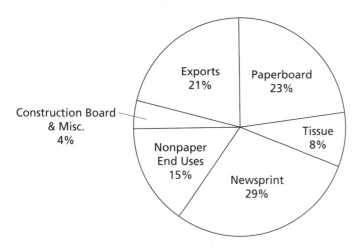

Uses for Recycled Newsprint

- Paperboard 23%
- Tissue 8%
- Newsprint 29%
- Nonpaper End Uses 15%
- Construction Board & Misc. 4%
- Exports 21%

Figure 5–6 Of the nearly 5 million tons of newsprint recovered in this country in 1990, over half went into the production of newsprint and paperboard. Roughly one-fifth was exported.

tute include scrap created by stamping out envelopes and the unprinted trim waste of printers. Because pulp substitute was never printed, it requires no de-inking and is the most highly valued source of secondary fiber. It is used to make fine printing and writing papers as well as tissues.

A second category of wastepaper is **high-grade de-inking** paper, which includes green-bar computer paper, printed scrap, sorted white office papers, and old textbooks. Although this wastepaper requires the removal of ink, the absence of clay coating and problematic inks and varnishes makes it an acceptable addition to the pulp mix in the production of fine papers and tissues.

Old newspapers (ONP) comprise a third category of wastepaper and are used in the manufacture of newsprint, paperboard, and tissue (see Figure 5–6). Like high-grade de-inking paper, ONP is also free of clays, varnishes, and a wide range of inks, and therefore is a very useful source of fiber. Ironically, however, the repulping potential of ONP is far from being realized. In fact, in 1989, 58 percent of America's total supply of newsprint was being imported, while 68 percent of its waste newspaper was either buried or incinerated. Puzzling circumstances such as this are contributing to the growth of governmental action to stimulate ONP recycling.

Old corrugated paperboard is another excellent source of secondary fiber because concentrations of it can be found at retail and wholesale establishments and it contains comparatively few contaminants. In 1991, 57.2 percent of corrugated waste was recovered and predominately used to produce new corrugated paperboard, linerboard, and kraft towels.

Still another category of wastepaper that can be recycled is office paper, which includes paper used in laser printers and photocopiers. However, the heat used to fuse the toner/ink to the paper's fibers makes this ink very difficult to remove; therefore this category of wastepaper is primarily used to make tissue.

Magazines were once of no value in recycling because their high clay content was considered a contaminant. However, a recent development in the de-inking process, *flotation,* actually benefits from the presence of clay, so a mixture with a 30–70 ratio of magazine and newspaper stock is now commonly used. First instituted in America in the late 1980s, flotation promptly increased the value of coated wastepaper and raised the volume of recovered magazines from 2.7 million tons in 1991 to 7.3 million tons by 1992.

The final category of secondary fiber is termed **mixed paper** and it is comprised of paper grades that are not presorted; examples include direct mail, catalogs, and directories. The role of mixed paper is limited to the creation of products in which bulk is more valued than qualities like brightness and strength. Boards used to make cereal boxes, packing boxes, and clay-coated folding cartons are examples of such products.

Preconsumer and Postconsumer Wastepaper

From the standpoint of certain governmental regulating bodies and consumer groups, not all sources of secondary fiber are the same and, for this reason, further clarification is appropriate. For example, some secondary fiber is created during the papermaking process itself. When a break occurs in the web of paper being formed on the paper machine, the machine usually continues to operate while the break is re-

Preconsumer Waste
- Obsolete inventory
- Returned stock that was not printed
- Butt rolls
- Envelope cuttings
- Printing and bindery trim
- Damaged rolls

Postconsumer Waste
- Old newspapers, magazines, office paper, books
- Old corrugated containers and paperboard
- Over-issue magazines (unsold)

Table 5–3 Preconsumer wastepaper categories as defined by the EPA are shown above. It should be noted that these criteria for pre- and postconsumer waste at present apply only to paper purchased by the federal government.

paired and the paper formed in the meantime is diverted to a large room beneath the machine where it rapidly accumulates in a huge pile. This paper, termed broke, is then repulped and becomes part of another day's production. Because the mill then creates paper from reclaimed stock, this new paper has traditionally been considered recycled. Still other disputed sources of waste stock are damaged and obsolete mill inventories, and unprinted waste produced by envelope converters and printers. The argument is made that wastepaper from these sources never reached the consumer and has always been reclaimed anyway. Therefore, environmental and consumer groups have pressed for a distinction between these sources of recycled paper, *preconsumer* waste, and the discarded paper that would end up in landfills if special efforts to recover it were not made, *postconsumer* waste (see Table 5–3).

The Environmental Protection Agency has defined **preconsumer** (or manufacturing) **waste** as being "generated after the completion of the papermaking process but recovered from the waste stream before use by the final customer. . . ." It is important to recognize that this definition excludes mill broke, which occurs before the "completion of the papermaking process"; in the view of the EPA, fiber created by repulping mill broke does not qualify as recycled. This exclusion is a point of dispute with some paper industry officials who feel that any paper waste that is repulped is also being recycled. They argue that if mill broke were not repulped, it would be relegated to the waste stream, and thereby add to the landfill problem.

Figure 5–7 Bales of printed waste-paper sit in the warehouse of a paper mill prior to their being pulped and de-inked. *Courtesy of Cross Pointe Paper Corp.*

Postconsumer waste, according to the EPA, consists of "those products generated by a business for consumer use which have been separated or diverted from solid waste for the purpose of collection, recycling, and disposition" (see Figure 5–7). Because postconsumer wastepaper has historically been carried to landfills, it is the prime target for widespread recycling programs.

In response to these concerns, several mills are now labeling their recycled products as being composed of certain percentages of preconsumer and postconsumer waste. Increasingly, governmental requirements go beyond requiring a certain percentage of recycled fiber and mandate that a minimum percentage of postconsumer waste be in specified paper grades.

Economic Factors Affecting the Market for Recycled Paper

Although there is little opposition to the concept of recycling, certain economic realities impact at several points upon its actual execution. One such reality is the relationship of supply and demand. As an increasing number of municipalities establish curbside collection programs, the supply of old newspapers (ONP) has exceeded the capacity of the paper industry to process it, thereby resulting in a plummeting ONP market. In 1995, some mills were paying as high as $200 a ton, but by 1996 the average price had fallen to $20 a ton. Some towns that once sold their ONP for $35 a ton have since been forced to pay $35 a ton to have it hauled away.

The capacity of the paper industry to process wastepaper has lagged behind the growing supply for several reasons. First, the paper mills are generally *located in remote areas* near the forests and were built to process virgin fiber. Converting these existing facilities to process secondary fiber is rarely profitable, thereby requiring the construction of new plants closer to cities where the majority of the wastepaper mass is found. Building these new pulping mills with de-inking capacities requires huge amounts of capital, usually between $40 and $85 million. Such a project also requires two or three years of conceptualization before a new facility can begin production.

Another economic factor that restricts the volume of paper being recycled concerns *relative material cost*. Paper made from virgin fiber usually has less than 10 percent of its total cost go toward raw material acquisition. In contrast, collecting and transporting old newspapers can comprise 20 percent of the cost to make recycled paper. Once the fiber is acquired, however, the operating costs for producing recycled paper are lower than for the manufacture of paper with virgin fiber (refer back to Table 5–2). Clearly, economic realities will always be a factor in determining the role that recycled fiber plays in the paper industry.

In addition to ecological and economic concerns, another issue that has an impact on the market for recycled paper is the mills' ability to use increasing amounts of secondary fiber in the manufacture of bonds, books, covers, and other grades used in printing without compromising desired characteristics such as brightness and strength (see Table 5–4). While it is true that 200 of the nation's 500 paper mills depend primarily upon secondary fiber, the vast majority of these 200 mills produce tissue, corrugated cardboard, or paperboard.

Making Paper From Secondary Fiber

Producing paper from wastepaper products is more complex than might be imagined, largely because of the potential diversity of the raw material's contents. When a pulp mill accepts a load of loblolly pine wood chips, the mill can make some accurate assumptions about the chemical contents of that load. However, a load of wastepaper could contain both groundwood and chemically pulped fiber, different types of coating pigment, various dyes, varnishes, and adhesives, as

Wastepaper Category	Process	Finished Product
Pulp substitute	Pulping	Communication grades Tissue
De-inking	Fine paper de-inking (washing, flotation)	Communication grades Tissue
Newspapers	De-inking	Newsprint Folding cartons
Mixed paper	Pulping, screening, cleaning	Packaging Molded products
Corrugated	Pulping, cleaning, screening	Corrugated medium Liner board Kraft towels

Table 5–4 The pattern in paper recycling is for fiber from one paper product to be used to make a product with equal or lower demands on characteristics such as brightness or strength. For this reason, fiber from old newspaper is not used to produce bond or offset paper.

well as a few paper clips, staples, and rubber bands. And that list does not take into account the vast range of ink pigments, solvents, and additives that could be represented. As has already been mentioned, homogeneity is a key factor in the desirability of a load of wastepaper. Clearly, the task of removing contaminants is easier if the nature of those contaminants can be controlled. For this reason, repulping mills usually accept only a specified type of reclaimed paper product; some buy only old newspapers, others only old corrugated containers, and still others only computer printout papers. When a shipment of appropriate wastepaper arrives at a repulping facility (see Figure 5–8), it first goes through an **inspection** for undesirable materials that may result in partial or total rejections. The most critical inspection occurs at mills that produce fine papers for the printing industry where brightness and press performance are crucial. Because groundwood fibers are unacceptable in bond, book, text, and cover paper, any old newspaper and corrugated products must be eliminated from the mix. Envelopes with plastic in their windows and papers with glues and other adhesives (called *stickies*) also need to be sorted out. With such a wide array of contaminants to watch for, it is amazing that some mills have been able to reduce their rejection rate to a mere 1 percent, a success owed largely to their working only with conscientious suppliers.

After inspection, the wastepaper is ready for **repulping,** a process roughly equivalent to throwing paper and water

Figure 5–8 A forklift loads bales of wastepaper onto a large conveyor belt that will carry the stock up and into the pulper. *Courtesy of Cross Pointe Paper Corp.*

into a giant kitchen blender (see Figure 5–9). Once the wastepaper has been reduced to individual fibers, the slurry is ready for the complex process of contamination removal or cleaning. The number of cleaning stages and their sequences varies greatly and is determined by three considerations—the nature of the *source of fiber* (old newspaper, corrugated board, high-grade de-inking), the *chemistry of the inks*, and the *type of finished paper* that the pulp will produce. Understandably, if the product will be paperboard or tissues, there is less need for brightness than if it were to be book paper. Because communication papers (e.g., fine printing and writing papers) place the most rigorous demands for brightness, uniformity, and performance, the process of de-inking these grades will be examined in some detail. Again, this process varies from one mill to another in the equipment and chemistry that are used, as well as the sequence of the process stages.

The goal of the **de-inking** process is two-fold: to *separate* the contaminants from the fiber stock and *remove* them from the stock. Although clays, latexes, adhesives, and waxes are common contaminants, the most common are printing inks,

Figure 5–9 The whirling blades of the pulper reduce the combination of waste paper and water into a gray pulp that will be de-inked, cleaned, and bleached. *Courtesy of Cross Pointe Paper Corp.*

which themselves comprise a wide range of chemical properties such as simple letterpress inks to more complex formulations such as those used in ultraviolet-curing printing systems. However, from the viewpoint of those who de-ink wastepaper, there are only two fundamental classifications—soft inks and hard inks.

Soft (greasy) inks are the oil-based inks used in most printing. These inks are not so difficult to separate from the fibers, but excessive agitation can cause them to remix with the fibers and reduce brightness. In sharp contrast, a higher agitation level is needed for removing, dispersing, and floating away the *hard* inks found in electron-beam and ultraviolet-cured printing.

While still in the pulper, the swirling action and friction among the fibers has already freed some of the ink and placed it into a suspension. This action is enhanced when the fibers absorb water and swell up, causing some of the ink particles to fall away. The next step is to dilute the stock to a 4 percent consistency and pump it onto the **coarse screening** stage which separates out larger contaminants such as staples, chunks of plastic, and lumps of fiber when these impurities fail to pass through the holes of the screen plate. Coarse screening is commonly supplemented with **centrifugal cleaning** that eliminates small, but heavier-than-fiber, substances such as sand and dirt (see Figure 5–10).

After screening and cleaning, the stock may be thickened to 11 percent consistency in preparation for **bleaching.** The chemistry to bleach the fibers varies among de-inking facilities, but the process usually consists of multiple stages. A further dilution to 1.5 percent prepares the bleached pulp

Figure 5–10 After being de-inked the pulp is diluted and further cleaned of contaminants in centrifugal cleaners that remove heavier-than-fiber particles. *Courtesy of Cross Pointe Paper Corp.*

for a **fine screening** which removes small particles of ink and dirt. The pulp is chemically treated to give the ink particles an affinity for water that assists in their separation from the pulp when the water moves through screens. Either pressure screens, side-hill screens, or a combination of both may be used to further clean the pulp (see Figures 5–11).

Until the late 1980s, *screening* and *washing* were the primary means of removing ink. However, an increase in the use of unconventional (hard) inks and polymeric inks/toner has created the need at many mills for **flotation** cleaning of the fibers. Flotation, a process developed in Europe, now allows the recycling of paper containing ink/toner from photocopiers and laser printers. In flotation, air is injected into the pulp and the rising bubbles carry ink particles to the surface where they are skimmed off (see Figure 5–12).

After flotation, water is drained off to thicken the pulp for **dispersion,** kneading the stock to scatter any traces of remaining ink so that they will not be seen by the naked eye. At this point, the fibers are as clean as they are going to get, and that can be very clean; some recycled sheets approach a 90 brightness rating.

After the cleaning process, between 70 and 80 percent of the wastepaper that entered the pulper remains for making paper. The remainder, consisting of *fines* (fibers that are too short or weak), coating, filler, and inks, forms a sludge that must be disposed of somehow (see Figure 5–13). Once accepted at public landfills, this sludge may contain concentrations of contaminants that are potentially harmful to the water table. For this reason, many repulping operations

Figure 5–11 Side-hill washers spill a diluted pulp mix over angled screens that allow ink particles to fall through the screen and separate from the pulp, which collects at the bottom for further cleaning. *Courtesy of Cross Pointe Paper Corp.*

have established their own landfills. It is ironic that 20 percent of the paper retrieved from the waste stream reenters in a more potent form. At present, research for alternatives to burying this sludge is focused on exploring waste-to-energy incineration and using waste materials to create useful products.

Last, the elimination of fines points up the fact that cellulose fibers cannot be reused forever. Although there is disagreement as to exactly how many recyclings are possible (many experts say two or three), it is acknowledged that, even if all paper were recycled, a supply of virgin fiber would still be needed to replace the lost fines. Therefore, despite its

Figure 5–12 Air bubbles rise to the surface of dilute pulp, carrying with them small ink particles. The bubbles then spill over the circular well in the middle of the vat and leave behind cleaner pulp. *Courtesy of Cross Pointe Paper Corp.*

importance, recycling can have only a limited effect on both timber harvesting and landfilling.

Printing on Recycled Paper

In 1971 America was introduced to the three chasing arrows of the paper recycling symbol (see Figure 5–14) that represented a new awareness of responsibility for the environment. However, in an effort to be consistent with the spirit of conservation, many paper mills in the 1970s introduced recycled bond, text, and book papers that—to be polite—were inadequate. Today recycled paper still has an image problem for many people in the printing industry as a result of those early products. However, during the last 20 years the paper industry's technology for and expertise in making quality communication grade papers from recycled stock has grown tremendously.

Printers traditionally judge a paper stock by two criteria—*price* and *performance*. Twenty years ago, many recycled papers fared poorly in both categories for a number of reasons. First, several mills tried to introduce recycled grades into their line of products using their existing facilities. Since then, mills have upgraded their technology. Second, because most paper mills were located in remote areas near large forests, transportation costs for the new fiber sources raised the price of the finished product. However, the cost of fiber ac-

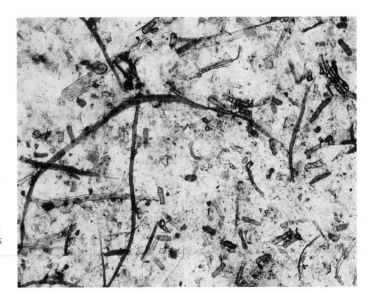

Figure 5–13 This microphotograph of recycled stock reveals the shortening of fibers that results from multiple pulping and refining operations. Fines are eliminated from the pulp to preserve strength and performance (75x enlargement).

Figure 5–14 White arrows with a black background indicate a product made with recycled fiber. Black arrows with a white background indicate a product that can be recycled.

quisition has declined as new pulping facilities have been built near population centers. Also, the availability of quality wastepaper has improved as large, coordinated efforts have been undertaken by the business sector to sort wastepaper for recycling. In fact, the Resource Information System, Inc. estimates that, in 1990, 28 percent of America's communication grade (printing/writing) wastepaper was recovered; the RISI has projected that, for 1995, the recovery rate will climb to 38 percent.

Still another reason for earlier recycled products often costing more than their virgin fiber counterparts was the huge investment in cleaning and de-inking technology. However, as more recycled paper is produced and sold, those costs are able to be shared by the larger volume. By 1992, recycled sheets commonly were 10 percent more expensive than comparable sheets made from virgin fiber, down from a 30 percent margin in the 1980s.

The poor performance of the earlier recycled sheets has also been addressed. Building new plants that have been designed to process secondary fiber instead of trying to adapt existing facilities has also allowed paper companies to provide a product with improved performance in the pressroom and in the finishing department.

Many printers report that they are able to get excellent results from recycled paper, but occasionally only after making some modest modifications in their operations. The Graphic Arts Technical Foundation offers some precautions to printers using recycled paper:

1. Be certain that your customer is aware of the optical qualities of the requested sheet. Recycled papers are available in everything from #1 premium coated to uncoated groundwood.
2. When printing large or expensive jobs on an unfamiliar stock, submit samples to the ink company and ask for an appropriate ink.
3. If the physical and optical properties are satisfactory, there should be no more problems when printing recycled paper than when printing virgin stock.
4. When printing, use the minimum amount of fountain solution to keep the plate clean.
5. Quick-release blankets will also reduce piling tendencies on the blanket.
6. Remember that a bad experience with a recycled paper only applies to that stock. Many other recycled papers are excellent.

The Market for Recycled Paper

The three arrows of the paper recycling logo can represent the three main components of successful recycling: *supply* of acceptable waste fiber, the *capacity to produce* quality paper from it, and a *demand* for these products. In the 1970s and 1980s, one or two of these factors may have been in place at the same time, but seldom were all three. When there was consumer demand, the production technology and facilities were not yet in place so printers often found recycled paper difficult to acquire and sometimes disappointing to work with. Then, when the product got better, the resultant growing demand for high-quality recycled paper sometimes overwhelmed the supply of adequately "clean" wastepaper.

However, the picture looks brighter for a closing of this three-part loop: material supply, production, and demand. As the papermaking technology continues to improve and

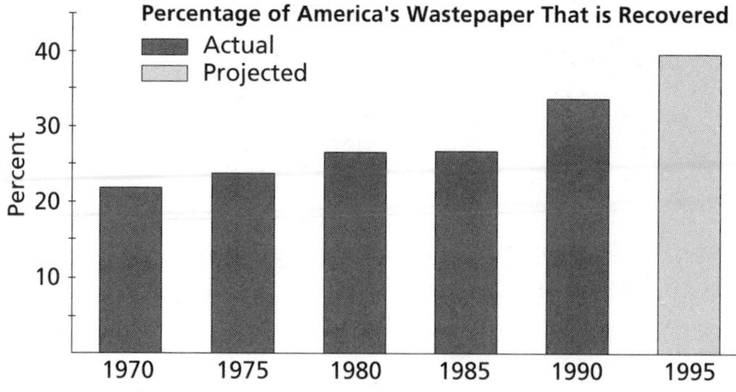

Percentage of America's Wastepaper That is Recovered

Figure 5–15 The five-year period between 1985 and 1990 witnessed a significant increase in the percentage of America's recovered wastepaper.

as ink and toner manufacturers make their products easier to remove from wastepaper, the product line and performance quality of recycled sheet will continue to improve. Then, the demand for recycled products will grow, thereby creating a greater demand for secondary fiber. This will also provide greater incentives for the corporate world and other consumers to sort out wastepaper from the waste stream. For example, the AFPA reports that 1992 U.S. paper and paperboard recovery topped 33 million tons, more than double the 1970 total; this trend is predicted to continue.

Many printers are hopeful that the current interest in recycled paper is a fad that will wane, thereby sparing them from having to adapt to its challenges. However, all indications contradict this viewpoint. First, the supply of recovered paper is growing steadily (see Figure 5–15). Second, the paper industry is responding to this growing supply with major investments in de-inking facilities and the technology to produce recycled sheets that are only getting better. If, with the cooperation of ink manufacturers, recycled paper continues to improve and cost less, its desirability can only be expected to increase—both in the public and private sectors.

Questions for Study

1. What large-scale forces have created the need to address the issue of resource recycling?
2. Examine the validity of the two most commonly heard rationales for recycling paper.
3. How does America compare to other industrialized nations in terms of the amount of refuse generated and in the means of disposing of it?
4. Compare the costs of papermaking with virgin and secondary fiber.

5. What roles have the federal and many state governments assumed in the matter of recycled paper? What goals appear to have been established?
6. List and explain the four basic factors that determine the value of a material for recycling.
7. List the seven categories of secondary fiber and the products that they usually go into.
8. Distinguish between preconsumer and postconsumer waste as defined by the EPA. List common sources of fiber that constitute each.
9. Why does the EPA not qualify mill broke as a source of recycled paper fiber?
10. List the contaminants that might cause a recycling mill to reject a load of wastepaper.
11. List the contaminants that must be removed from recycled pulp before it is ready for the paper machine.
12. Characterize the two basic types of ink that a recycling mill deals with.
13. Discuss washing and flotation as techniques for de-inking.
14. What factors prevent a 1,000 pound bale of wastepaper that is repulped from producing 1,000 pounds of paper?
15. Why do many printers wish to have nothing to do with recycled paper?
16. What has changed in the recycled paper business since the 1970s?
17. Why is it more of a challenge to produce bond or offset paper from postconsumer waste than it is to produce paper towels or paperboard?
18. What steps can printers follow to smooth the transition to using recycled paper?
19. What factors are involved in "closing the loop" of paper recycling? What is the prognosis for this process?

Key Words

homogeneity
ONP
mixed paper
de-inking
dispersion
pulp substitute

old corrugated
preconsumer waste
coarse screening
flotation
high-grade de-inking
postconsumer waste

centrifugal cleaning
inspection
mass
repulping
fine screening
degree of contamination

6 Physical Characteristics

AFTER STUDYING THIS CHAPTER, THE STUDENT SHOULD BE ABLE TO:

■ Distinguish among brightness, gloss, whiteness, and fluorescence. Include test methods in your response.

■ Explain the ways of imparting opacity to paper and how each method impacts the sheet's physical characteristics.

■ Distinguish among tear, burst, tensile, surface, and folding strength. Include test methods in your response.

■ Explain how the degree of beating received by the pulp influences the finished sheet.

■ Distinguish between acid and alkaline-sized sheets relative to both manufacture and physical characteristics.

■ Explain why grain direction and dimensional stability are important in printing.

■ Define caliper (as opposed to bulk) and explain how it influences sheet stiffness.

When examining the **physical characteristics** of paper, we are looking at a long list that includes color, opacity, brightness, gloss, bulk, stiffness, porosity, smoothness, weight, absorbency, permanence, folding strength, surface strength, tear strength, tensile strength, and burst strength. A sheet of paper can be produced to excel in any one of these qualities and, as a by-product, the sheet will also embody one or more of the other characteristics. But the bad news is that no quality can be built into a sheet of paper without also sacrificing one or more of the other qualities. In short, the printer literally cannot have everything. This fact of life requires that the paper buyer be aware of what can and cannot be expected when seeking a sheet to meet a particular set of requirements. This knowledge is clearly a prerequisite for a realistic approach toward paper and minimal disappointment in its use in printing and finishing operations.

Paper's physical characteristics determine a particular sheet's appropriateness for a particular application. Just as a two-seat sports car is impractical for family transportation, a paper that would be ideal for a corporate letterhead would likely be a disaster if used to print a full-color magazine or catalog. To avoid selecting the wrong paper for any situation, people involved in stock selection must have an awareness of the important characteristics that are needed and understand how to select a sheet that embodies them.

These physical characteristics are relative; that is, nearly all paper has them to some degree. For example, consider the ability to resist tearing (tear strength) of a $1 bill and a facial tissue; the currency has many times more resistance than the tissue, but they both have at least *some* tear strength. The point to understand is that the amount of a particular characteristic a certain sheet possesses is what determines its fitness for a given application. For this reason, a paper's characteristic needs to be measured quantitatively so that comparisons can be made. To accomplish this task, testing devices, the procedures for using them, and the standards for presenting their results have been developed. This chapter will define the most significant of paper's characteristics and briefly explain the test methods for assessing them.

Finally, it is crucial that people who order or use paper understand the complex interrelationships that exist among the characteristics. In short, a sheet can be manufactured to have a high level of any characteristic desired by the customer—but an abundance of any one characteristic necessarily comes at the expense of one or more others. This give-and-take relationship requires that people using paper understand what is being given up when they ask for a sheet with one or more attributes. To address this need, the chapter will also examine the means that paper mills have of imparting attributes to paper and also explain the side effects that necessarily result.

The Interactions Between Light and Paper

Among the most obvious physical characteristics to the untrained eye is the way that paper interacts with the light that falls upon it. What happens here will determine the sheet's color, brightness, gloss, and opacity.

When light strikes any surface (including paper) there are three phenomena that can occur: the light can be *reflected, absorbed,* or *transmitted.* With nearly all paper, each of the three phenomena takes place to some degree depending upon the nature of the sheet.

Reflection is a measure of the amount of light that strikes the paper and then bounces away without penetrating the surface. The more light that is reflected from the surface, the brighter the sheet appears. As is commonly observed, lightly colored objects reflect the sun's rays better than darker objects, which tend to absorb light. Dark objects contain pigments that absorb light, so reduced pigment content produces greater reflection.

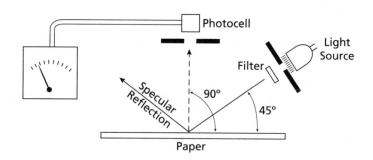

Figure 6–1 A brightimeter emits a beam of blue light that strikes the paper sample at a 45-degree angle. To avoid measuring specular reflection (gloss), the light reflected at 90 degrees is measured.

Brightness

A sheet's **brightness** is a measurement of the amount of light reflected, rather than absorbed, by a surface. Paper is made brighter by bleaching the pigments that are naturally found in pulp and in coating materials. A *brightness tester* (or brightimeter) shines a prescribed amount of a specified wavelength of light onto a sheet and then measures the diffuse reflection. As shown in Figure 6–1, the light strikes the paper at a 45-degree angle while the measurement is taken at a 90-degree angle to the paper. By measuring diffuse reflection, the sheet's specular or direct reflection (gloss) is not being measured. The specified wavelength of the light is 457 nanometers, which is in the blue portion of the visible spectrum.

One way to increase a sheet's overall reflection is to apply **fluorescent** dyes to the surface. These pigments appear to reflect a greater volume of visible light than struck them originally. The principle at work depends on a light source, such as the sun or a fluorescent tube, emitting a wide range of light waves of several different wavelengths. The human eye is sensitive to (capable of seeing) only some of these wavelengths, which comprise what is called the **visible spectrum.**

Outside of this visible spectrum are ultraviolet (UV) light waves which cannot be seen. However, when these light waves strike fluorescent dyes, the light energy is transformed into a form of energy that stimulates electrons to jump to a higher energy level than is normal for them (see Figure 6–2). As they return to their regular energy level, they lose energy in the form of visible light. In this way, a surface can appear to reflect more visible light than struck it in the first place. Clearly, this phenomenon can make a sheet of paper appear brighter than mere bleaching could achieve, providing that the light source is emitting ultraviolet light.

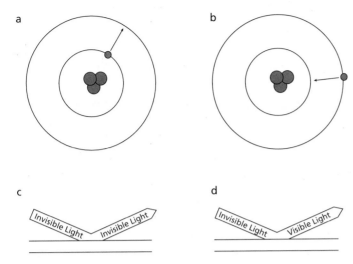

Figure 6–2 Fluorescence is the transformation of invisible ultraviolet light into visible light. When fluorescent pigments are bombarded with UV light, electrons are temporarily excited to a higher level (orbit) as seen in illustration (a). When these electrons fall back to their normal level, they give up their extra energy in the form of emitted light (c). Electrons that fall back in stages (like a ball bouncing down the steps), emit light with lower energy and longer wavelengths, thereby making it visible. Illustration (c) shows invisible ultraviolet light striking regular paper with no effect. Illustration (d) shows ultraviolet light striking paper containing fluorescent dyes and being transformed into visible light, adding to the overall brightness of the sheet.

Whiteness

Whereas brightness measures the quantity of reflected light, **whiteness** measures both quantity and uniformity of reflection across the visible spectrum—the percentages of red, blue, and green light that are reflected. If a sheet reflects 85 percent of the red, green, and blue light, it is balanced in its reflectance and will appear as neither a blueish cold-white nor a yellowish warm-white. In contrast, if a sheet reflects 85 percent of the red and green light and 90 percent of the blue light, that sheet will have higher brightness, but will rate lower in whiteness (see Table 6–1).

Color

The concept of **color** results from the light absorbing nature of pigments. The phenomenon of color is explained in detail in Chapter 13, but a brief overview is presented here.

Colored paper is made by adding pigments in the form of dyes to the pulp. These pigments absorb some portion of the

	Percent Reflection			
Paper	Blue	Green	Red	Appearance
A	85	85	85	Highest whiteness
B	90	85	85	Highest brightness
C	60	70	70	Lowest in both

Table 6–1 The reflection levels of three sheets of paper have been measured and they are compared for brightness and whiteness. Sheet A reflects light uniformly; sheet B reflects the most light, but not uniformly; sheet C is neither highly reflective nor uniform; this sheet would appear gray with a yellowish cast.

light that strikes them and reflect the remainder. Natural white light consists of equal amounts of red, blue, and green light. This composition can be demonstrated by using a prism to refract these three components into separate bands of color (refer ahead to Figure 13–1). Certain pigments absorb certain amounts of these components and this action creates the phenomenon of color.

For example, if a sheet's pigments absorb all of the blue and red light, then only the green light is being reflected to the eye and the sheet appears green. By the same token, if all of the green and blue light is absorbed, the surface would appear red because only the red light is reflected to the eye (see Figure 13–4). Pigments that absorb all of the three main components of light reflect none of it with the result that they appear black. And, predictably, a surface that absorbs no light but instead reflects all of it will appear white, the color of the natural light that struck it. For this reason, the addition of any pigment that will impart color to a sheet of paper will also lower the sheet's brightness. Color is the result of light absorption; the darker the color, the more light is absorbed and the less is reflected. A black sheet with a dull surface may reflect only 5 percent of the light falling onto it. A dark blue sheet might reflect 20 percent, a light blue sheet 60 percent, and a coated white sheet up to 95 percent.

A sheet's color can be measured with either a spectrophotometer or a colorimeter. The **spectrophotometer** is useful in creating a visual record of a color by producing a curve that measures the reflectance percentage for all wavelengths in the visible spectrum (refer ahead to Figure 12–14).

A **colorimeter** provides numerical readings for aspects of a color—hue, saturation, and lightness—and is often used to assess a sheet's whiteness. Because color testing devices are more commonly used with ink than with paper, the colorimeter and spectrophotometer are described in Chapter 12.

Gloss

Related to brightness, but not the same, is the sheet's **gloss** or its "shine." Gloss is not a measurement of how much light is reflected back, but rather how much of the overall reflection is *specular,* rather than *diffuse,* reflection. The difference between specular reflection and diffuse reflection was explained in more detail in Chapter 4; however, a brief explanation is offered here. A surface can bounce back light in either a uniform pattern (specular reflection) or randomly (diffuse reflection).

The uniformity of the surface determines how the light rays will be reflected. A smooth, mirrorlike surface will produce specular reflection while an irregular surface will bounce back the light at irregular angles, producing diffuse reflection (refer back to Figure 4–1).

Imagine the neglected finish of an old white automobile. Because of its color, almost all of the sun's rays that strike the old paint will be reflected away, giving the finish a high level of brightness. However, the car will not shine until a layer of wax is applied, filling in the low places of the rough surface. The smoothness that results after buffing will not reflect more light than before, but will reflect the same amount of light at a uniform angle, thereby changing diffuse reflection into specular reflection. So, whereas brightness is the result of pigments, gloss is the product of a smooth surface. And, whereas brightness is a measurement of the degree of overall reflection, gloss is a measurement of how much of the overall reflection is specular.

Interestingly enough, a very smooth sheet that is totally covered with black ink would have very low brightness and yet high gloss. The sensation would be like that of seeing yourself in the finish of a black car's waxed finish; although the dark color is absorbing most of the light, that which is bouncing off the smooth finish is specular reflection. By the same token, a highly-bleached sheet of blotter paper with its very rough surfaces would have high brightness, yet no gloss because the high reflection is all diffuse.

Because brightness measures a sheet's diffuse reflection while gloss measures specular reflection, a **gloss meter** functions something like a brightness meter except that the angle of incidence and the angle of reflection are the same. For most papers this angle is 15 degrees, but exceptionally high

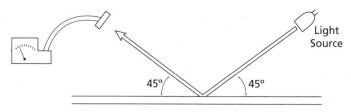

Figure 6–3 A gloss meter (inset) shines a beam of light onto the paper sample at a 45-degree angle and measures the amount of light that is reflected at a 45-degree angle. *Courtesy of Testing Machines Inc.*

gloss sheets may measure more accurately at 70 degrees. Figure 6–3 shows how a gloss meter measures the amount of light reflected at the same angle at which the surface was struck. A gloss meter also can be used to measure the gloss level of a printed ink film.

Paper's Opacity

Light that does not bounce off the surface can either be absorbed into the sheet or transmitted through the sheet and out the other side. A sheet of paper that transmits a significant amount of light is easy to see through and is said to lack **opacity.** Such a sheet is poorly suited for carrying a printed image on both sides, and, because most printing is two-sided, opacity is an important characteristic.

As might be imagined, one way to reduce the amount of light that is transmitted is to increase the amount that is reflected. A second tactic could be to absorb as much as possible, but because high absorption is the result of dark pigments, this is not possible with white paper. This problem is worsened by the fact that cellulose fibers are basically transparent. However, because these fibers are denser than air, they can scatter the light moving through the sheet because of an action called **internal reflection.**

Based on a law of physics, internal reflection is what occurs when a beam of light moves into the interior of, in this case, a sheet of paper, passing in and out of the numerous individual fibers that it encounters. Because tiny pockets of air exist between these fibers, the light moves through a fiber, then through air, then another fiber, and so on until it has passed through the entire sheet and out the other side. This alternating between the two different densities—air and fiber—changes the light's speed and direction. When the light enters the denser fiber, it is refracted; that is, its speed is reduced and it is made to veer off at a certain angle. Once the light pops out of the fiber, it regains its original speed and direction (see Figure 6–4). However, in this process of moving from one medium of a certain density to another medium with a different density, a loss of some of the light occurs and this scattering action is the key to making white paper more opaque.

Whatever portion of light entering the sheet is scattered by internal reflection does not pass through to the other side; in short, the more light that is scattered, the less light is transmitted. The important point to understand is that this scattering is the result of **interfaces,** the term applied to each instance of internal refraction. The more interfaces in a given sheet, the more opaque that sheet becomes.

The opacity of a sheet is measured by placing the paper sample over a white background and illuminating it. A measurement is then made of the amount of light reflected from the sample's surface. A second reading is then taken of the reflectance, except this time the background is black. The second reading will be lower than the first because any light that passed through the sheet the first time was reflected back through it by the white background; during the second measurement the transmitted light was absorbed by the black background. The ratio between the two readings is the sheet's opacity (see Figure 6–5).

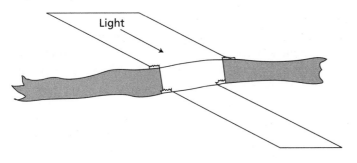

Figure 6–4 When light strikes a cellulose fiber, some scattering of light takes place as the surface is penetrated. The light is refracted and slowed as it moves through the fiber. Upon leaving the fiber and re-entering the air within the sheet, light regains its original direction and speed, but loses still more light during this exit.

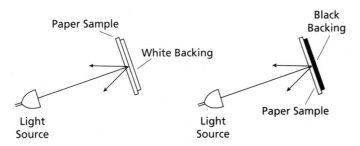

Figure 6–5 An opacimeter first measures the light reflected from the paper sample with a white backing. The white backing is then replaced with a black one and another reading of reflection is taken. *Courtesy of Testing Machines Inc.*

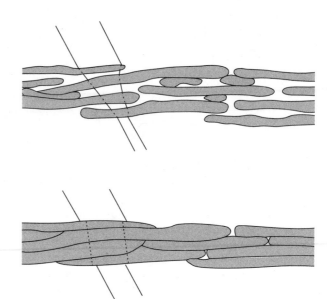

Figure 6–6 When calendering is increased, the caliper of the sheet is reduced because of the closer bonding of individual fibers. This loss of porosity reduces the number of interfaces and the sheet loses opacity.

A method of comparing the opacity of two sheets of paper is simply to place them alongside one another with a printed page beneath them. The sheet that is the more difficult to read through is the more opaque.

Making Paper More Opaque

There are several factors that determine the number of interfaces within a given sheet of paper and these factors can be controlled by the papermaker. The first factor that determines a paper's opacity is its *caliper*, or thickness, because high caliper means that the fibers are separated by tiny air pockets that ensure numerous interfaces. On the other hand, if the sheet has been compressed by the pressure of the calendering operation, then the fibers are pressed more tightly together, forming a denser sheet with fewer individual surfaces and, therefore, fewer interfaces and less opacity (see Figure 6–6). This relationship between calendering and fiber density brings up the first trade-off that the buyer of paper must make: all other factors being equal, an increase in smoothness (achieved through calendering) necessarily results in a decrease in opacity. In addition, because increased beating of the fibers also causes them to bond more tightly, the strength that this procedure provides will again be at the expense of the sheet's opacity. Therefore, sheets with a comparatively rough surface will have greater opacity

Figure 6–7 The platelike shape of calcined kaoline clay can be seen in this 5,000-power enlargement. When air pockets get between the plates, the number of interfaces increases. *Courtesy of Mead Central Research.*

than if they had been glossy. Other side effects of decreasing the caliper are lower ink absorption and decreased stiffness.

A second method of increasing the number of interfaces is to increase the *number of fibers* in the sheet. Although an increase in the number of fibers will allow a glossy sheet to also have good opacity, it will automatically raise the weight of the paper and cannot be considered useful where a lightweight sheet is required, such as an item to be mailed.

The third way to improve opacity is to add **filler** to the sheet. The addition of finely ground *clay* to the pulp will increase opacity, not because clay refracts the light more than does fiber, but because the clay particles are much smaller and provide so many more surfaces (and therefore interfaces) than the fibers alone. The form of clay most commonly used in papermaking is *kaoline* (KAY-oh-lin) because it is plentiful (mostly found in Georgia), can be effectively bleached, and can be ground to a tiny particle size. When kaoline clay is refined by being heated to a sufficient temperature (calcined), its particles become platelike (see Figure 6–7). This flat shape makes *calcined* (KAL-signed) kaoline more effective at improving paper's opacity than unrefined kaoline clay.

Clay fillers can be augmented by the inclusion of *titanium dioxide* particles which, in addition to being smaller, have a greater density and are more refractive than clay. This higher index of refraction makes titanium dioxide a very ef-

Figure 6–8 A form of PCC that imparts a high level of opacity to the sheet is the scalenohedral crystal shape. *Courtesy of Mead Central Research.*

fective filler, but its high cost prevents its use in most papers and clay is easily the most common filler found today.

An increasingly popular filler is *calcium carbonate*, which occurs naturally in the form of chalk. In the past, calcium carbonate was not a primary filler material because it was chemically incompatible with rosin-sized paper. However, the recent popularity of alkaline-sized paper has allowed the benefits of calcium carbonate fillers to become realized; these benefits include high levels of both opacity and brightness. *Ground calcium carbonate* (GCC) has been the traditional filler material, but a recently developed form called *precipitated calcium carbonate* (PCC) generates higher brightness levels than were previously possible. The production of PCC can be controlled to generate crystals of various shapes and sizes (see Figure 6–8), and a combination of crystal types is generally thought to impart the optimum mix of physical properties (brightness, bulk, opacity, porosity, and tensile strength) to the sheet.

But regardless of the material that is used, the amount of filler content that goes into a sheet will necessarily displace its weight in fiber. For example, if 10 pounds of filler were added to the mix intended to produce 50-pound paper, then 10 pounds of fiber would have to be removed in order for the finished product to still be 50-pound paper. Therefore, if two different sheets of 50-pound paper were made—one with no filler and the other with 10 pounds of filler—the latter would contain less fiber. This reality has a profound effect on the performance of the sheet. Remember that it is the bonded fibers that provide the sheet with its structure and, when some of these fibers are displaced by their weight in filler particles, the sheet's tear, fold, tensile, burst, and surface strengths suffer. This reduction in strength is also due to filler particles getting between the fibers and interfering with the fiber-to-fiber bond.

A distinction should be made between **show-through** and **strike-through** because, although similar to the novice, they are fundamentally different and result from different causes. Show-through results from poor opacity; light is simply passing through the sheet and carrying the image printed on the other side with it. Strike-through, however, is the visibility of an image through a sheet because of ink penetrating the paper. To illustrate the distinction, a blank sheet of paper can be laid over a page of black type. If the type beneath it cannot be seen, the sheet has good opacity. If, on the other hand, the same sheet is printed and its images can be read from the other side, strike-through is occurring and the blame lies with the sheet's porosity, not its opacity.

Summing Up Opacity

In short, the buyer of paper must know that even though there are three ways to increase paper's opacity, there also will be predictable side effects with each of them. If the bulk or caliper is kept high by using minimal calendering, then (a) the smoothness will decrease, (b) the gloss will decrease, (c) the ink absorbency will increase, (d) the stiffness will increase, and (e) the sheet will be more spongy and perhaps less suited to gravure or letterpress printing. If the basis weight is increased by using more fiber, then (a) the postage cost may increase, and (b) stiffness and strength will increase. If the amount of filler content is increased, then (a) the sheet will become weak, (b) the caliper will decrease because filler takes up less volume than the same weight of fiber, and (c) smoothness will increase because the particles will tend to fill in the valleys of the surface.

Smoothness

The **smoothness** of a sheet's surface can affect the quality of ink transfer and drying time, as well as gloss. A sheet's level of surface smoothness is primarily achieved through calendering or the polishing of a coating layer. *Smoothness testers* usually assess a sheet's smoothness by measuring the amount of time (in seconds) required for a given volume of air to leak between the sheet and a flat metal plate that is

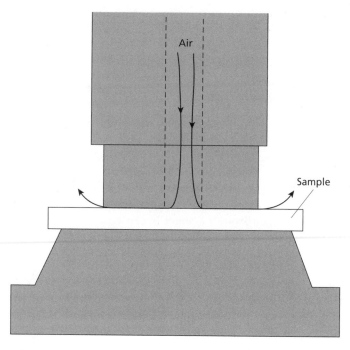

Figure 6–9 This device can test a paper sample for smoothness as well as porosity and softness. To measure smoothness, air pressure is exerted downward against the clamped paper sample. The amount of air that escapes is then measured. *Courtesy of Gurley Precision Instruments.*

held against it. The lack of levelness in a sheet's surface allows more air to pass across it; therefore, the lower the volume of air passage, the smoother the sheet is rated (see Figure 6–9).

Several devices are made for testing paper's smoothness by measuring air leakage. However, a high level of smoothness produces a high reading with some of the instruments and a low reading with others, so any smoothness reports should include the brand of the testing machines.

The Dimensions of Paper's Strength

A paper's *strength* has several dimensions. Its resistance to being punctured, torn, pulled apart, and ruptured are all very separate physical characteristics that often do not reside in the same sheet to a comparable degree. Appropriately, each type of strength has its own method of testing.

Tear Strength

Tear strength measures a sheet's resistance to tearing after an initial cut has already been made. The fiber length of the sheet is the primary factor in determining tear strength, although the fiber-to-fiber bond also has an influence. Tear strength is a very desirable attribute for web press operators who wish to avoid web breaks during the run.

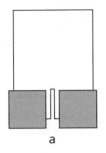
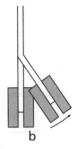
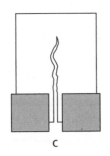

a   b   c

Figure 6–10 Tear strength is measured by clamping a pre-slit strip of paper on two sides (a) and applying a lateral force (b). The amount of force being applied at the moment of tearing is recorded.

The Elmendorf tearing tester is widely used to measure the force required to get the sample to tear. As shown in Figure 6–10, the sample is held by two clamps separated by a small gap. In this gap, the sheet is cut with a knife. Force is then applied to the sheet when one of the two clamps begins to rotate away from the stationary clamp. The pressure being applied at the moment of tearing is the sheet's tear strength.

Burst Strength

Envelopes and other forms of packaging must have adequate **burst strength,** which is the resistance to rupturing from a force that is uniformly applied to one side of the sheet. The test for burst strength consists of clamping the sample with a circular ring; the portion of the paper within the ring is then pressed against by a rubber diaphragm that expands upward as liquid pressure is increased from below. The amount of pressure being applied at the moment of the paper's rupturing is the sheet's burst strength (see Figure 6–11). Because the Mullen tester is the standard instrument used for assessing a sheet's burst strength, this value is commonly referred to as its Mullen strength. As was explained earlier in the chapter, there is a positive correlation between burst strength and both the pulp's fiber length and degree of beating.

Some typical burst strengths of common paper products are as follows: newsprint, 6; magazine cover, 22; business envelope, 30; grocery bag, 41; and a new piece of currency, 70.

Tensile Strength

In the paper industry, **tensile strength** is the force required to snap a 1-inch wide strip of paper. Tensile strength is of some interest to a printer with a web press that pulls a continuous length of paper from a roll and wishes to avoid web

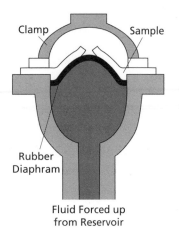

Clamp   Sample

Rubber
Diaphram

Fluid Forced up
from Reservoir

Figure 6–11 A Mullen tester (left) measures a sheet's burst strength by holding down the sample with a ring-shaped clamp and then increasing liquid pressure against a rubber diaphragm that presses against the bottom of the paper. The amount of pressure required to burst through is noted. *Courtesy of B.F. Perkins.*

breaks that occur when the paper cannot withstand the pull. However, a sheet's ability to avoid web breaks is even more closely related to its tear strength than its tensile strength. Many packaging applications rely heavily on good tensile strength.

Tensile strength testers simply clamp onto both ends of the sample strip and apply a slowly increasing pull against the paper until it breaks. The greatest pressure, measured in pounds, that the sample withstands is its tensile strength. The grain direction of the sample must be noted because paper's tensile strength is always stronger when the grain direction is parallel to the direction of the pulling force.

Tensile-at-the-Fold

As was explained in the section on fold strength, some sheets crack at the fold and show thin white lines within solid image areas. To assess a sheet's resistance to cracking, a **tensile-at-the-fold** test dries a sheet, folds it, and then tests its tensile strength.

Folding Strength

Printed products that must withstand being folded and unfolded repeatedly without breaking at the fold should have high **folding strength** (or *folding endurance*). An example of such a product is the ubiquitous road map. Folding strength is measured by a device that clamps one end of the test sample and rotates it a prescribed number of degrees (either 180 or 135 degrees, depending on the machine used) in both directions until it breaks at the fold (see Figure 6–12). The

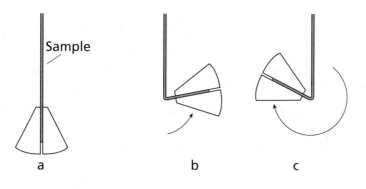

Sample

a   b   c

Figure 6–12 Folding strength is measured by clamping the sample (a) with a mechanism that rotates to the right (b) and then to the left (c) at least 90 degrees from the vertical plane. The number of rotations required to snap the paper is noted.

number of complete rotations achieved before breaking is the sheet's folding strength.

Folding strength benefits most from extended beating and long fibers. The folding strength of some paper products is as follows: newsprint, 40; magazine cover, 150; and currency, 3,000.

Folding Quality

Promotional materials that fold where a solid printed image has been printed must resist cracking at the fold or the ink film will crack with the paper's surface and reveal a white line where the paper's interior becomes visible (see Figure 6–13). The sheet's ability to fold without cracking is its **folding quality.** Although folding quality is related to fold strength, other factors that play a definite role are the sheet's weight, moisture content, and the substance thickness of the coating (if any). Heavy sheets are more likely to crack at the fold, as are sheets with low moisture content and heavy coating layers; collectively, they can create a very high risk sheet.

One technique for assessing folding quality is merely to apply a solid ink film to several sheets of paper, then fold them and gather the resulting signatures to form a stack one-half inch or more. An examination of the spine allows the observer to note the amount of white streaking.

Surface Strength

The degree to which the surface of a sheet of paper can resist rupturing when being pulled by a force perpendicular to that sheet is its **surface strength.** Paper with inadequate surface strength is likely to pick during printing and the greatest risk of picking occurs during lithography because of its use of high tack inks.

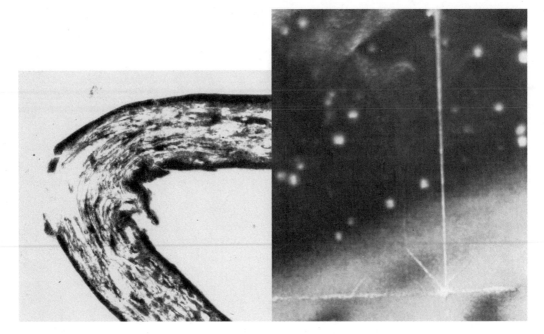

Figure 6–13 The white lines on the poster detail at right did not appear on this poster until the poster was folded, thereby cracking the coating layer as well as the ink layer. The 200x microphotograph at left shows the effect that folding has on both the fibers and coating. *Courtesy of Mead Central Research.*

Many printers have used an unfamiliar paper and encountered picking problems during the press run that could have been avoided through preliminary testing for surface strength. A common technique is to use the *Dennison wax test,* which uses sticks of wax with calibrated adhesive levels that are melted on one end and pressed down against the paper. When cool, the stick is pulled upward sharply. The sticks with greater adhesiveness carry higher numbers and the sheet's assigned surface strength is the highest numbered stick that the paper is able to resist. For example, if a sheet first ruptures with a number 12 wax stick, it is referred to as "OK11."

Although the wax test is not complicated to use, it is best suited to uncoated papers and other tests are often used when a coated sheet needs to be tested. One method uses the ICT tester which applies a solid ink with a known tack level to 1 × 8 inch strips of the paper being tested. The speed of the two surfaces—ink and paper—increases at a sharp rate during the printing from zero to fast enough to generate a great pull of the ink on the paper's surface. After the quick printing, the strip is removed from the tester and the distance from the beginning of the printed image to the first instance of picking is measured. Sheets with the greatest distance have the greatest surface strength (see Figure 6–14).

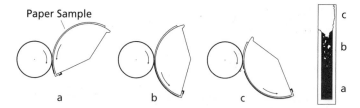

Paper Sample

Figure 6–14 The ICT test for surface strength applies a solid image to a strip of paper at increasing speeds. At point A the speed is comparatively slow; at B it is faster; and at C, it is so fast that serious picking has already occurred. The printed strip is then examined. Note that no picking occurred at A, some occurred at B, and the surface ruptured at C.

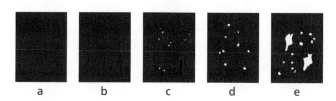

Figure 6–15 The I.P.I. pick test prints five solid areas with different tack levels to the test sheet. In this simulation the sample's surface withstood the ink's pull in images A and B, showed evidence of picking with C, and picked more seriously with D, and began to rupture with E.

The International Print Ink (I.P.I.) test applies five inks of increasing tack levels across the width of the same sheet. The sheet is then examined to see which tack levels were withstood (see Figure 6–15). To picture how this test works, simply imagine a press with five inks sitting alongside one another in the ink fountain.

Four-color jobs printed on one or two-color lithographic presses often do not pick until subsequent passes. The most common cause is that the dampening solution of previous passes through the press "softened" the sheet and lowered its surface strength. To test for this phenomenon, the *I.P.I. wet test* can be used to dampen the sheet before the application of the five inks.

Achieving Strength During the Beating Stage

Most aspects of paper strength are achieved during the beating stage. Occurring after the pulping and bleaching stages, beating is the process of grinding and flattening the fibers as well as fraying them at the ends. Beating serves to strengthen the fiber-to-fiber bond in two ways: (1) as the fibers are flattened, they lose their tubular shape and mat much more closely together; (2) as the fibers are roughened at their ends, they become frayed and produce tiny hairlike extensions, called *fibrils,* that intertwine with fibrils of other fibers and lock together, thereby improving the fiber-to-fiber bond (see Figure 6–16).

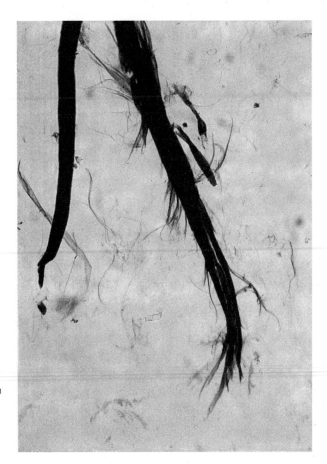

Figure 6–16 A 200x enlargement of softwood fibers reveals the fibrillation that occurs as the result of the beating operation. The interlocking of fibrils increases a sheet's strength in most categories.

To understand the necessity of the beating process, a look at the cellulose fiber in its natural state is important. Originally designed to move water through the tree, they are similar to a drinking straw in both shape and function. Imagine trying to form a sheet with pieces of tiny drinking straws; their tubular shape impairs their contact with one another and produces an overall sheet surface that is uneven.

The beating stage is very important because the amount of refining that the pulp receives will profoundly affect the finished sheet in several ways. First, the sheet will become much stronger in nearly all respects. Folding strength, tensile strength, and burst strength increase significantly in proportion to the amount of time the fibers are beaten. Therefore, paper that is made for bags, maps, or other items that receive rough handling will require extra time in the beater. Interestingly enough, extra beating actually causes tear strength to decrease; there are two factors involved in this oddity: (1) tear strength depends primarily on the length of the fiber; and (2) the beating process necessarily cuts the fibers while flattening and fraying them.

As might be expected, there are numerous other side effects to the beating process. As the individual fibers are flattened, they cumulatively produce a flatter and thinner sheet. This effect is compounded by the fibers' packing more closely together. The result is that, as beating is increased, the sheet's caliper (or bulk) decreases. This compacting means fewer air pockets between the fibers and results in the sheet's becoming less porous. As porosity decreases, so does the sheet's absorbency, meaning that ink holdout will improve.

Because porosity creates the interfaces needed for internal scattering of light, increased beating time also will result in decreased opacity. An example is glassine, the transparent product of extensively beaten fibers. Although flattened fibers will mat together to form a smoother surface, further smoothing on the calender stacks will be difficult because the less porous a sheet is, the less potential it has to be smoothed during the calendering operation. A "hard" sheet (comprised of flat, hard-packed fibers) will not yield to the polishing action and will tend to retain its surface contours. This rigidity means that a hard sheet will lack the micro-smoothness that calendering can create in a softer sheet and, therefore, may impair printing quality in the letterpress and gravure processes where smoothness is important.

A Summary of Beating

To summarize, significantly increased beating will improve a sheet's folding strength, burst strength, tensile strength, and ink holdout. At the same time it will lower the sheet's caliper, opacity, porosity, and perhaps printability. Tear strength increases with moderate beating, but this trend is reversed with increased beating.

Sizing—Internal and Surface

As was explained in Chapter 3, paper is internally sized to reduce its absorption of water. **Internal sizing** is crucial for paper used in lithography to prevent excessive penetration of the dampening solution and potential weakening of the sheet that can cause picking on subsequent passes through the press. A paper that has received only moderate sizing is said to be *slack-sized*, while sheets with heavy sizing are called *hard-sized*.

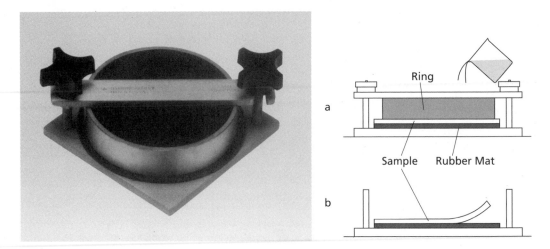

Ring

a

Sample   Rubber Mat

b

Figure 6–17 The Cobb tester (inset) clamps a metal ring over the paper sample and pours a liquid into the ring. After a prescribed wait, the liquid is poured out, the ring removed, and the sample is weighed. *Courtesy of Gurley Precision Instruments.*

Several techniques are available for assessing a sheet's resistance to water penetration and a few are described here. One test consists of placing water-soluble dye particles on a small paper sample that is then floated on water. The time required for the water to pass through the paper and begin to dissolve the dye is recorded; 45 seconds would be considered good sizing.

A second technique involves connecting electrodes to two ends of a sheet of paper and allowing a solution that conducts electricity to contact both sides of the sheet. As the sheet absorbs the solution, it becomes increasingly conductive; the time required for the sheet's resistance to reach a specified level is the measure of its sizing level.

Other techniques weigh the paper sample, expose it to water for a specific period of time, and then weigh it again. Because the increased weight is attributed to water absorption, a minimal increase is indicative of a high sizing level (see Figure 6–17).

The Effects of Sizing

As was explained in Chapter 2, cellulose fibers are very absorbent; in fact, they are so absorbent that the ink of a felt-tip pen will feather out quickly when placed against a paper towel or sheet of newsprint. To compensate for this absorbency, paper must be sized and made more resistant to liquids, not only to improve the sharpness of the printed image but also to prevent the sheet's fiber-to-fiber bond from being weakened by the fountain solution of offset lithography. There are two types of sizing and most printing grades of paper receive both; they are *internal sizing* and *surface siz-*

ing. These two types of sizing are fundamentally different in (a) purpose, (b) the materials used, (c) the stage of paper-making at which they occur, and (d) the effects that they have on the sheet's other physical characteristics.

Internal sizing occurs in the beating stage where individual fibers are coated to impart some resistance to liquids. Traditionally, rosin, a sticky substance derived from softwood trees, is added to and mixed with the bleached pulp. With the aid of a precipitant, the rosin coats the individual fibers and renders them less likely to absorb liquid. In the case of writing paper, this liquid is ink; in lithographic printing, the prime source of concern is that enough fountain solution could enter the sheet to weaken the fiber-to-fiber bond and allow the sheet to rupture against the pull of high tack inks during subsequent press passes.

If moisture-proofing were not such a necessity, it might be omitted because of the host of ill effects that it causes. The rosin itself is not the problem; the harm is done by the aluminum sulfate (papermaker's alum) that is added to precipitate the rosin onto the fibers. At this time, tiny particles of aluminum sulfate also precipitate onto the fibers and, when the sheet is formed, find their way between fibers and impair the fiber-to-fiber bond. As a result, the sheet's fold, surface, tensile, burst strengths and stiffness are all decreased. The aluminum sulfate particles also make the paper acidic and greatly reduce its permanence. The only positive side effect is that opacity is modestly improved because the spaces created between the fibers increase internal reflection.

Alkaline-Sized Paper

The fact that rosin sizing produces an acidic sheet that may deteriorate badly after only 50 years has caused librarians and, therefore, book publishers to demand greater permanence from internally sized paper. For this reason, **alkaline sizing** has become a major issue with the paper and printing industries.

Alkaline sizing is the alternative to the traditional rosin sizing that produces a sheet that contains aluminum sulphate and an acidic pH of 4.5 to 6. Alkaline sizing, in contrast, produces a sheet with a pH value of 7 or 8. A material's pH (potential hydrogen) value indicates its acidity or alkalinity on a 0 to 14 scale, with 7 being neutral. Values below 7 indicate increasing acidity, while values above 7 indicate increasing alkalinity. Because the pH scale is logarithmic with a base of 10, a pH value of 4 is 10 times more acidic than a 5 pH and 100 times more acidic than a 6 pH. Clearly, a subtle change in pH value has a major impact on a sheet's chemistry and nowhere is this impact more evident than in its permanence.

Although rosin sizing has been used since the early 1800s, it became very common after 1850, but books and other documents made with it were found to lack the permanence of books made without it centuries earlier. For example, the pages of the 500-year-old Gutenberg Bible still appear white and strong in sharp contrast with the brownish, brittle pages of rosin-sized books printed in the twentieth century. Surveys in the 1980s of America's major libraries found between 10 and 50 percent of their collection could not be used without risking damage to their pages. In fact, the New York City library alone has 2.5 million books, the equivalent of 35 miles of shelves, that are slowly turning brown and becoming brittle.

As a result, libraries must now spend hundreds of millions of dollars over several years to microfilm the estimated three million brittle books that are deemed worthy of this procedure. Librarians have been joined by both authors and publishers in seeking more permanent paper for book publishing. Policies have been adopted at both state and federal levels to require that documents be printed on alkaline paper. Still another sector that is recognizing the value of alkaline paper is business and industry; in America alone over one million paper documents are created every minute and it is predicted that the accumulation of stored documents will reach 46 trillion by 1994. Nearly 90 percent of these documents will be paper.

Since early in the twentieth century, paper chemists have known that aluminum sulfate (alum) was the primary cause for paper's being self-destructive, as the alum combines with sulfur dioxide in the atmosphere over time to raise the acidity of the sheet and break down the long molecular chain of cellulose. However, the paper industry did not want to abandon rosin sizing because of the enormous cost and technical pitfalls that are inherent in such a major production change.

As late as 1980, only 10 percent of America's groundwood-free paper was alkaline, but market demand in the 1980s resulted in a 30 percent share by 1990. Because more mills are making the conversion, the growing experience with making alkaline paper has overcome nearly all of the early technical problems and mills are finding several benefits of the new chemistry. First, the synthetic polymers that replace rosin as the primary sizing agent allow over twice as much filler to be added to a sheet of paper, which lowers material costs because filler costs less than the bleached fiber that it displaces in the mix.

Second, calcium carbonate can be used as a filler in alkaline pulp, whereas clay is the dominant filler in rosin-sized pulp. Calcium carbonate (chalk) is naturally brighter than bleached clay, so brighter pulp is a welcomed side effect. Material costs are further lowered because calcium carbonate's brightness reduces the need for the expensive titanium dioxide that is used to brighten rosin-sized sheets. Third, alkaline papers also have greater fold, burst, tensile, and surface strengths because of the absence of the aluminum sulfate particles that interfere with the fiber-to-fiber bond. Fourth, the calcium carbonate filler greatly improves opacity. Last, although the shift in pH value from the 5 of a rosin-sized sheet to a pH of 7 with an alkaline sheet usually requires some adjustment in the fountain solution of an offset press, printers who make the necessary adjustment report better printability than they had achieved with rosin-sized paper.

Surface Sizing

Not until the sheet has been formed and dried on the paper machine does the **surface sizing** process occur. While still on the paper machine, the surface of the sheet receives a coating of starch to augment the internal sizing. Writing papers are surface sized to retard the spreading of ink across the surface and to reduce surface rupturing when erased. Printing papers that have been surface sized are less likely to lose surface fibers due to the pull of high tack inks. This increase in surface strength is not without its side effects, though. The starch will cause the paper to become stiffer and the further cementing of the surface fibers will cause the sheet's burst strength and tensile strength to increase. Unfortunately, the starch fills in the gaps between the surface fibers and, by reducing the interfaces, reduces opacity. Also, because the starch makes the sheet stiffer and more brittle, fold strength is reduced.

A Summary of Internal and Surface Sizing

In summary, if paper receives internal sizing with rosin, it becomes more resistant to liquids; however, its permanence as well as the fold, tensile, surface, and burst strength all decrease, while the opacity improves somewhat. Surface sizing improves the sheet's surface, tensile, and burst strength, but decreases its fold strength and opacity.

Improving Permanence

A paper's ability to resist deterioration through aging is the measure of its **permanence.** The two factors that most impair permanence are (1) the amount of lignin in the sheet and (2) its acid content or pH level. *Lignin,* the natural cement that holds cellulose fibers together in wood, is the agent that causes newsprint to yellow and weaken after perhaps only a matter of days upon exposure to light. *Groundwood* papers are often one-third lignin and are the least permanent variety of the printing papers. A common example of paper with a high groundwood content is *newsprint,* the substrate of newspapers and telephone books. Several magazines consist of groundwood paper that has a white coating on both sides; the coating not only improves brightness but also extends permanence past several years. *Chemical pulping,* followed by oxygen delignification and bleaching, is the best way to remove the lignin from wood fiber, and when done properly can produce paper that will last 200 years. *Cotton and linen fibers,* however, are naturally lignin-free and can produce paper that can accurately be called permanent; the exception is when the source of the fiber is old rags or clothing that has been heavily laundered and bleached, resulting in paper made from already deteriorated fibers.

As was just discussed, a second threat to permanence is the paper's *acidity,* usually caused by the aluminum sulfate used during internal sizing. This problem can be reduced through alkaline sizing. It should be noted that alkaline and pH neutral are not synonyms for *permanent,* despite their commonly being used interchangeably. The term *alkaline* merely indicates that the pH of the sheet is 7.0 or higher, while the term *permanent* paper is accurately reserved for a sheet that is chemically stable and long-lasting regardless of its pH. The only sheets that can truly be called permanent are made totally from either good quality cotton or linen.

Printers wishing to find alkaline papers can refer to Grade Finder publications; double stars indicate those brands that contain no groundwood and have a pH value of 7.5 or higher.

Porosity

A sheet's **porosity** concerns the ease with which a gas or liquid can make its way through the fibers. Paper is inherently porous because it consists of a network of fibers with tiny air pockets between them. The air volume of a sheet of printing paper can range from 20 to 70 percent, depending on the ratio of long and short fibers, the filler content, the amount of calendering received during manufacture, and whether or not it is coated. Short fibers fit together more closely than long fibers; filler material fills some voids; calendering makes the sheet more dense; and coating seals the sheet's surface to some extent. All of these factors reduce porosity.

A sheet's porosity will certainly influence its absorbency of freshly printed ink, thereby impacting the ink's drying rate. Also, paper that will be gummed (such as envelopes) should not be so porous that it allows excessive penetration or strike-through. In the feeder system of a sheet-fed press, lightweight paper with excessive porosity can interfere with the suction-cup pickup and produce double feeding. However, heat-set presses require paper to have sufficient porosity to allow moisture to escape the sheet fast enough to avoid blistering.

Paper porosity is tested by forcing air under a specified pressure against the sample and measuring the time required for a given volume of air to escape through the sheet. More than one porosity testing device is manufactured so care must be used to note which one was used to produce the rating because the readings generated are not comparable (refer back to Figure 6–9). Also, some devices are more sensitive at measuring highly porous sheets, while others are better with very dense sheets.

The Influence of Calendering on the Sheet

Another way that the manufacturing process can influence the nature of the printed sheet is in the degree of **calendering** that the paper receives. Calendering consists of sending the paper between heavy cast metal rollers that smooth and level both sides of the sheet. Calendering is performed after the final drying on the paper machine. The pressure of the rollers against the paper increases surface uniformity; the high spots are simply pressed down and the entire sheet is compressed to some degree. The calender stacks will probably be the only smoothing operation that uncoated papers receive. Coated papers, on the other hand, will likely have this on-machine calendering supplemented with the polishing of the *coating layer.* That is, after the paper has been calendered on the paper machine, it is coated, and then taken to the *supercalender* machine which polishes the pigments in the coating layer. The result is a sheet with even greater microsmoothness and, therefore, an increased gloss level. However, the friction that produces the gloss also can slightly discolor the coating materials which, in turn, can lower the brightness.

Although a smoother sheet is the primary goal of calendering, there are some negative side effects which should be somewhat predictable by now. First of all, when the sheet is compressed, the fibers are packed more closely together. This density results in fewer interfaces and opacity falls. In addition, the greater density means that porosity and absorbency decrease, which improves ink holdout. The increased density also means that caliper is reduced. A supercalendered sheet can lose 25 percent of its caliper.

The coating that is applied prior to the supercalendering operation is used to improve the brightness, gloss, opacity, and ink holdout of the paper. The only negative effect of coating is a decrease in fold strength. However, as the percentage of adhesive in the coating mixture is increased, the sheet's performance will suffer in several ways. Chapter 4 discusses this relationship in more detail.

The Effects of Grain Direction on the Sheet

A characteristic of a roll of paper that cannot be controlled by the manufacturer is its **grain direction.** The general alignment of the individual fibers on the screen of the paper machine is the result of the forward motion of the screen combined with the sideways shake that also occurs. For this reason, the fibers' overall alignment will always be in the direction that the sheet was traveling on the paper machine. Therefore, while still on the roll, the grain direction will always be around the roll, rather than across it. After the paper on the roll has been converted into sheets, however, the grain direction cannot be seen by casual observation because a 19×25-inch sheet could have been slit and sheeted from the roll as either *grain long* or *grain short* (see Figure 6–18).

Although the grain direction of a sheet of paper is not obvious to the eye, any one of several quick tests will reveal "how the grain runs." It is important for the sheet-fed printer to acquire paper with the proper grain direction because this affects a number of other physical characteristics: folding quality, tensile strength, stiffness, and dimensional stability (this last one will be defined later in the chapter). The first three characteristics may influence the way that the finished piece will perform and the latter two may determine which edge of a press sheet will become the gripper edge; that is, which edge will go into the press first.

Looking first at fold strength, it should be noted that paper (especially in the heavier weights) folds much more easily and lies flatter if the fold is parallel to the grain direction, rather than against it. Therefore, a book's signature should

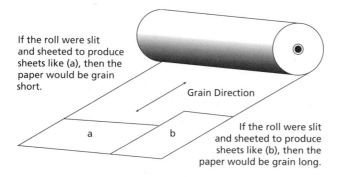

If the roll were slit and sheeted to produce sheets like (a), then the paper would be grain short.

Grain Direction

a   b

If the roll were slit and sheeted to produce sheets like (b), then the paper would be grain long.

Figure 6–18 Note how a sheet's being grain long or grain short is determined by how it was sheeted from the original roll.

be printed so that the final fold will be with the grain. This result will be accomplished by coordinating the imposition of the images on the plate, the paper's grain direction, and the lead edge of the sheet. When heavier papers must be folded against the grain, they often must first be scored or the finished piece will not lie flat after folding.

Paper is stiffer in the grain direction and this fact can influence whether the lead edge of the press sheet is the short or long dimension. The main consideration here is that if the sheet's grain is parallel to the direction of travel through most presses, the resultant stiffness will reduce jams and improve runnability. A second consideration of grain direction and stiffness involves printing an item that is designed to stand up, such as a display unit or table tent. With these examples, the image should be placed onto the paper so that the grain direction of the finished unit will be vertical to reduce sagging.

A Quick Look at Dimensional Stability

Dimensional stability is the degree to which a sheet of paper can retain its size despite changes in its moisture content. In truth, *paper is not dimensionally stable* because it is primarily composed of cellulose fibers which are *hygroscopic,* that is, very sensitive to changes in their environment's humidity. Just as a wooden door or drawer will pick up moisture and swell during a humid month, so will a sheet of paper. The amount of expansion will vary due to a variety of factors, but the most significant two are (1) grain direction and (2) density.

The analogy that was made earlier between a cellulose fiber and a drinking straw might prove useful in understanding the role that *grain direction* plays in dimensional stability. Because cellulose fibers are hygroscopic, if a sheet of paper

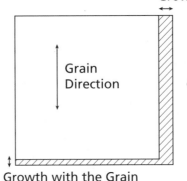

Growth Across the Grain

Grain Direction

Shaded Areas Indicate Growth Due to Moisture

Growth with the Grain

Figure 6–19 As paper picks up moisture, it grows more *across* its grain direction than it does *in* the grain direction.

is placed in a very humid environment, the fibers will begin to absorb moisture until the moisture content of the sheet matches its environment. The absorption of moisture causes the fibers to expand and this growth will be transferred to the entire sheet—but not to the same extent in both dimensions.

For example, if a 20 × 20-inch sheet of paper were placed into a room with a high relative humidity, the width might expand ¼-inch, while the height would expand only ¹⁄₁₆-inch. The reason for the difference in growth is grain direction. Recall the similarity of cellulose fibers to flattened drinking straws. As they pick up moisture, they begin to swell and expand considerably from side-to-side, but they do not grow in length nearly so much (see Figure 6–19). Therefore, because the individual fibers expand in one direction more than another (width more than height), the sheet itself will expand more in one direction than another. The dimension of the greater expansion will be perpendicular to the grain direction. The significance of this relationship is explained further in Chapter 7.

A second factor in determining a sheet's dimensional stability is its **density,** a measure of how tightly packed the fibers are to one another. Density varies in proportion to the beating of the fibers and the calendering of the sheet. Heavy beating and/or heavy calendering results in a very dense paper that lacks porosity so, when the fibers expand, they will immediately push against one another and force the overall sheet's expansion to accommodate their growth. On the other hand, in a less dense sheet with greater porosity, the fibers can undergo significant expansion before they push against one another; therefore, as the density of the sheet is increased, the dimensional stability is decreased.

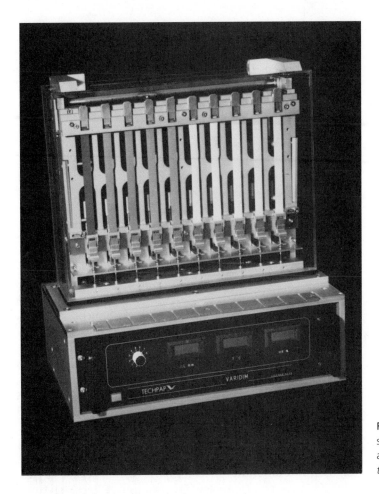

Figure 6–20 An expansimeter hangs sample strips from different sheets in a humidity controlled chamber. *Courtesy of Testing Machines Inc.*

The dimensional stability of several paper samples can be measured and compared with an *expansimeter*. This device hangs test strips of paper of a known length in a humidity controlled environment; the strips are kept straight with a slight downward pressure. The humidity is then altered for a predetermined amount of time and any dimensional changes that result are noted when the strips are remeasured (see Figure 6–20).

Moisture and Paper

Moisture content is a very critical aspect of papermaking. Not only will a change here affect the sheet's dimensions, but most of the other physical characteristics of the sheet, as well. Paper can be manufactured with high or low moisture contents, but, as has consistently been the case throughout

Figure 6–21 A sword hygrometer's blade is inserted into a pile of paper to sense the relative humidity for which the paper is balanced. The indicator shows a relative humidity of 54.9 and also a temperature of 70.4° F. *Courtesy of Rotronic Instrument Corp.*

this chapter, either extreme will have both good and bad effects. For this reason the paper mill aims at what it considers the best compromise for a given sheet's probable end use. A quick look at the effect of moisture content reveals that those physical characteristics that relate to the fiber-to-fiber bond will improve as moisture content is increased. Therefore, a higher moisture level will see the burst, surface, and tensile strengths increase. Also, a high moisture level will make the sheet less brittle and more pliable, resulting in less stiffness, along with greatly improved folding quality and tear strength.

Obviously, the moisture content of a sheet of paper right after its manufacture will rapidly change when that sheet is exposed to a different environment. For an example of this, consider a car trip from New Orleans to Los Angeles. Along the way, a road map that stood up well to use in humid Louisiana may break apart at the folds after the car has been in the dry Arizona desert for only a day or so.

Measuring Moisture Content

The moisture content of a pile of paper can be measured with a *sword hygrometer*—a device with a bladelike tube that can be inserted into the pile to measure the electrical conductivity of the air between the sheets. A wire connects the handle of the blade to a device that displays the relative humidity for which the pile of paper would be balanced (see Figure 6–21).

Caliper and Bulk

As has already been demonstrated, a sheet's thickness influences several important characteristics such as opacity, stiffness, runnability, and foldability. Therefore, it is crucial that printers view a sheet of paper as a three-dimensional object and learn to accurately measure its caliper (thickness). **Caliper** is the thickness of a single sheet of paper, usually expressed in one-thousandths of an inch (or points). For example, a sheet of Sub. 20 (or 20 pound) bond paper is usually .004-inch or four *points* in caliper.

Caliper is greatly influenced by not only the substance weight, but also by the presence of filler and coating as well as the degree of calendering that the sheet received. The interplay of these variables makes it impossible to predict caliper on the basis of substance weight or vice versa. For example, a sheet of Sub. 80 cover stock has a caliper of 10.5 points, while a heavier sheet of Sub. 100 coated cover stock is actually thinner, at 9.5 points. At the same time, Sub. 70 coated book stock is 3.5 points thick, while a Sub. 70 book with an eggshell finish is 6 points thick—an increase of 71 percent.

Paper caliper can vary across the sheet, so several measurements must be taken to generate an average. The greater variation usually is found across, rather than with the grain. Also, variation is greater in heavyweight sheets designed primarily for strength. Caliper is expressed in thousandths of an inch—usually called points or mils. For example, Substance 20 bond and Substance 50 book paper are approximately .004 inches or 4 points.

Because of the compressible nature of paper, micrometers with extra wide contact points are required, as is a specified .5 kg/cm^2 applied pressure (see Figure 6–22).

Measuring stock caliper accurately is crucial for book publishers who are concerned with **bulk,** the thickness of a number of sheets or pages of paper. The *bulking number* for a stock is the number of sheets in 1 inch after they have been squeezed between plates with 36 pounds of pressure per square inch (psi) for 30 seconds. The bulking number, multiplied by two, gives the *pages per inch* (ppi) for that stock. The ppi can also be calculated by dividing the four-sheet caliper (in points) into 8,000 or the single-sheet caliper (in points) into 2,000. For example, if a Sub. 50 offset sheet has a caliper of .004-inch or 4 points, then one inch will contain 500 pages (2,000 ÷ 4).

To calculate the bulk of a given number of pages, multiply the number of pages in question by the sheet's caliper. For example, 480 sheets of Sub. 80 coated book paper with a .004-inch caliper would occupy 1.92, or nearly two inches. The ability to predict bulk is crucial in knowing how large

Figure 6–22 The base of a paper micrometer must be wide enough to not compress the paper and provide an inaccurate reading. This sample appears to be 5.17 thousandths of an inch thick. *Courtesy of Testing Machines Inc.*

141

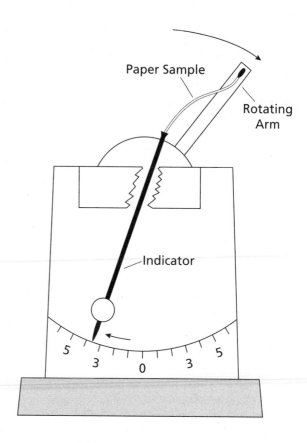

Figure 6–23 A stiffness tester (left) connects a test strip of paper to the top of an indicator arm and a rotating arm. The movement of the rotating arm places a lateral stress on paper; its resistance to bending is noted where the indicator arm points to the scale. *Courtesy of Gurley Precision Instruments.*

the spine of a book's cover must be or how many booklets can be placed into shipping containers.

Stiffness

A sheet's **stiffness**—its ability to resist being bent—can be important to the press operator as well as to the end-user. The caliper of a sheet of paper largely determines its stiffness and, therefore, the sheet's runnability through both the press and high-speed finishing equipment. In fact, if the caliper of a sheet were doubled, it would become *eight times* stiffer. Printers wishing to maximize stiffness, therefore, should choose low-calendered (antique) uncoated sheets with low filler content. Unfortunately, increasing numbers of customers want four-color process printing on glossy sheets that must also be lightweight to reduce their postage cost. In addition, printers must be aware of stock caliper if the job includes a postal reply card or picture postcards because the U.S. Postal Service requires a minimum caliper of .007-inch for cards that it handles.

Several devices are made to assess paper's stiffness, but most work on the basic principle of applying a lateral pressure to one end of a sample and measuring how far the other end of the sample moves some sort of rigid indicator (see Figure 6–23).

Two sheets of paper can be visually compared for stiffness by cutting them to the same size and simply pushing them over the edge of a table or other flat surface. The sheet that sags less is the stiffer of the two. Because grain direction influences a sheet's stiffness, be certain that the grain directions of the test sheets are parallel.

Substance Weight

When a bond sheet is referred to as "20 pound," the **substance weight** (or *basis weight*) is being given. Chapter 9 explains substance weight in detail, but a brief explanation is that it is the weight of one ream (500 sheets) of that sheet in a specified size. In the metric system, a sheet's weight (grammage) is measured in grams per square meter of a single sheet.

When weighing a sheet, any one of several scales can be used to compute the substance weight from only a single sheet that can be as small as 2×4 inches. The most modern devices are electronic and provide digital displays. Substance weight is important for customers who will be mailing printed materials.

TAPPI Test Standards

The Technical Association of the Pulp and Paper Industry (**TAPPI**) has developed standards for most techniques used to test the properties of paper. These standards establish the specifications and procedures that are necessary before data from testing labs around the country can be comparable. Companies that market paper testing devices will usually reference the particular TAPPI test to which a given machine has been designed to conform. It should be noted that TAPPI often has more than one method for assessing a single property. A good example of this diversity of technique concerns testing a paper's sizing. The test that places a piece of paper carrying particles of a water-soluble dye and measures the time before the dye melts in TAPPI Standard T–433. The test that weighs the sample before immersing it in water for a specified time period and reweighing it is TAPPI Standard T–491. Several others exist for sizing testing, so a person reading the test data must be made aware of the TAPPI test that was used. Also, testing instruments are designed to conform to a particular TAPPI test. Relative to paper smoothness, the Digi-Bekk and Gurley testers conform to TAPPI T–47, while the Sheffield three-column model conforms to TAPPI T–538. From an international perspective, there are several other organizations that have developed standards for paper testing, but TAPPI tests are by far the most widely referenced in America.

Printability Testers

Multipurpose devices called *printability testers* are available from more than one manufacturer. Although the list of tests that can be performed varies among the machines, a sampling of their abilities includes testing for both wet and dry picking, smoothness, ink setoff, abrasion, backtrapping, dot gain or loss, and ink transfer during lithographic, gravure, or flexographic printing.

Summary of Physical Characteristics

The primary lesson of this chapter is that the buyer of printing paper must be aware that when a certain physical characteristic is sought in paper, there is a necessary trade-off to be made. By imparting the desired characteristics to the sheet during manufacturing, other characteristics will be significantly altered. The buyer must be aware of the cause-and-effect relationships and understand what necessarily will be sacrificed and to what degree any trade-offs will occur. At this point, the buyer can prioritize what will be needed from the paper and place the order accordingly. After all, only this awareness of a given paper's capabilities can prevent a sheet from being used inappropriately (i.e., sending letterpress paper through an offset press) or asking more of a sheet than is realistic.

Questions for Study

1. What are the three phenomena that can occur when light strikes a surface? Which translates into brightness?
2. What is the difference between gloss and brightness? Can a paper have high gloss and low brightness? If so, how?
3. How do fluorescent dyes improve a sheet's brightness?
4. Describe how internal reflection improves a sheet's opacity.

5. What role does the number of interfaces play in improving a sheet's opacity? Itemize three ways of increasing the number of interfaces.
6. What are the two main reasons for beating pulp?
7. Why does heavy beating decrease tear strength while improving fold, tensile, and burst strength?
8. How does beating affect a sheet's caliper?
9. How is internal sizing different from surface sizing?
10. How does alkaline sizing compare with rosin sizing relative to a sheet's strength and opacity?
11. What two factors primarily determine permanence? How can they be controlled?
12. What is revealed by a paper's pH value of 5?
13. Can the grain direction of a roll of paper be predicted?
14. List three reasons why grain direction is important to a printer.
15. How would you define dimensional stability? What are the two main factors that affect it?
16. How thick, in thousandths of an inch, is 9 point paper?

Key Words

brightness
visible spectrum
colorimeter
opacity
filler
smoothness
tensile strength
folding quality
internal sizing
porosity
hygroscopic
hygrometer
stiffness
fluorescent

whiteness
gloss
internal reflection
show-through
tear strength
tensile-at-the-fold
surface strength
alkaline sizing
grain direction
density
caliper
physical characteristics
spectrophotometer
gloss meter

interfaces
strike-through
burst strength
folding strength
permanence
dimensional stability
expansimeter
bulk
TAPPI
color
calendering
moisture content

7 Using Paper in Printing

AFTER STUDYING THIS CHAPTER, THE STUDENT SHOULD BE ABLE TO:

■ Identify the causes for curled paper.

■ Explain how paper's moisture content can produce wavy edges or tight edges and how these conditions affect the sheet's performance on the press.

■ Describe how a sheet's grain direction affects a sheet's press performance and dimensional stability.

■ Compare and contrast coating pick, fiber pick, and blocking-in-the-pile.

■ Distinguish among linting, dusting, powdering, piling, and scumming.

The chapter on physical characteristics established that a number of variables exist in the manufacture of a given sheet of paper and that any variation will affect the nature of the finished sheet. It was also pointed out that no sheet can be superior in all respects. Because of the complex nature of papermaking, the printer must expect a wide range of performance levels when different kinds of paper are used. Sometimes problems will occur that can be traced to a defect in the paper's (1) manufacture, (2) transportation or storage, or (3) use on the press. In most instances, the problem can be solved by an adjustment by the press operator, but occasionally the situation is irreversible. This chapter is intended to acquaint the reader with some of the most common problems involving paper in the pressroom and provide insight into their causes and possible solutions.

Structural Curl

The first broad category to be examined involves curl—a condition whereby a sheet of paper is not flat (see Figure 7–1). Because curl can occur during manufacturing, handling, or printing, there are several categories. **Structural curl** occurs at the paper mill through too much beating of the fibers or nonuniform drying of the formed sheet. Papers that are coated-one-side (**C1S**) can also pick up a curl with the grain immediately after the coating process. This

Structural Curl

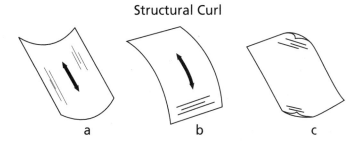

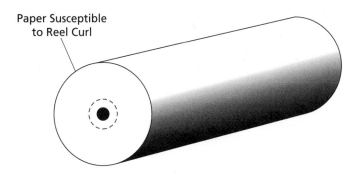

Figure 7–1 Structural curl can occur parallel, perpendicular, or even diagonally to the sheet's grain direction. In all instances, it is a permanent condition.

Figure 7–2 Paper that is sheeted from the indicated portion of this roll will have the curl of the reel "set" into them.

happens because the uncoated side has less protection from the surrounding humidity and will pick up moisture faster than the coated side; therefore, the uncoated side will swell at a greater rate, producing a curl toward the coated side. This can also occur with papers that are coated-two-sides (**C2S**) when the amount of coating varies from one side to another or any time when the two sides have been finished differently (e.g., degree of supercalendering). Sheets with structural curl cannot be corrected and should be immediately reported to the paper merchant by the printer.

Reel Curl

A second type of curl that originates at the mill is **reel curl** or *wrap curl*. Unlike structural curl, which is built into the sheet during manufacture, reel curl takes place after the finished paper has been wound onto the roll.

While in the roll, the paper near the core is obviously wound around a small circumference and this curve becomes part of the paper (see Figure 7–2). To avoid producing paper with reel curl, the paper mill should only sheet paper to a certain distance from the roll's core. The cutoff point is determined by the weight of the paper, the diameter of the core inside the roll, and the length of time the paper has

Figure 7–3 As moisture comes in contact with the top of this sheet, the fibers near the surface absorb the moisture, swell, and cause that portion of the sheet to grow. This is why paper will always curl away from the wet side. As the moisture moves to the bottom of the sheet and the top begins to dry, the curl will reverse itself.

Wet Side Where Fiber Growth Occurs

been left on the roll; reel curl is more pronounced with heavyweight paper, narrow cores, and long periods on the roll. Like structural curl, reel curl should not pass a mill's inspection and be sold to a printer. Sometimes, however, paper with reel curl can make its way to the printer if the inner portion of a roll is sold to a paper *jobber* rather than be designated as broke and repulped. The jobber may then slit and sheet the small roll and sell the resulting sheets of paper. The primary problems that structural and reel curl present the printer are found in the feeder system of a sheet-fed press and the appearance of the final job because neither problem disappears during printing.

Paper Curl That Results From Printing

There are two types of curl that can develop during printing: **offset curl** and **back-edge curl.** If lightweight paper is printed on an offset press with very little ink coverage, **offset curl** (or moisture curl) can be the result (see Figure 7–3). Because the nonimage area of a plate is always carrying dampening solution, as the amount of ink that is transferred to the paper decreases, the amount of moisture it receives increases. This moisture will first cause the sheet to expand across the grain and away from the printed side and then dry with the curl toward the printed side. The lithographic printer can minimize offset curl by (a) reducing the amount of dampening fluid, (b) avoiding paper with poor dimensional stability (see Chapter 6), and/or (c) using the largest press sheet possible. (For the reader who is not familiar with printing technology, Chapter 14 explains the principles involved in lithography as well as the other printing processes.)

Back-edge curl (also known as *tail-end hook*) is a problem that can occur only on a sheet-fed offset press. It is a combination of high tack inks on the blanket and solid areas being printed near the trailing edge of the sheet. When both situations exist, the tacky ink does not release the sheet from the blanket until the paper is finally stressed to the point that it

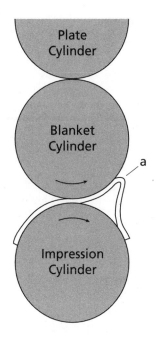

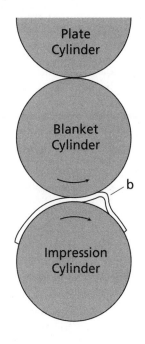

Figure 7–4 Back-edge curl occurs as the sheet is pulled away from the ink-covered blanket roller (point A). If this release point were back at point B, back-edge curl would not occur.

picks up curl from the release angle (see Figure 7–4). The result is always a curl away from the printed side and reflective of the direction of travel through the press. Several factors can minimize back-edge curl: (a) the degree of tack in the ink used; (b) the placement of solid areas on the plate; (c) the stiffness of the paper as determined by caliper, basis weight, and grain direction; and (d) press speed.

How Moisture Affects Paper

The second broad category of problems that occur when printing on paper involves its reaction to changes in moisture. As was explained in the chapter on physical characteristics, paper is a *hygroscopic* substance—that is, it readily absorbs and gives off moisture to conform to its environment. This characteristic not only will cause the dimensions of the sheet to change somewhat, but will also affect its fold, tear, burst, and surface strength, along with the caliper and stiffness. These hygroscopic properties are also of prime concern to the press operator as well because they are a major cause of printing problems. To understand how changes in moisture content can be so significant to the printer, one must first understand the concept of relative humidity and the way that printers measure the moisture content of both the paper and the surrounding air.

The Impact of Relative Humidity on Paper

When a roll or a carton of paper is made at the mill, it may ultimately find its way to a printing firm anywhere from Florida to Nevada—locations with obviously different relative humidities. A sheet that is unwrapped in a Louisiana plant in August may find itself surrounded by air that has an 80 percent **relative humidity** (R.H.) and, as a result, will start to pick up moisture until it is stabilized with the room. If the same sheet were shipped to a Las Vegas printer, it could find itself exposed to a pressroom relative humidity of 25 percent, whereupon it would promptly begin to lose moisture on its way to equilibrium with the surrounding air (see Table 7–1). Therefore, a sheet that is too moist for one plant may be too dry for another.

By the same token, the relative humidity of the same pressroom can vary greatly from month to month, creating the same problem. For example, a Pittsburgh printing company could have a 70 percent R.H. in August and then find the furnace-heated air of February to be 10 percent R.H. These shifts in humidity can be a chronic source of aggravation to the printer. However, because a paper mill cannot

	J	F	M	A	M	J	J	A	S	O	N	D
Atlanta	41.0	44.8	53.5	61.5	69.2	76.0	78.8	78.1	72.7	62.3	53.1	44.5
Boston	28.6	30.3	38.6	48.1	58.2	67.7	73.5	71.9	64.8	54.8	45.3	33.6
Columbus	26.4	29.6	40.9	51.0	61.2	69.2	73.2	71.5	65.5	53.7	42.9	31.9
Denver	29.7	33.4	39.0	48.2	57.2	66.9	73.5	71.4	62.3	51.4	39.0	31.0
Detroit	22.9	25.4	35.7	47.3	58.4	67.6	72.3	70.5	63.2	51.2	40.2	28.3
El Paso	42.8	48.1	55.1	63.4	71.8	80.4	82.3	80.1	74.4	64.0	52.4	44.1
Los Angeles	56.8	57.6	58.0	60.1	65.7	69.1	70.5	69.9	66.8	61.6	56.9	63.0
Miami	67.2	68.5	71.7	75.2	78.7	81.4	82.6	82.8	81.9	78.3	73.6	69.1
Minneapolis	11.8	17.9	31.0	46.4	58.5	68.2	73.6	70.5	60.5	48.8	33.2	17.9
New York City	31.5	33.6	42.4	52.5	62.7	71.6	76.8	75.5	68.2	57.5	47.6	36.6
Phoenix	53.6	57.7	62.2	69.9	78.8	88.2	93.5	91.5	85.6	74.5	61.9	54.1
Pittsburgh	26.1	28.7	39.4	49.6	59.5	67.9	72.1	70.5	63.9	52.4	42.3	31.5
Richmond	35.7	38.7	48.0	57.3	66.0	73.9	78.0	76.8	70.0	58.6	49.6	40.1
San Francisco	48.7	52.2	53.3	55.6	58.1	61.5	62.7	63.7	64.5	61.0	54.8	49.4
Seattle	40.1	43.5	45.6	49.2	55.1	60.9	65.2	65.5	60.6	52.8	45.3	40.5

Table 7–1 The average monthly temperatures (°F) of these American cities over a 30-year period reveal the wide range of conditions that an open pile of paper would be exposed to inside a pressroom. In March, a Miami pressroom would likely need little or no heating, but the freezing outside temperatures in Minneapolis would probably require heating the pressroom enough to lower the relative humidity and cause tight edges in the paper. Because paper from the same production run at the mill could be opened in any city and month, each pressroom must have the climate control capability to provide the 45–55 percent R.H. with which the paper is compatible.

know its paper's geographic destination or the relative humidity of a given pressroom on a given day, it makes more sense for paper to be made with a constant moisture content and for printers around the country to control the climate of their pressrooms by using air conditioning in the summer and humidifiers in the winter. With climate control, the pressroom can avoid wide swings in humidity from morning to afternoon and throughout the year (for an explanation of relative humidity, see Appendix A).

To be technically correct, relative humidity is a measurement that only can be applied to air; however, printers will often refer to the relative humidity of paper. What they are actually describing is the *equilibrium* that will develop between the paper and air with that relative humidity. For example, if a paper's moisture content is such that it will neither absorb nor release moisture when placed into a room with a 45 percent R.H., then the paper is said to have a 45 percent R.H. The degree to which a pile of paper is in equilibrium to the atmosphere of, say, a pressroom can be determined by using a paper hygroscope, an instrument consisting of a swordlike blade that is inserted into the pile and an attached meter that indicates the R.H. at which the paper would be in equilibrium.

Wavy Edges in Paper

The ability to measure the moisture balance between a skid of paper and its pressroom is crucial because of the hygroscopic nature of paper. It was noted earlier that when a single sheet of paper is placed in a room with an R.H. it is not balanced with, the sheet will either take on moisture and expand or give up moisture and shrink. But what would happen if the sheet were not uniformly exposed to the air? Take, for example, one of the middle sheets in a 4-inch pile of paper. That sheet would have 2 inches of paper above and below it and, therefore, would only be in contact with the air at its edges. In this instance, if the surrounding air is more moist than the paper it is balanced for, the middle sheets will absorb moisture and expand only *at the edges*. Ordinarily, the swelling of fibers would cause the sheet to grow across its full length and width, but, if the large central part of the sheet remains unaffected by the air, the outer portions will be held back and not allowed to actually become longer. To accommodate their fibers' growth, the edges will necessarily become wavy (see Figure 7–5). Hence, **wavy edges** are the result of moisture absorption by only the outside portion of a pile of paper and do not occur when an entire sheet is ex-

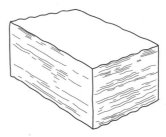

Figure 7–5 Wavy edges begin at the outside of the pile of paper where fibers contact the surrounding air, absorb moisture, and swell. The longer that paper is exposed to moist air, the farther toward the center will the waviness move.

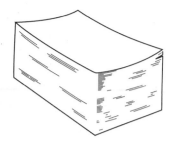

Figure 7–6 Tight edges result in saucer-shaped sheets as the outside edges of a pile of paper lose moisture to dry surrounding air.

posed; for example, the top sheet of a pile will not develop wavy edges.

Wavy edges are a real problem for the press operator because only the perimeter of the sheet is wavy, while the majority of the sheet is flat. When a wavy-edged sheet is forced between the cylinders of a press, the curled edges will be flattened into wrinkles. This effect is most pronounced during offset lithography because the planographic nature of the plate and the blanket are very unforgiving of any deviations from being perfectly flat. Wavy-edged paper can also present problems when precise trimming is needed or when close register multicolor printing is done near the press sheet's edges.

The problems of excessive humidity in the pressroom are not limited to sheet-fed press operations; web press operators must protect their rolls of paper, as well. As roll ends absorb moisture, they can swell, become spongy, feed with uneven web tension, and misregister or wrinkle. When rolls of coated paper absorb excessive moisture, they can blister under heat drying. Finishing and converting operations can also be hindered by high humidity because moisture absorption causes a sheet to lose stiffness and potentially hurt its runnability in folding and other high-speed operations.

Wavy edges can result in the winter when paper that has been chilled in transit or storage is unwrapped inside a warm pressroom. The outside edges of the cold paper will chill the surrounding air, thereby lowering its capacity to hold moisture and raising the relative humidity to 100 percent. At this point, surplus moisture will condense on the edges of the paper and quickly be absorbed. The only way to prevent this headache is to give the paper adequate time to warm up before unwrapping it.

Tight Edges in Paper

When a skid of paper is exposed to air that is significantly drier, the shift in moisture content goes in the opposite direction. The outside portions of the sheets shrink and **tight edges** (or *shrunken edges*) result (see Figure 7–6). This tightening of the edges will produce a sheet with raised corners and, like wavy edges, the result can be wrinkling, inaccurate trimming, and faulty registration near the outer part of the press sheet. But these are not the only problems. If the sheet's R.H. falls below 35 percent, static electricity can develop, resulting in jams in the feeding system of sheet-fed presses, folders, and other equipment. Tight edges are very commonly found in northern states' pressrooms during winter months. Furnaces without humidifiers are the usual cause because they can lower the pressroom's relative humidity to as low as 5 percent.

Because paper mills usually produce paper that is balanced for 40–45 percent R.H., the printer's best approach to avoiding the host of problems associated with moisture exchange is simply to *control the climate of the pressroom*. When this is not possible, the next best approach is to minimize the time between when the paper is exposed to the air (unwrapped) and when it is printed. Bear in mind that rewrapping skids with regular paper will not prevent moisture exchange between the skid of paper and the surrounding air.

The Effects of Moisture Absorption During Offset Printing

As was explained in the chapter on physical characteristics, a sheet of paper that picks up moisture will grow more across the grain than in the direction of the grain. This trait is important to the offset press operator due to the offset blanket's application of moisture onto the sheet. Imagine how the grain direction will affect a two-color job that will be run on a one-color offset press. Obviously the sheet will be sent through the press for the first color, and, after the ink dries, it will be sent through again for the second color. However, because the sheet absorbed dampening fluid as well as ink from the blanket, it expanded in the delivery pile and, even if allowed to dry over the weekend, will remain a little larger than it was before it was printed. Because paper has a characteristic called *hysteresis,* it never shrinks back to the exact size it was—even when it returns to the original "relative humidity." This phenomenon means that paper will give up less moisture on its way down from a high relative humidity than was absorbed on its way up (see Figure 7–7).

To the press operator, hysteresis spells trouble in the form of misregister. For example, if the sheet has expanded after being printed by the first plate, then the image on the second

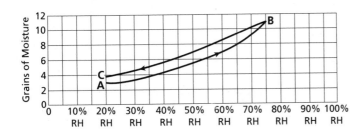

Figure 7–7 This chart traces the dimensional changes of a sheet of paper as it starts with a low relative humidity (point A), absorbs moisture (point B), and is then dried to its original relative moisture (point C). Observe that the sheet does not return to its original moisture level, and therefore original size, when the original relative humidity is achieved.

153

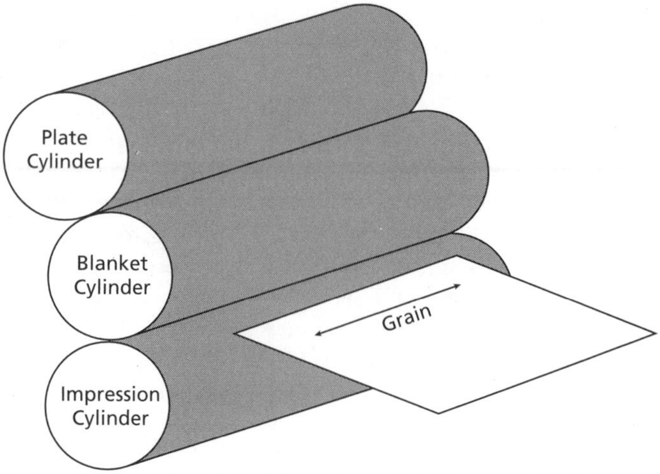

Figure 7–8 In this drawing, the press sheet is being fed so that the sheet's grain direction is parallel to the rollers. Another way of phrasing this is to say that the grain direction is perpendicular to the direction of travel.

plate is now too small to register with the first plate's image. The solution here is to "stretch" the image on the second plate by increasing the packing of the plate and blanket cylinders, which will have the effect of expanding the second image relative to the paper in the dimension that wraps around the plate. Adjusting the packing has no effect on the image in the across-the-plate direction. This procedure can allow the image on the second plate to keep pace with the growth of the image on the paper, providing that the sheet was printed with its grain direction parallel to the press rollers (see Figure 7–8). The press sheet's grain direction is important because a sheet's growth is significantly greater across the grain than with the grain. If the grain direction is parallel to the rollers, the sheet's growth will be greater from the lead edge to the trailing edge than from side-to-side and it is the growth in this head-to-tail dimension that packing adjustments can counteract.

Grain Direction Relationships With the Press Sheet

The person who places the order for paper and specifies the desired grain direction will do so to maximize production efficiency. An example of this principle is how the *grain direction* will relate to the press rollers. It was pointed out in an earlier chapter that paper is stiffer in the direction that the grain runs. Therefore, when a press sheet's grain direction is parallel to the direction of travel, meaning perpendicular to the press rollers, greater stiffness occurs with the usual result of fewer jams. But when a job calls for multiple passes through the press, this improved runnability must be sacrificed for better register control as just described. To com-

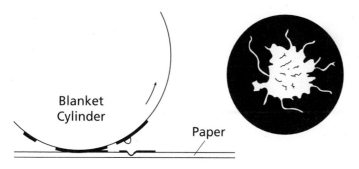

Figure 7–9 When high-tack ink meets paper with inadequate surface strength, the ink film may not always split as it comes out of the nip between the blanket and plate cylinders. When the pull of the ink exceeds the fiber-to-fiber bond of the sheet, the sheet's surface ruptures and a cluster of fibers clings to the blanket. Examining the ruptured sheet with a magnifying glass will reveal a very irregularly-shaped hole with an uneven floor.

pound matters further, the press sheet's grain direction is also of concern to the bindery department for a couple of reasons: (a) it can reduce jams in the folder and other equipment; and (b) it greatly affects folding quality. Because no way exists to please all of the people all of the time, a conference with the production manager is the best policy to follow when a question of priority arises.

Problems with Picking

A broad range of difficulties can face the press operator when paper comes in contact with ink, and a number of these problems concern the paper's surface strength. The first condition to be examined is **picking**—the rupturing of the surface due to the pull of the ink film. In letterpress printing these small particles come off onto the plate and in offset printing they attach themselves to the blanket. In both cases, the transfer will be evidenced on successive sheets until the press is stopped and the particles are removed. There are two types of picks: fiber pick and coating pick.

Fiber pick occurs when *clumps* of surface fiber are lifted from the rest of the sheet by the ink film (see Figure 7–9). Fiber pick usually occurs with large solid areas. When these fibers transfer onto an offset blanket, they promptly become wet and begin to reject ink, creating white areas in the solid areas of successive sheets. To identify fiber pick, the press operator can simply move through the delivery pile until the first instance of the blemish is found. Close examination will reveal a shallow hole from where the fibers were pulled; this rupture will be irregular in shape and have a rough floor. Fiber pick can be caused by paper with inadequate fiber-to-fiber bond or ink with excessive tack. In either case, the problem can usually be minimized by reducing the ink, slowing the press, or both.

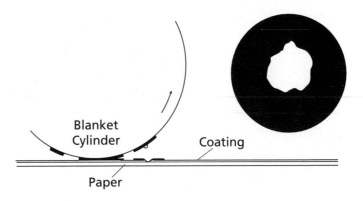

Figure 7–10 Coating pick is similar to fiber pick except that the paper's weakness is not in its fiber-to-fiber bond, but rather in the adhesive level of the coating. As a result, a close examination of coating pick will find a more uniformly defined hole with a smoother floor than is found with fiber pick.

Blanket
Cylinder

Coating

Paper

Coating pick is similar to fiber pick, but it is exclusive to coated papers and is due to inadequate adhesive in the coating mixture. To identify coating pick, once again find the sheet where it occurred and examine the rupture under a magnifying glass (see Figure 7–10). If the outline of the hole is not very ragged and the floor is smooth, then it is the coating—not a clump of fibers—that has been pulled away by the ink film. As is the case with fiber pick, the best solution is to reduce ink tack, slow the press, or both. Due to the high tack ink that they use, lithographic printers see more fiber and coating pick than their counterparts in letterpress or gravure printing.

Blocking-in-the-Pile

A problem that is often mistaken for coating pick is **blocking-in-the-pile.** Whereas picking takes place at the moment that the plate or blanket comes away from the paper, blocking-in-the-pile occurs in the delivery system of the press as the sheets are drying. What occurs here is a heavy layer of ink attaches itself to the sheet above it and pulls off bits of coating (see Figure 7–11). To diagnose blocking-in-the-pile, compare the top of a sheet with the back of the sheet above it. The dilemma can usually be eliminated by not allowing the delivery pile to get so tall and thereby not allowing the weight to build up. Laying down a thinner ink film, using a faster drying ink, or substituting more absorbent paper can also help.

Linting, Dusting, and Offset Spray Problems

Under the broad category of unclean paper is a host of ways that extraneous particles can find their way onto the blanket or plate and interfere with print quality. What they have in

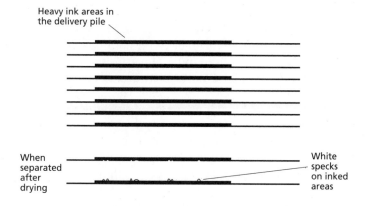

Heavy ink areas in
the delivery pile

When
separated
after
drying

White
specks
on inked
areas

Figure 7–11 In this representation of blocking-in-the-pile, note how the white specks on the solid areas have come from the bottom of the sheet directly above.

common is that they both require the press to be stopped for otherwise unnecessary washups.

One of these problems is **linting,** a situation that can occur on a sheet with adequate surface strength. Whereas picking is the separation of clumps of fibers from the sheet, linting is the lifting of an occasional fiber that did not really become part of the sheet's surface structure. Although not as severe at first as picking, during a long run the accumulation of these individual fibers can severely hurt print quality.

Dusting is the picking up of loose particles that were created by a dull slitter wheel or cutting blade and scattered between the sheets. In this instance the fault lies not with the paper's structure, but rather with its subsequent handling, either at the mill or later. A similar situation can develop if an unclean press is allowed to deposit excessive offset spray onto printed sheets as they enter the delivery pile. The spray contains white particles designed to prevent the set-off of a printed image onto the back of the sheet above it. If these sheets are then given a subsequent pass through the press, the ink can lift these particles from the sheet and onto the blanket or plate. Here, the problem is not with the paper at all.

Powdering and Piling Problems

Powdering and *piling* are two annoyances for the offset printer that are similar in origin. **Powdering** occurs when the ink film pulls some filler particles from an uncoated sheet or coating particles from a coated sheet. **Piling** is also a lifting of mineral particles, but it is the dampening solution—not the ink—that is the cause. Still another headache can occur if the sheet is not adequately moisture-resistant (e.g., either starch-coated paper or coating adhesive that is too water-soluble), the clay particles can be washed from the paper and onto the blanket. With both powdering and piling, the accumulation of mineral particles onto the blanket has an abrasive effect on the plate and can wear the film from the nonimage areas of the plate. This, in turn, can result in ink getting onto nonimage areas—a condition known as **scumming.**

There are still other developments that can arise in certain printing situations: splitting, blistering, embossing, and delamination are among these but, because they are characteristics of heat-set drying, blanket-to-blanket web printing, or some other specific circumstance, they will not be examined here. However, an adequate number of printing problems have been discussed in this chapter to make the point that the type of paper used and its quality are critical considerations when maximum press performance and the quality of the final product are sought.

Summary

At this point some basic statements can be made about the requirements for printing paper.

1. Paper must be flat.
2. Grain direction can affect paper's runnability and dimensional stability.
3. The moisture content must be compatible with the relative humidity of the pressroom.
4. Adequate surface strength is required to prevent picking.
5. Paper with dust, lint, or powder on the surface will print badly.
6. Paper used in offset lithography must be adequately moisture-resistant.

The problems examined in this chapter are commonly attributable to some characteristic of the paper. Problems that are usually traced to the ink, an interaction between the ink and paper, or another factor are examined in Chapter 14.

Questions for Study

1. List the three main categories for problems that can occur with paper.
2. How can structural curl result? Can this condition be remedied?
3. How is reel curl different from structural curl?

4. Offset curl and back-edge curl can both occur on the press, but how are they different? What can be done to avoid each?
5. Define relative humidity. How does it relate to the printer?
6. What causes wavy edges?
7. Which is worse, a pressroom with a relative humidity that is too dry or one that is too wet?
8. What effect can hysteresis have on a job that will have to make multiple passes through a press?
9. Would a printer concerned about close register on a three-color job that will be run on a single-color press want the grain to be parallel or perpendicular to the direction of travel? Why?
10. What is the difference between fiber pick and coating pick? What is the cause of each?
11. What causes blocking-in-the-pile and what are three remedies?
12. How are linting, dusting, and powdering different? Which can cause scumming and why?

Key Words

structural curl
offset curl
wavy edges
picking
blocking-in-the-pile
piling
C2S

back-edge curl
tight edges
fiber pick
linting
powdering
reel curl
relative humidity

hysteresis
coating pick
dusting
scumming
C1S

8 Classifications of Paper

AFTER STUDYING THIS CHAPTER, THE STUDENT SHOULD BE ABLE TO:

■ Identify the five broad classifications of printing papers.

■ List the categories of bond papers and explain how grade and fiber content influence price.

■ Describe how carbonless bond works.

■ Describe the following categories of bond paper: manifold, MICR, safety, ledger, vellum, and translucent.

■ List the categories of book paper and identify their key characteristics.

■ Compare and contrast bristol and cover papers.

■ List the types of paper that are considered utility sheets.

In previous chapters, it was pointed out that the manufacture of paper can be controlled to produce a wide range of products. For the printer, this is a mixed blessing; the good news is that there are many categories of paper from which to choose, but the bad news is that the printer must make an informed choice from a wide range of alternatives. The use of an inappropriate sheet can result in extra expenses and production headaches for the printer, as well as dissatisfaction for the customer.

However, it is also the case that a thoughtful selection of a sheet by classification, category, weight, color, and texture can propel a printed piece into a new dimension of beauty and appeal. It is this skillful blending of ink and paper that can distinguish the adequate from the excellent and the missed deadline from the punctual. In short, a thorough knowledge of paper is fundamental for success in the arenas of both production and design.

In broad terms, paper can be structured into five classifications. They are *business, book, bristol, cover,* and *utility*. Each classification contains several categories of different papers, each made to meet a specific printing need.

Business Papers

A logical place to begin is with the **business papers** classification because these sheets are so common in everyday life.

As the name suggests, these papers are commonly found in the business world and within this grouping there is a diverse collection of stocks.

Bonds

Bond papers comprise the first category of business papers to be examined. Although the name comes from its originally being used to print stocks and bonds, it is primarily thought of today as the sheet for letterhead stationery. This end use obviously requires that the sheet be able to handle pen-and-ink writing, typing, erasure, and folding. Because letterhead stationery is expected to carry not only business communications, but also reflect a positive image of the company or agency sending it, several aesthetic considerations are available and bonds are often purchased with concern for color, texture, durability, stiffness, and rattle (the sound made by a sheet when held at a corner and shaken). Because bond paper is usually printed on offset duplicators, it is given a high degree of moisture resistance through both internal and surface sizing. This prevents the sheet from absorbing too much of the offset blanket's dampening solution, which usually dominates the blanket due to letterhead stationery's light ink coverage. In this way, bond is similar to offset paper, which will be discussed later. However, because bond paper is usually printed on one side and with one or two colors of ink (as with letterheads, forms, and invoices), the sheet has very little filler and will not have the opacity or dimensional stability of an offset stock.

When buying bond paper, decisions will have to be made on both the sheet's *fiber content* and *quality grade*. Table 8–1 indicates the range of choices available and their cost comparisons. Sulfite bond is comprised of wood fibers that were

	Sulfite	25% Cotton	50% Cotton	75% Cotton	100% Cotton
Premium #1	1.94	2.04	2.37	2.71	3.29
#1	1.15				
#4	1.08	1.61			
#5	1.00				

Table 8–1 This chart shows the availability of both sulfite and cotton bond in the various grades. Also shown is a cost comparison in which the #5 sulfite sheet is the basis of comparison for all other grades. That is, if a given quantity of #5 sulfite bond costs $1.00, that same quantity in, for example, a #1 sulfite sheet would cost $1.15.

chemically pulped to remove nearly all of the lignin and then bleached. Within this group are several grades that vary in quality and price. At the bottom is the No. 5 grade sulfite bond, sometimes referred to as *utility bond,* and as the name implies, this sheet is the least expensive and is used for invoices, statements, fliers, and some letterheads. By moving up to higher grades, a buyer can obtain a sheet with greater whiteness, strength, rattle, and purity. The cost can also be expected to increase, but probably not the press performance. Many printers are surprised to learn that the more expensive sheets give them more trouble in both runnability (moving through the press without jamming or misfeeding) and printability (accepting the printed image). The reason for this apparent contradiction is that the primary differences among the grades concern the sheet's appearance and feel, not its performance on the press. Recent studies in which press operators ran various grades of unmarked bond papers have confirmed what previously had been widely suspected. A glance at the chart will reveal the absence of No. 2 and 3 grades, which once existed before the law of supply and demand saw these middle grades become extinct.

At the top of the sulfite bond column are the Premium No. 1 and No. 1 grades. Nearly all mills that produce a line of bond paper will offer a No. 1 sheet. Less popular is the Premium No. 1, which costs more than other grades.

Once called *rag bonds* because of their historic fiber source, **cotton bonds** will have at least some cotton in them. Cotton content is expressed as a percentage, with the remaining percentage being chemical wood pulp. The addition of cotton to a sheet brings with it two advantages—*permanence* and *strength*. Bond that is made of 100 percent cotton fiber contains no lignin or aluminum sulfate to cause the sheet to discolor or become brittle with age. This quality level is obviously necessary for legal documents, wills, stocks, bonds, certificates, and other items requiring long-term (or archival) permanence. The sheet's high strength is because cotton fiber is roughly four times as long as wood fiber. There are some trade-offs to be made, however. A move from a No. 5 sulfite sheet to one made of 100 percent cotton will see an increase in permanence, erasability, brightness, and cost; however, the long fiber and greater refining of the cotton sheet will impart to it some of the runnability problems discussed earlier with respect to sulfite grades.

As Table 8–1 reveals, a buyer can select from four degrees of cotton content, but the lower quality grades that exist for sulfite bonds are gone. As can also be seen, the most expensive of all bonds would be a mill's Premium grade 100 percent cotton bond.

Not shown on the chart is a specialty bond that exists for typists who are error-prone: **erasable bond** allows typing mistakes to be easily removed with minimal disruption of the fibers. The sheet carries a substantial coating of *starch* so that the typing is applied onto and removed from the starch surface instead of the fibers.

One characteristic of all No. 1 grade bonds is that they carry a watermark to identify their manufacturer; this is also true of sheets with a cotton content, where the watermark reveals the percentage of cotton fiber in the sheet. Some companies or agencies have the stock for their letterhead stationery custom-made by a mill to feature their own name or insignia as a watermark. The investment in a custom watermark makes a point about how important some companies consider the "first impression" that letterhead stationery can create.

In the same vein, many firms and associations are concerned about the *color* and *finish* of their letterhead stationery. In addition to a plain or **wove** finish is the **cockle** finish, which resembles the handmade sheets of yesteryear (see Figure 8–1). Premium No. 1 bonds very often have a cockle finish. A point of interest here is that the cockled finish of handmade paper was the result of nonuniform air drying that was overcome with the introduction of the drum drier. The cockled finish is achieved by resorting to a rather primitive method of drying that turns out to be more expensive because it is slow. Other options exist that can give a letterhead a classic look. One is a **linen finish,** achieved by embossing a texture into the sheet. A second is the **laid finish,** which suggests the look of handmade paper by simulating, through a combination of watermarking and embossing, the imprint of the laid wire screen used to form sheets at the vat (refer back to Figure 1–6). As might be expected, any sheet that receives additional finishing "off the paper machine" (after the roll has been removed from the machine) will reflect this extra treatment in its price.

The substance weights that are available with bond paper are Sub. 13, 16, 20, and 24. Of these, Sub. 20 is used most often. Chapter 9 will explain substance weights in detail.

Business-forms bond (or *forms bond*) is made especially for the printing of continuous business forms on web presses. Because it is made for web printing, business-forms bond is not classified as a bond, but is mentioned here be-

Figure 8–1 These close-up photographs reveal three finishes that are commonly available in bond papers. From left, they are (a) wove, (b) cockle, and (c) laid.

cause the word bond is in its name. The common substance weights are Sub. 12, 15, and 20. Business-forms bond needs to be fairly lightweight to produce clear carbon copies.

Many multipart forms have traded in carbon paper for the carbonless paper method of making copies. **Carbonless** bond (also known as *chemical transfer paper*) works on the principle of using pressure to rupture capsules carried throughout a sheet's backside coating. Once broken, the capsules release a dye that becomes visible when it contacts the receptive chemical coating on the top side of the sheet below it (see Figure 8–2). Obviously, this form of image transfer works only when the sheets within a set are facing in the correct order.

Two kinds of carbonless capsules exist: cluster and mononuclear. *Cluster* capsules are thin-skinned and can be accidentally broken with modest pressure, so they are mixed with larger starch particles called stilts that absorb small pressures to prevent accidental rupturing (see Figure 8–3). A newer development are *mononuclear* capsules that require no stilts because they have tougher skin.

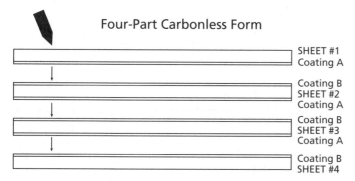

Four-Part Carbonless Form

SHEET #1
Coating A

Coating B
SHEET #2
Coating A

Coating B
SHEET #3
Coating A

Coating B
SHEET #4

Figure 8–2 Carbonless forms allow copies to be generated when pressure from a pen or typewriter (represented by the heavy arrow) causes tiny capsules coated onto the backside of sheet one to rupture and release a dye that causes the coating on the top side of sheet two to turn dark. As this pressure continues through the sheets of a multipart form, so does the transference of the image. The coating with the capsules (a) must always be in contact for the receiver coating (b).

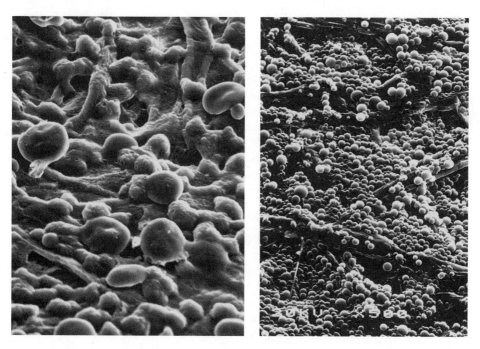

Figure 8–3 On the left is cluster coating on carbonless paper. The larger objects are starch stilts and the smaller ones are the actual capsules. At right are mononuclear capsules. Both photographs are at 500 power. *Courtesy of Mead Central Research.*

This wide range of products under the broad label of "bond" is necessarily due to the equally wide range of businesses and applications in the world. Imagine a nationwide wholesaling firm that needs to send out a price change notice to its hundreds of dealers. For this use, a low-grade sulfite sheet will perform very well. However, the same company may use a 100 percent cotton sheet for the correspondence from the office of its president.

With this diversity of product line, bond papers clearly play an important role within the printing industry. Matching up the correct sheet with each application requires a solid knowledge of this wide spectrum of choices.

Xerographic Papers

A second category of business paper is **xerographic paper** made especially for use in electrostatic photocopiers. As the capabilities of photocopiers have increased to the point that they are actually replacing offset duplicators in some printing facilities, the volume of xerographic papers sold has ballooned. For the commercial and in-plant facilities that split the work between offset duplicators and photocopiers, mills are producing sheets that are compatible with both printing processes; these are called **xerographic dual-purpose.** Whereas offset printing requires a well-sized sheet, electrostatic copiers place a severe test on a sheet's ability to resist curling after being heated. This heat melts and sets the image-producing powder/toner. Because the first photocopiers required more heat than modern models, this resistance to curling is less of a factor than was once the case.

Like the bonds, xerographic paper is available in three grades: No. 1, No. 4, and No. 5; however, the xerographic dual-purpose sheets are only available in No. 4 and Premium No. 4 grades.

Duplicator and Mimeograph Papers

Two other types of business papers that are often mistakenly lumped together are the **duplicator** and **mimeograph** sheets. These machines were once very commonly used in schools, churches, and offices, but they have largely been lost to the photocopier industry. Both duplicator and mimeograph paper are available in No. 1, No. 4, and No. 5 grades and three weights: 16, 20, and 24 pound.

Lightweight Papers

Within the group of business papers are three types of lightweight sheets. The first is **manifold**—a thin, smooth sheet usually found in conjunction with carbon paper. The W–2 forms that are used with income tax preparation are examples of manifold paper. It is available in weights that range from 8 to 10 pounds, in white as well as colors, and in two grades: No. 1 and No. 4, with the first being watermarked.

Onion skin is similar to manifold in both general appearance and use, but it is more durable and better looking due to its cotton content, which can range from 25 to 100 percent. Onions play no role in supplying the fiber content of onion skin paper; the name comes from the sheet's resemblance to onion skin—thin and transluscent. This paper is graded only by cotton content and is available in weights ranging from 7 to 10 pounds.

The third lightweight sheet is **airmail**—a sheet intended more for pen-and-ink and less for printing than the other two lightweight sheets. It is available in 9, 13, and 16 pound weights.

Ledger Paper

At the opposite end of the weight spectrum for business papers is **ledger.** Designed for recordkeeping, ledger is similar to bond, except that it is much heavier and usually has been supercalendered to harden the surface for good erasing qualities and to stand up to repeated handling over a long period of time. Ledger paper is available with cotton contents of 25, 50, 75, and 100 percent as well as an all-sulfite content. The common weights are 24, 28, 32, and 36 pounds.

The demand for ledger paper has decreased greatly, as increasing numbers of businesses move to electronic bookkeeping.

Specialty Papers

Three specialty papers will be examined together because they have some similarities with one another. The first is **MICR** (Magnetic Ink Character Reading) paper, which is receptive to the magnetic ink used on banking documents, such as checks and deposit slips. MICR paper is produced to comply with specifications developed by the American Banking Association. It is available in two grades: No. 1 and No. 4.

Another stock used in banking is **safety paper,** on which checks and certificates of title are usually printed. Safety paper derives its name from a preprinted background image that makes erasing or otherwise altering written or typed images impossible. Sometimes safety paper has been preprinted with an invisible ink that produces a voiding message when bleach or ink eradicator is used in an attempt to chemically remove a written message. Safety paper must also have MICR characteristics.

The third sheet is similar to MICR paper, but in a different way than is safety paper. **OCR** (Optical Character Recognition) is paper on which typed material can be fed into a scanning device that will read the characters and allow manuscripts to skip the rekeyboarding stage of computer typesetting. Understandably, the primary demand placed on OCR paper is that it must be completely free of any visual impurities that might cause the optical scanner to misread the images.

Three categories of paper used by nearly everyone are tablet, writing, and envelope. **Tablet** is one type of paper that is usually packaged in small quantities to be purchased by the ultimate consumer. Within this category exists a wide range of quality levels, with the low end being substantially or entirely made from groundwood fiber. Tablet paper usually has rules printed across it as guides for writing.

A higher-quality level is found with **writing paper,** because this category includes the boxed stationery found in gift stores. Writing paper is made to accept pen ink well and to erase well, but does not reflect a concern for printability. The same priorities apply to **envelope stock.** Because the primary concerns of the manufacturer are opacity, burst and fold strength, and the ability to be gummed and moistened with limited curl, printability is somewhat slighted. Although many bonds and writing papers come with a matching envelope, envelope stock often has been manufactured separately and is less like its contents than appearance would suggest. Envelope papers are available in 20, 24, 28, and 32 pound weights. The larger, stronger envelopes are often made from special kraft paper and are not in the business papers classification. The available colors for large envelopes are white, brown, and gray; these grades can run up to 40 pounds in weight.

Only two specialty sheets remain in this examination of business papers. **Vellum** paper is a strong, high-quality sheet that is made to imitate parchment. In fact, many paper catalogs list this grade under the heading of "parchment." The most common application for vellum paper is in printing diplomas and similar certificates.

It should be noted that the term vellum is commonly used to describe the rough, eggshell-like finish of many grades of stock, including bond, bristol, tracing, and label papers. This can be a little confusing, but just remember that when vellum is used in connection with another category such as in "vellum bristol" or "vellum bond," the term is describing the finish.

The final category of business papers is **translucent bond** (or *translucent paper*), a sheet used primarily by architects, engineers, and others to make master copies from which ozalid and blueprint copies can be made. The common weights are 11, 13, 15, 18, and 20 pounds and it is available with fiber contents of all-sulfite and 25 percent cotton.

Book Papers

The second large classification of papers to be examined is **book paper,** produced to serve the giant publishing industry, but used extensively by commercial printers for smaller jobs. The subcategories of book papers are divided by physical characteristics (such as coated and uncoated); the printing process (as with offset, letterpress, and gravure sheets); and specific application (such as bible and text papers). The following list shows the primary categories of book papers.

> **Offset**
> > Coated one-side
> > Coated two-sides
> > Uncoated
> > Opaque—bible
> **Text**
> **Letterpress**
> **Rotogravure—gravure**
> **Book publishers**
> **Label**

The items in this list are not exclusive of one another. That is, offset paper can be printed by letterpress; text paper is nearly always run on offset presses and, because it is uncoated, could be viewed as an uncoated offset paper. So, although this classification called book papers is a complex one, it represents such a large volume of the overall printing industry that its members need to be understood.

Offset Paper

Offset paper will be the first group studied because lithographic (offset) printing is the giant of the industry. In fact, offset paper has become almost synonymous with the heading "book paper," and the two are often used interchangeably. This grouping offers a very wide range of grades,

	Gloss	Dull	Matte	Embossed	Brightness Level
Premium No. 1	X	X	X	X	86–92
No. 1	X	X	X	X	85
No. 2	X	X	X	X	83
No. 3	X	X		X	81
No. 4	X	X	X		76–77
No. 5	X	X			69–70

X—indicates availability.   Brightness figures represent the percentage of light reflection.

Table 8–2 Coated offset paper is available in as many as six grades and four basic finishes. The brightness level figures indicate the percentage of light reflectance.

finishes, and weights from which to choose. The common feature of all offset paper is that it has received both surface and internal sizing and must offer good surface strength to withstand the pull of high-tack offset inks. As is the case with all book papers, offset stock is designed to be printed on both sides, and for this reason contains filler additives to provide the necessary opacity. A Sub. 40 (or 40 pound) supercalendered offset sheet will contain, on average, 28 percent filler; this is in sharp contrast to a bond sheet that may be only 3 percent filler. Much offset paper is **coated** to enhance ink holdout and sometimes to add gloss. Within the range of coated offset sheets, there are four basic finishes. These are *gloss,* with a gloss range of 55 percent or higher; *dull,* with a gloss range of 21–54 percent; *matte,* with a gloss range of 0–20 percent; and *embossed,* in which one of several definite textures is built into the sheets' surfaces.

Table 8–2 presents the availability of the six grades and four finishes of coated offset paper. Also shown are the brightness ranges for each grade, which demonstrate that the higher grades provide the greatest whiteness. It might also be noted that the No. 4 grade sheets may contain some percentage of groundwood and that No. 5 sheets will likely contain some groundwood. These lower grades are appropriate for low-cost, high-volume printing jobs such as catalogs; whereas, the top four grades are commonly used in annual reports and other printed pieces where high quality is the primary concern. The basis weights of coated offset paper range from the 34, 50, and 60 pound sheets in the No. 5 grades to Premium No. 1 sheets that come only in the heavier 70, 80, and 100 pound substance weights.

As was noted earlier, all paper sold for offset printing must have satisfactory surface strength and opacity; the main trade-offs in moving from higher to lower grades involve brightness, gloss, and cleanliness.

One of the advantages of the offset method of printing over letterpress is that it does not require a smooth surface in order to deliver high-quality printing, such as fine-screen halftones. Offset printing flattens the sheet between the blanket and impression cylinders just as the ink transfer is occurring. Because the sheet is made smooth at the moment of printing, offset paper need not be coated to effect a good image transfer. Uncoated offset paper is available in a wide range of weights, colors, and finishes, and because it is not coated and has received less calendering, it will feature greater bulk and opacity than its coated counterparts in the same substance weight.

For applications where good opacity is important, especially in lightweight sheets, **opaque offset** is a popular choice. These papers are a favorite in catalogs or other pieces where lighter weights are needed to reduce mailing rates, but the two-sided printing requires extra filler content to provide opacity. Opaque offset comes in Premium No. 1, No. 1, No. 2, No. 3, and No. 4 grades and basis weights from 30 through 100 pounds. The best combination of being light in weight while maintaining high opacity is **bible paper,** a specialty sheet that not only is lighter than opaque offset, but provides even greater opacity.

Text Paper

Among the most interesting categories of all the book papers is **text,** which originally got its name from being used to carry the text or body of books and other quality publications. Today, its broad spectrum of colors and selection of rich finishes makes it a favorite of designers working on high-quality advertising pieces, menus, greeting cards, booklets, announcements, or programs. Whether made with a felt or embossed finish, text papers can impart a distinguished appearance to a printed piece (see Figure 8–4).

Because they are used in such a wide range of applications, high-grade text papers are not only made to run well on the press, but also handle embossing, hot-foil stamping, and die-cutting operations, as well.

Text papers can also be ordered with a *deckle edge* on one of the four sides. This is the rough, "torn" look that is sometimes used to impart a handmade look to invitations, greeting cards, and announcements (see Figure 8–5). Text papers

Figure 8–4 The deep pattern of this text sheet was achieved through embossing. Several embossed finishes are available with text papers and their matching cover weights.

are commonly found in 60, 70, and 80 pound basis weights and in several finishes, including felt, antique, and many embossed textures.

Other Categories of Book Paper

Whereas offset paper is produced by scores of mills and in over 200 brands, only around a dozen mills supply the rather small **letterpress** market. Sheet-fed letterpress machines require a smooth surface for halftone printing, but can usually use a high-grade offset sheet with satisfactory results. A significant market does exist for *web-letterpress* paper, a stock usually ordered by publishers who use mostly No. 4 and No. 5 grade coated papers that contain groundwood fiber. Letterpress printing on uncoated stock is primarily limited to newspaper and paperback book publishers using one of several grades of groundwood stock.

Rotogravure paper represents a very significant slice of the paper sales pie but, like letterpress paper, is manufactured only by approximately a dozen mills nationally. The paper is made to run on rotogravure presses, which require a smooth finish. In the case of sheet-fed gravure, many printers have found that higher-grade coated offset sheets perform satisfactorily. As is the case with letterpress paper, rotogravure is nearly always coated. An exception to this is gravure stock made of groundwood fiber.

The book publishing industry requires paper with unique performance standards, and for this reason, there are mills that manufacture stock to meet these requirements, called **book publishing** paper. One such standard is *bulking thickness,* the exact caliper of an individual sheet, which is critical in hardbound books because a book's total number of pages must fit perfectly into the already-made hard-bound cover.

Figure 8–5 This dark text sheet features a deckle edge on one side.

Bulking is also a concern to publishers who want to keep large books from becoming so thick and unwieldy as to adversely affect sales. At the other extreme, books with few pages, such as the yearbook of a small high school or a children's book, may be printed on a high bulk stock to make them look larger. High bulk is achieved by using minimal calendering, improving the stiffness and runnability of the paper. The trade-off is that the paper's ink holdout suffers when the polishing of the surface is reduced.

Two other requirements of book publishers are that their papers have high opacity and good performance. This last term means that the stock must fold well, be able to lie flat when the book is open, and have good tear strength.

One of the main differences between book and magazine publishing is the degree of gloss usually found in the products. Whereas, magazine publishers often need at least moderate gloss to give snap to the halftones and colors, book publishers avoid gloss to the point of using matte-finish stocks to reduce glare and eye strain. Basically, there are three groups of book publishing paper: coated, uncoated chemical fiber, and uncoated groundwood fiber. Stocks from all three groups are compatible with offset printing.

Label paper (often called "litho label") is manufactured specifically to meet the rigorous demands placed on having to accept printing on one side and gumming on the other. Usually coated on only one side, label stock must be able to maintain structural stability and flatness while being printed, varnished, die-cut, and/or laminated to a wide range of objects and materials, such as glass bottles, cardboard boxes, and vinyl recordings. Label paper can be cast coated, gloss coated, or uncoated, and is usually available in

No. 1 and No. 2 grades. The range of common weights includes 60, 70, 80, and 100 pound substance weights.

Bristol Papers

Several kinds of papers fall under the umbrella term **bristol**, which comes from the city in England where they were first made. Basically, bristol means a heavyweight paper with a caliper of at least .006-inch. Examples of subcategories are *index* bristols (from which index cards are made), *postcard* bristols, *posting* bristols (used with mechanized bookkeeping machines), and *tough check* bristols (used for signs, extra heavy covers, and window displays). *Vellum* bristol is the name given to an extra firm paper with a vellum finish. The other finish for bristol paper is *smooth.* Just because they are heavy, bristols are not to be thought of as inferior paper; they are a good quality product usually made by laminating multiple rag content sheets until the desired thickness is achieved. The thinner bristols can be made at one time on a special kind of paper machine called a *cylinder machine* because it can form thicker sheets than is practical with the Fourdrinier machine.

Cover Papers

Cover papers are a classification usually comprised of extra heavy offset and text stocks because covers are often required to match their contents as well as protect them. For this reason, several offset and text papers come in "matching cover," which means that designers have a wide range of beautiful colors, finishes, and textures from which to choose when selecting a cover paper; some even have a deckle edge (refer back to Figure 8–5). At one end of the range of choices is the cast coated sheet with its high gloss, while at the other end are the uncoated, deeply embossed textured sheets. Cover papers are used by creative designers in scores of applications besides being on the outside of publications. Menus, advertising pieces, greeting cards, business cards, invitations—wherever their weight, caliper, and finish are needed to impart richness and sophistication, cover papers can be used effectively.

The popular weights of cover papers are 50, 60, 65, 80, and 100 pounds. Some cover stock, such as the cast coated sheets, are marketed not by substance weight, but rather by caliper, which is measured in *points.* One point equals a thousandth of an inch so that a 9 pt. cover sheet is .009-inch thick.

A popular cover paper with graphic designers is the **duplex** cover, a sheet that appears to be one-color on the front and a second color on the back. This effect is achieved simply by laminating two different cover papers and, when used creatively by the designer, the visual effect can be very striking. Covers for menus or other products that will receive substantial handling are often printed on *plastic-laminated* or *pyroxlin-coated* covers because of their resistance to grease or general soiling. Some cover stocks are embossed to simulate the look and feel of cloth or leather.

Utility Papers

Finally, there is the **utility papers** classification, wherein all the remaining categories find a home. This classification is sometimes referred to as specialty, miscellaneous, or functional papers. A brief description will be given to most of them. **Newsprint** is a paper made chiefly of groundwood fiber (75–95 percent) with some chemical fiber added to improve the sheet's strength. Although comparatively inexpensive, newsprint lacks the brightness and permanence of book papers. **Tagboard** or *tag* is an especially strong, rigid paper made from long fibers such as chemical wood pulp and jute and is calendered to smooth and harden the surface. It is used in applications demanding long life or rugged use, such as shipping tags. Tagboard is available in two colors (white and manila), in three weights (100, 125, and 150), and in three grades (Premium, No. 1, and No. 3).

Gummed papers constitute a type of label paper in which the adhesive is activated by water. The ungummed side can be uncoated or coated, depending on the finish desired. **Pressure-sensitive papers** are a special label paper in which physical contact, rather than water, activates the adhesive.

Blanks are a very heavy cardstock that are marketed by the number of plies the sheet consists of rather than by substance weight or thickness (for example, six-ply card blank). Printing blanks can be coated on either one side or both sides. **Railroad board** is also a heavy, multiply paper that can be coated or uncoated, but it comes in only colors.

A limited market exists for **synthetic papers** in the printing industry. A synthetic paper can be entirely formed from a man-made material or it can be a blend of man-made and cellulose fibers. Rayon, nylon, and acrylics are examples of synthetic materials that can be used with cellulose fibers to improve the finished sheet's fold strength, tear strength, and tensile strength. Unlike cellulose fiber, these fibers do not have a natural adhesion to one another, so a binder is required. The most common synthetic papers used in printing, though, are the polyolefin polymers such as polypropylene and polystyrene. Unlike the matting together of fibers, these sheets are polymers extruded into pellets and then flattened and stretched into sheets.

Because these totally synthetic sheets are very resistant to tearing, moisture, and exposure to ultraviolet light, they have several outdoor applications, such as banners and landscape nursery tags. The fact that totally synthetic sheets have no grain direction also makes them appropriate for large maps that must receive several folds. A salad dressing manufacturer specifies synthetic paper for labels on jars that sit in the produce section of supermarkets and must withstand misting. Synthetic sheets also bring high opacity levels at low calipers and weights.

Nonetheless, synthetic papers are not popular with all printers because printing on them requires a special effort. Their nonabsorption of ink usually makes for comparatively slow drying and their sensitivity to high temperatures prohibits heat-set printing. Hot stamping and embossing can also require great care. Lastly, synthetic sheets neither recycle nor incinerate as easily as do conventional papers. When these potential drawbacks are added to the fact that synthetic papers cost significantly more than paper made from cellulose fiber, it seems apparent that their use will be very limited to situations where their attributes are of primary importance.

Paper Merchants' Catalogs

Printers generally acquire their paper from paper merchants' catalogs rather than directly from the mills. These wholesalers have catalogs that identify their inventory of paper stocks by classification or category, brand name, weight, size, color, and finish. The printer should be warned that a collection of these catalogs may have little consistency among them in the way the categories are arranged. Table 8–3 is a listing of the headings from the catalogs of six different paper merchants in the same city. For example, note that text papers are a separate group in the books of wholesaler F, but are combined with cover stocks by wholesalers A and C; meanwhile, wholesalers B, D, and E place them with the uncoated book papers. Also, catalog E shows headings for "uncoated offset" and "coated offset," while catalogs A and C use the terms "uncoated book" and "coated book" to identify

Paper Co. A

Uncoated Book
Coated Book
Label & Blotting
Text & Cover
Boards, Index, Bristol
Bonds, NCR, Safety, Mimeo, Duplicator, Ledger,
    and Thin
Envelopes, Bond Envelopes, Cards,
    Announcements, and Stationery

Paper Co. B

Text-Uncoated Book
Offset-Opaque
Coated Book
Cover
Safety
Thin Paper, Onionskin
Bond
Duplicator, Mimeograph
Ledger
Gummed, Blotting
Blanks, Tagboard
Envelopes
Announcements, Wedding, Cards

Paper Co. C

Coated Book
Uncoated Book
Text and Covers
Bristols, Blanks, Board
Gummed, Blotting
Writing Paper
Envelopes, Announcements
Miscellaneous

Paper Co. D

Uncoated Book, Text
Coated Book
Cover
Bond, Mimeo, Duplicator,
    Ledger, Safety, Thin

Carbonless, Carbon paper
Bristol, Index, Tag, Blanks, Board
Gummed, Pressure Sensitive, Label
Envelopes, Announcements

Paper Co. E

Text, Plain Book
Uncoated Offset
Coated Offset
Uncoated Cover
Coated Cover
Bonds, Safety
NCR Paper
Kraft Wrapping, Chipboard, Tapes
Ledger, Mimeo, Duplicator, Thins
Index, Bristols, Tagboard, Blotting
Boards, Blanks
Pressure Sensitive, Gummed
Announcements, Tags, Stationery
Envelopes

Paper Co. F

Uncoated Offset and Opaque
Text
Coated Book
Cover
Mimeo, Duplicator, Safety, Carbonless, Thin
Sulphite and Rag Bond
Ledger
Bristol, Board, Blanks, Blotting
Gummed and Pressure Sensitive
Envelopes and Announcements
Cut Cards and Tags

Table 8–3 The section headings from six different paper merchants' catalogs within the same city are shown to reveal the lack of uniformity that exists. A person buying paper from these six wholesalers would first need to become familiar with each catalog's structure.

	Bond	Book	Bristol	Carbonless forms	Cover	Duplicator	Groundwood	Index	Ledger	Mimeograph	Parchment	Tablet	Tag	Text	Xerographic
Annual reports		•			•				•					•	
Booklets		•												•	
Brochures		•			•				•		•			•	
Business forms				•											
Catalogs		•													
Church bulletins	•					•				•					•
Diplomas	•										•			•	
Door hangers			•		•		•								
Dust jackets		•			•										
Greeting cards		•			•					•					
Index cards							•								
Letterheads	•														
Magazines		•													
Menus			•		•		•	•	•		•		•	•	
Newsletters	•	•								•					
Newpapers							•								
Paperback books		•					•								
Posters		•			•								•		
Postcards								•	•						
Price Lists	•	•								•					•
Classroom handouts						•									
Signs			•											•	
Tablet books							•					•			
Telephone books							•								
Yearbooks		•													

Table 8–4 This chart reflects some of the more common applications of various types of paper. In no way, however, does it attempt to establish guidelines or to limit the utilization of any of these categories.

many of the same lines of stock. Therefore, the printer may need to adapt to more than one system when working with a number of wholesalers. Otherwise, the diversity of the catalog structure can be confusing.

Summary

This chapter has not attempted to present every type of paper that is manufactured. To get some idea of the scope of the product range, it is interesting to note that the *Competitive Grade Finder* contains over 1,000 listings in its Group Classification Index and the range of colors and weights available swells the total choices to the tens of thousands. What this chapter has done is describe the five basic classifications and their more common categories and subcategories. An attempt has also been made to offer some insight into what can be expected of these sheets and how they are commonly used. Table 8–4 shows some popular uses for a number of sheets described in this chapter. It should be noted, however, that because creative people are not bound by convention, it is common to see a sheet used in a most unorthodox, but nonetheless effective, application. Working with such a kaleidoscope of colors, textures, weights, and other physical characteristics can be one of the most challenging and rewarding aspects of the printing industry.

Questions for Study

1. Why does bond paper lack the opacity of a comparable sheet of offset stock?
2. What is the fiber content of 25 percent rag bond?
3. Differentiate among cockle, laid, and linen finishes.
4. Why is it crucial that carbonless papers be kept in a certain order within the set?
5. Why must duplicators and mimeograph papers have different physical characteristics?
6. What do MICR and OCR papers have in common?
7. What are the six categories of book papers?
8. Name the four finishes available with coated offset stock.
9. What is the relationship between grade and brightness levels of coated offset paper?
10. Within the publishing field, what are the primary differences between the paper used for books and the paper used for magazines?

Key Words

business papers
erasable bond
laid finish
carbonless
onion skin
safety paper
book paper
text paper
label paper
newsprint
bond
linen finish
cockle finish
xerographic paper
ledger
OCR

offset paper
deckle edge
bristol papers
tagboard
cotton bond
wove finish
business-forms bond
manifold
MICR
translucent bond
opaque offset
book publishing paper
duplex paper
blanks
xerographic dual-purpose
duplicator

mimeograph
airmail
tablet paper
writing paper
vellum
envelope stock
bible paper
rotogravure paper
letterpress paper
gummed papers
utility papers
pressure-sensitive papers
railroad board
synthetic papers

Buying Paper

9

AFTER STUDYING THIS CHAPTER, THE STUDENT SHOULD BE ABLE TO:

■ Define substance weight, ream weight, and M weight.

■ Define basic size and explain its role in performing paper calculations.

■ Use the equivalent weight formula to calculate the ream weight for a given size sheet.

■ Explain how the volume of an order affects the price per pound.

■ Use a paper merchant's catalog to calculate the cost of a specified order of paper.

■ Explain how a larger sheet size or more sheets can actually lower the price per pound.

■ Explain the secondary paper market and its role in the industry.

■ Explain how TQM influences the papermaking process.

■ Demonstrate how to maximize the yield of small sheets by a larger sheet when grain direction is a concern.

■ Calculate the weight of a roll of paper and convert its weight to linear feet.

The nature of the printing business requires access to a wide range of paper sizes and weights. Add to this the variety of types of paper such as bond, text, offset, book, and the numerous press sizes found in printing firms and the result is a very diverse and complex array of paper from which to choose. Any printer ordering paper must specify not only the quantity, but also the weight, manufacturer, brand name, color, category, and finish. An example of an order might be "6,000 sheets of 19 × 25-inch, 50 pound, Alum Creek[1] Sabre cream offset dull finish." Although, at first, this appears cumbersome, each element in the above description represents a decision that the printer has made and an instance of control that has been exercised over the appearance or performance of the order. So, the good news is that this wide range of choices allows for a great deal of versatility in the finished product, but the bad news is that it also creates a fairly complex system of packaging and marketing for the paper buyer to deal with.

Substance Weight

A quick look at a package or carton of paper will reveal that paper is assigned a "weight." Figure 9–1 shows a label from a package of 8 ½ × 11-inch, bond paper. The S20 identifies the paper as being "20 pound" bond, but exactly what is it that weighs 20 pounds? One sheet of the paper? A full pack-

181

Figure 9–1 This package label contains much information for people who can understand terminology like "S20."

Figure 9–2 The "weight" of paper is a reference to the weight of a certain number of sheets in a particular size.

age? The fact is that 20 pounds is the *substance weight* or *basis weight* of the paper and therefore is the weight of one ream (500 sheets) in the basic size. To help picture this, examine what is shown in Figure 9–2, a pile of 500 sheets of a certain size. If this were weighed and found to be 20 pounds, the paper would be labeled 20 pound paper. Even though one may be holding only a piece of paper torn from the top sheet in that pile, that piece would still be called 20 pound paper despite the fact that its actual weight might be less than an ounce. So, regardless of the size or quantity of paper, the substance weight is a measurement of *the weight of 500 sheets of that paper cut in the basic size.*

Basic Size

Generally speaking, different categories of paper have different **basic sizes.** At the time that these basic sizes were agreed to by the paper manufacturers, these dimensions were also among the most commonly used. Although that relationship is not always the case today, there does exist a certain logic to the basic sizes shown here.

Business papers (bond, mimeo, ledger)	17 × 22 inches
Book, offset, text papers	25 × 38 inches
Cover papers	20 × 26 inches
Index stock	25 ½ × 30 ½ inches
Utility papers (newsprint, manila, tagboard)	24 × 36 inches
Blotting paper	19 × 24 inches

To gain some insight into why different types of paper have different basic sizes, consider the *end result* of the paper as well as the *size of the press* that will probably be used to do the printing. For example, letterhead stationery is usually made from bond paper and commonly run on an offset duplicator which can accommodate sheets up to 11 × 17 inches. Looking at the sizes involved here, a connection can be seen among the 8 ½ × 11-inch finished size, the 11 × 17-inch press sheet, and the 17 × 22-inch basic size, with each being twice the size of the previous sheet. So, if a printer were to buy 17 × 22-inch bond paper, by cutting it in half he would have two 11 × 17-inch press sheets, which would equal the maximum size of the press; also, each press sheet would be able to produce two 8 ½ × 11-inch finished letterheads with no waste (see Figure 9–3). Although paper is available in a range of sizes besides the basic size, the substance weight of the paper is always a reference to the

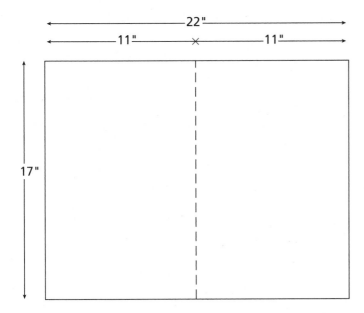

Figure 9–3 A sheet measuring 17 x 22 inches will produce two 11 x 17-inch sheets with no waste.

weight of a ream of that paper if it had been cut in the basic size. Knowing this fact makes it possible to figure the weight of any quantity of sheets in any size. To illustrate this point, imagine ordering 500 sheets of 11 × 17-inch Sub. 20 bond paper. Because paper is usually priced by the pound, before the cost of the order can be predicted, the weight must first be found. Before doing any arithmetic, ask what the substance weight is measuring—or, in other words, "what weighs 20 pounds?" The answer is, of course, one ream (500 sheets) in the basic size (17 × 22 inches).

If the order were for a ream of 17 × 22-inch paper, the weight would be 20 pounds (as shown in Figure 9–4) but because the 11 × 17-inch purchase size is one-half the basic size, the weight would be cut in half as shown and the order would come to 10 pounds. By the same logic, if an order were placed for 500 sheets of 22 × 34-inch Sub. 20 bond paper, then the weight of the order would be 40 pounds because the purchase size is twice the basic size, as shown in Figure 9–5. To continue in this vein, the weight of 500 sheets of 8 ½ × 11-inch Sub. 16 pounds paper would be 4 pounds (16 ÷ 4 = 4) because 8 ½ × 11 inches is one-quarter of the basic size.

Equivalent Weight Formula

An examination of a paper merchant's catalog quickly reveals that there are several paper sizes available that do not have this incremental relationship to the basic size. To cite

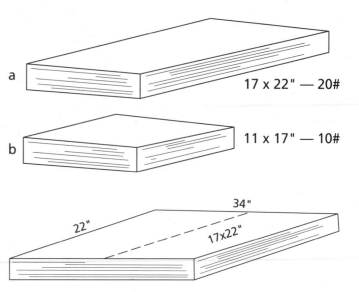

Figure 9–4 When the size of a ream of paper is cut in half, so is the weight of a ream. In other words, the weight is directly proportional to the sheet size.

Figure 9–5 A ream of 22 x 34-inch Sub. 20 bond paper weighs 40 pounds because it is twice as large as the 17 x 22-inch basic size for bond.

examples of this, bond paper can be ordered in sizes such as 19 × 24, 17 × 28, 24 × 38, 28 × 34, 17 ½ × 22 ½, 22 × 25 ½, and 22 ½ × 35 inches. These nonincremental sizes obviously make the process of figuring the basic size more difficult because they are not half, double, or a quarter of the basic size, but they can still be calculated by means of simple proportion. By using what is called the **Equivalent Weight Formula** (or the Comparative Weight Formula), the ream weight of any size sheet can be found. This formula works on the principle that the same ratio that exists between the *desired size* and the *basic size* will also exist between the *ream weight of the desired size* and the *substance weight.*

Figure 9–6 shows two piles of Sub. 20 bond that have been cut to different sizes. Pile A is a ream of 17 × 22-inch paper and, because 17 × 22 inches is the basic size for bond, the pile will weigh 20 pounds. Pile B is also a ream, but because its area in square inches is *one-half* the basic size, its weight is easily figured to be 10 pounds. Even though this can be done mentally, an analysis of the procedure reveals the key: when the area is reduced by 50 percent, the weight is also reduced by 50 percent. Expressed as an equation, we have the Equivalent Weight Formula:

$$\frac{\text{area of the basic size}}{\text{area of the desired size}} = \frac{\text{substance weight}}{\text{weight of the desired size}}$$

By placing the appropriate numbers into the equation, allowing *x* to represent the unknown weight of the desired

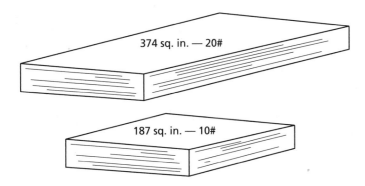

Figure 9–6 In this instance, the areas in square inches of the sheets are compared and the weights of the piles will reflect this area-to-area ratio.

size, and proceeding with the solution, the formula yields the same answer that logic had provided.

$$\frac{17 \times 22}{11 \times 17} = \frac{20}{x} \quad \rightarrow \quad \frac{374}{187} = \frac{20}{x}$$

$$\rightarrow \quad 374x = 3740 \quad \rightarrow \quad x = \frac{3740}{374} \quad \rightarrow \quad x = 10 \text{ pounds}$$

In the above problem the Equivalent Weight Formula was used to calculate a problem simple enough to be done mentally. However, this formula becomes more valuable in figuring the weight of some of the nonincremental sizes listed earlier; these are sizes that do not lend themselves to mental calculations. For example, instead of 11 × 17-inch paper, 28 × 34-inch can be used.

$$\frac{17 \times 22}{28 \times 34} = \frac{20}{x} \quad \rightarrow \quad \frac{374}{952} = \frac{20}{x} \quad \rightarrow \quad 374x = 19,040$$

$$\rightarrow \quad x = \frac{19,040}{374} \quad \rightarrow \quad x = 50.9 \text{ pounds (wt. of 1 ream of}$$

28 × 34-inch 20# bond, rounded to 51 pounds)

M Weight

Observe that the above answer is labeled as being the weight of one ream of the paper being ordered. Understandably, this is referred to as the ream weight. An even more useful item is the weight of 1,000 sheets, which is called the **M weight** (derived from the Roman numeral for 1,000) and is produced by doubling the ream weight. For example, paper that has a ream weight of 51 pounds would have an M weight of 102 pounds. The value of knowing the M weight becomes apparent when figuring the total weight of an order of paper. By multiplying the M weight by the number of Ms being ordered, the total weight will result.

For example, earlier in this chapter the Equivalent Weight Formula was used to find the ream weight of 28 × 34-inch Sub. 20 (or 20#) bond paper. The 51# ream weight was then doubled to give the M weight of 102#. Knowing this, the M weight (Mwt) can be multiplied by any number of "Ms" that might be needed; this conversion from number of *sheets* to number of *Ms* is done by simply moving the decimal point three places to the left. In this fashion, 6,000 sheets becomes 6 Ms; 14,500 sheets becomes 14.5 Ms; 2,176 sheets becomes 2.176 Ms; and 770 sheets becomes .77 Ms. So, if 21,200 sheets of the 20# 28 × 34-inch bond were needed, the calculation for the total weight would be expressed as

$$21.2 \text{ Ms} \times 102\# \text{ (Mwt)} = 2,162.4\#$$

Because paper is commonly sold by weight, the next and final step is to multiply the number of pounds by the price per pound. Assuming a price per pound of $.60, the price of the order comes to $1,297.44 (2,162.40 x $.60).

Calculating Order Weight and Cost

To work a problem all the way through, calculate the cost of order weight 8,600 sheets of 8 ½ × 11-inch Sub. 16 bond that costs $1.14 per pound. The first step in this process is to figure the ream weight. Logic reveals that because a ream of the paper in the basic size weighs 16# and the order size (8½ × 11 inches) is one-quarter of the basic size (17 × 22 inches), then the weight of a ream of 8½ × 11-inch paper would be one-quarter of the substance weight divided by 4 pounds. Although hardly necessary in this problem, the Equivalent Weight Formula could also be used here to produce the same answer:

$$\frac{8\frac{1}{2} \times 11}{17 \times 22} = \frac{x}{16} \rightarrow \frac{93.5}{374} = \frac{x}{16} \rightarrow 374x = 1,496$$

$$\rightarrow \quad x = 4 \text{ pounds (weight of 500 sheets of } 8\frac{1}{2} \times 11\text{-inch paper)}$$

Next, doubling the weight of 500 sheets (ream weight) will give the weight of 1,000 sheets (Mwt), which multiplied by the number of Ms in the order will reveal the total number of pounds in the order.

$$4\# \text{ ream weight} \times 2 = 8\# \text{ Mwt}$$
$$8\# \text{ (Mwt)} \times 8.6 \text{ Ms} = 68.8\# \text{ (total weight of 8,600 sheets)}$$

Finally, the number of pounds multiplied by the price per pound gives the total cost.

$$68.8\# \times \$1.14/\text{lb.} = \$78.43$$

Packaging of Paper

Paper can be purchased in either rolls or sheets for web and sheet-fed presses, respectively. This part of the chapter will deal with paper designed for the vastly more common sheet-fed presses. Basically, *sheets* of paper are packaged in two ways: on *skids* and in *cartons*. A skid of paper consists of a large stack of paper on a wooden pallet and that weighs around 1 ton. The paper will likely be *flagged* with small strips of colored paper protruding from the pile at 500 sheet intervals to aid in keeping track of usage. The entire pile will be wrapped as a unit with plastic, a moisture-resistant paper, or both. Paper ordered on a skid is easier to unwrap and move about as a unit than if it were to arrive in several cartons. Also, the uncomplicated form of wrapping, combined with the volume of ordering a ton or more of paper, generates a lower price. A disadvantage of skid-wrapped paper is that it needs to be used reasonably soon after being opened because the entire contents are then made vulnerable to becoming soiled, as well as to changes in the pressroom's humidity.

A printer ordering significantly less than a ton of paper or wanting to distribute its use over a longer span of time will likely order paper packaged in cartons rather than skids. The weight of a carton of paper depends upon the category, size, and basis weight of the order, but cartons are usually between 120 and 160 pounds to facilitate handling by one person. For bonds and other business papers, the contents of a carton are kraft-wrapped into 500-sheet packages, so a 3,000-sheet carton of paper would contain six ream-wrapped packages. Cartons that do not contain ream-wrapped packages will be flagged, as are skids, in 500-sheet multiples. A paper catalog will indicate the packaging of particular paper by indicating the number of sheets in a carton and if the sheets are also wrapped in 500-sheet packages. This wrapping of paper into packages and cartons allows for the use of a large order of paper to be spread out over a period of months because any unopened paper remains protected by the ream wrapping. However, the extra cost involved in this type of packaging is reflected in the price of the paper to the printer when compared to paper on skids.

Although paper is generally sold by weight, the way it is packaged and the size of the order determine the price per hundred pounds, which is called *price per hundredweight* and is abbreviated **cwt.** Paper merchants charge less per hundredweight on orders that are large enough to cross price thresholds within the merchant's pricing structure. These thresholds can vary somewhat from one merchant's catalog to another, but common thresholds are as follows:

less than a package (fewer than 500 sheets)
1 package
1 carton
4 cartons
8 cartons
16 cartons
5,000 pounds
10,000 pounds (on skids)
36,000 pounds (a carload)

How Volume Influences Cost Per Pound

As was mentioned earlier, the size of a paper order can influence the price per hundredweight (cwt) of the paper and ultimately influences the cost of the total price. Simply put, when a price threshold is crossed, the price per cwt changes and the *higher the volume of the order, the lower the price per cwt.* Figure 9–7 demonstrates this pattern. Looking in the broken package column, it can be seen that the price per cwt for Sub. 16 White Saginaw bond is $288.00, or $2.88 per pound, when fewer than 500 sheets are ordered. To fulfill

PRICES PER CWT.

	Bkn. Pkg.	Pkg.	1 ctn.	4 ctn.	16 ctn.
Hawkeye Bond					
White 16 lb.	177.30	118.20	100.50	88.05	81.60
20 lb.	168.15	112.20	95.25	83.55	77.40
Colors 16 lb.	190.95	127.20	108.15	94.80	87.75
20 lb.	181.80	121.20	103.05	90.30	83.70
Saginaw Bond					
White 16 lb.	288.00	192.00	168.00	146.90	135.40
20 lb.	275.70	183.83	160.80	140.60	129.60

Figure 9–7 The price per one-hundred pounds decreases as size-of-the-order thresholds are crossed.

an order of, for example, 350 sheets of this same stock, a merchant would first have to open a carton, remove a ream package, unwrap it, count out 350 sheets, and wrap them along with the remaining 150 sheets for which a buyer may not be found later. Because the merchant would rather simply sell an entire package and save the cost of counting, wrapping, and perhaps not selling the remainder, the price cwt is significantly lower (67 percent) if a full package is purchased. If a full carton is ordered, the price per cwt is even lower, and so on.

Imagine a paper stock that has a broken package price per cwt of $100, and therefore a price per pound of $1.00; listed below is some idea of what the price per pound might be if the same stock were ordered in larger volumes.

Broken package	$1.00 per pound
Full package	.67 per pound
Full carton	.58 per pound
4 cartons	.51 per pound
16 cartons	.47 per pound

Even greater savings are possible by ordering skids of paper directly from the mill. These price thresholds are 5,000, 10,000 and 36,000 pounds and they save money in two ways: (1) because of the mill-to-printer shipment, the paper merchant does not handle or warehouse the paper; and (2) skids are more economical packaging than cartons.

Using the Paper Merchant's Catalog

A basic tool of the paper buyer is the paper merchant's catalog. Because printers desire a variety of brands, sizes, colors, and weights, the catalog is designed to accommodate these variables as well as indicate grain direction, availability of shipment, and packaging data. The good news is that it is structured to allow for all of these choices; the bad news is that using a paper catalog requires a thorough understanding of its organization. The benefit of learning the catalog's ins and outs, however, is well worth the effort because it will result in acquiring the best sheet at the best price.

To learn this system, the assumption will be made that an order must be placed for 7,000 sheets of Sub. 20 19 × 24-inch white Northpark Bond, grain short. The first step is to turn to the catalog's index (Figure 9–8) and see if that merchant carries Northpark Bond and, if so, which page its data will be found on. Turning to that page reveals Figure 9–9, a page from a typical paper merchant's catalog. Directly under

BOND

Figure 9–8 Paper merchants' catalogs usually have an index for each section. Shown above is the index for a merchant's inventory of bond papers.

the brand name heading is found a description of the sheet. It is a watermarked sulfite sheet (contains no cotton); it comes in 14 colors and 1 finish (wove) and it will arrive in cartons that will encase 500-sheet packages. Also, it is noted that all seven of the warehouses listed at the bottom of the page stock this sheet. Moving down to the lower half of the page, we find that four substance weights are available and that the desired Substance 20 is among them.

Moving down the left column marked "size," we see that 19 × 24-inch paper is carried and that it is grain short; this merchant indicates the grain direction by underlining the

NORTHPARK BOND

Sulfite - Watermarked
Colors: White, Ivory, Tan, Gray, Green
Finish: Wove
Basic Size: 17 × 22
Grain Direction: Underlined

Substance wt.	Size	Mwt	Sheets/Pkg.	Sheets/Ctn.
13	1<u>7</u> × 22	26	500	5000
	17 × 2<u>8</u>	33	500	4000
16	1<u>7</u> × 22	32	500	4000
	17 × 2<u>8</u>	41	500	3000
	1<u>9</u> × 24	40	500	3000
	23 × 3<u>5</u>	68	500	1500
20	1<u>7</u> × 22	40	500	3000
	17 × 2<u>8</u>	51	500	3000
	1<u>9</u> × 24	49	500	3000
	23 × 3<u>5</u>	85	500	1500
24	1<u>7</u> × 22	48	500	3000
	23 × 3<u>5</u>	102	500	1500

PRICES PER CWT.

		Bkn. Pkg.	Pkg.	1 ctn.	4 ctn.	16 ctn.
White	Sub. 13	341.63	227.55	135.85	119.19	110.09
	Sub. 16	277.75	185.15	110.45	96.90	89.50
	Sub. 20 & 24	254.78	169.85	101.35	88.90	82.10
Colors	Sub. 13	358.72	239.12	142.64	125.15	115.60
	Sub. 16	291.64	194.41	115.97	101.75	93.98
	Sub. 20 & 24	266.25	177.50	105.90	92.90	85.80

Figure 9–9 A page from a typical paper merchant's catalog contains all the information needed to calculate the weight and cost of a given order for paper.

appropriate dimension. By moving from this size listing to the right until intersecting with the desired substance weight column, we find that the M weight for 19 × 24-inch Sub. 20 bond is 49 pounds. This is an important figure because it is the weight of 1,000 sheets (1 M) of the paper being ordered. Observe, also, that each carton contains 3,000 sheets (3 M); this is another piece of information that will soon be useful.

The first step in the calculation for the cost of the order is to figure its weight. Remember that *paper is sold by weight.* Because the order is for 7,000 sheets (7 M) and each M weighs 49 pounds, then 7 Ms × 49#/M = 343 pounds, the weight of the order. The next step is to multiply the number of pounds by the price per pound, but an examination of the pricing schedule located just above the center of the page reminds us that the price per cwt is determined by the volume of the order; that is, how many cartons are being ordered. Knowing that the order is for 7 M sheets and there are 3 M sheets in a carton, we proceed:

$$\frac{7 \text{ Ms}}{3 \text{ Ms}} = 2.334 \text{ cartons.}$$

Because the order will consist of more than one whole carton, but fewer than four cartons, the appropriate price category is "1 carton" and this means the price per cwt is $101.35. Moving the decimal point two places to the left provides the price per pound of $1.0135 (do not round this number), multiplied by the number of pounds being ordered gives the cost of the order:

$$343\# \times \$1.0135/\# = \$347.63$$

Expressed in a shorter form,

Ms (Ms being ordered) × Mwt (wt of an M) × $/# (price per pound) = Price of order

Of course, the appropriate price threshold for an order is first found by dividing the number of sheets being ordered by the number of sheets in a carton.

A second problem may help to reinforce this procedure. Consider finding the cost for 800 sheets of Sub. 16 23 × 35-inch green Northpark Bond, grain long. Multiplying the number of Ms (.800) by the Mwt (68#) gives the order weight of 54.4 pounds. Because the 800 sheets in the order are fewer than the 1,500 sheets in a carton, the order is clearly less than 1 carton. But because packages contain only 500 sheets and 800 is more than 500, the order does qualify for the package price. After converting the price per cwt to price per pound, multiply $1.9441/pound by 54.4 pounds and the resultant $105.76 is the price of the order.

Buying More Paper While Paying Less

As has been noted, the price per cwt, and therefore the price per pound, varies with the volume of the order. The paper buyer using this catalog to buy Northpark Bond would find five different price brackets for, say, white Sub. 13. Sometimes, when an order is close to the next price threshold, the prudent buyer of paper will *order more sheets than needed* to qualify for the lower price per pound of the next price bracket. Often, this move will result in actually buying more paper for less money. Consider the situation of needing 15,000 sheets of 17 × 22-inch Sub. 16 white Northpark Bond. Ordinarily, the steps would be as follows:

1. 15Ms × 32# (Mwt) = 480#
2. 15Ms ÷ 4 Ms (sheets/ctn) = 3.75 cartons (Because the order is for 1 or more cartons, but not 4 cartons, the 1 ctn price is appropriate.)
3. 480# × $1.1045 = $530.16

In this case, the perceptive buyer recognizes that the 3.75 cartons needed are very close to qualifying for the lower four carton price per pound.

To qualify for this price, the order naturally must be increased to four cartons or 16,000 sheets. The question becomes "Will the lower price per pound balance out increasing the number of sheets?" Pricing it both ways will provide the answer.

1. 16 Ms × 32# = 512#
2. $\dfrac{16 \text{ Ms}}{4 \text{ Ms (sheets/ctn)}} = 4 \text{ cartons}$
3. 512# × $.9690 = $496.13

Clearly, the opportunity to get an additional 1,000 sheets of paper and actually pay $34.03 less ($530.16 − $496.13) is worth taking advantage of, even if the extra paper is merely thrown out. The value of those extra 1,000 sheets is enhanced by figuring what they would cost if ordered separately at some future date. Because of the small size of the order, the higher package price bracket must be used.

1. 1.00 Ms × 32# (Mwt) = 32#
2. 1.00 Ms ÷ 4 Ms (sheets in a ctn) = .25 ctns (use 1 pkg price)
3. 32# × $1.8515/# = $59.25

When the value of the extra paper is added to the initial cost savings ($59.25 + $34.03 = $93.28), ordering a full four cartons is obviously a smart strategy.

Pricing Paper by M Sheets

Traditionally, paper merchants have priced their paper by hundredweight (cwt). The rapidly growing trend, however, is toward pricing paper per *thousand sheets* or "by the M."

MIDPOINT BOND (white)

Sulfite Bond
Watermarked
Finish- Wove
Packaged- Ream Sealed

17x22 BASIS	SIZE	M SHEET WEIGHT	SHEETS TO CTN	BKN	REAM	1 CTN	4 CTN	16 CTN
					M SHEET PRICE			
WHITE								
13	22x34	52M	2500	47.05	31.37	26.66	23.36	21.63
	24x38	64M	2000	57.91	38.60	32.81	28.76	26.62
	28x34	66M	2000	59.72	39.81	33.84	29.65	27.46
16	17x22	32M	4000	23.52	15.68	13.33	11.68	10.82
	17x28	41M	3000	30.14	20.09	17.08	14.97	13.86
	17½x22½	34M	4000	24.99	16.66	14.16	12.41	11.49
	19x24	39M	3000	28.67	19.11	16.24	14.24	13.18
	22x34	64M	2000	47.04	31.36	26.66	23.36	21.63
	24x38	78M	1500	57.33	38.22	32.49	28.47	26.36
	28x34	82M	1500	60.27	40.18	34.15	29.93	27.72
20	17x22	40M	3000	28.48	18.98	16.14	14.14	13.10
	17x28	51M	3000	36.31	24.20	20.58	18.03	16.70
	17½x22½	42M	3000	29.90	19.93	16.95	14.85	13.76
	19x24	49M	3000	34.89	23.25	19.77	17.32	16.05
	22x34	80M	1500	56.96	37.96	32.28	28.28	26.20
	24x38	98M	1500	69.78	46.50	39.54	34.64	32.10
	28x34	102M	1500	72.62	48.40	41.16	36.06	33.41

Figure 9–10 This particular paper merchant's catalog prices its paper "per thousand sheets" instead of "per hundred pounds."

This format makes computation one step easier because the number of Ms can be directly multiplied by the price per M to get the total cost. This process can be demonstrated in the exercise of ordering 16,000 sheets of 17 × 28-inch Sub. 20 Midpoint Bond White (see Figure 9–10). The procedure is as follows:

$$\frac{16 \text{ Ms}}{3 \text{ Ms/ctn}} = 5.33 \text{ ctns (use 4 ctn price)}$$

16 Ms × $18.03/M = $288.48 (cost for 16 M sheets)

After finding the appropriate price bracket, only one step remains—that of multiplying the Ms by the price per M. Note in this problem that no consideration was given to buying enough additional paper to qualify for the next price bracket because that would require buying too much to be cost effective.

A point that cannot be stressed too strongly is that—in all cases—the *prices listed are per M,* even when less than an M is being purchased. A case in point is the cost of buying 500 sheets of 17 × 28-inch Sub. 20 Midpoint Bond white (refer

Skylark Offset

FINISH: Gloss, Dull, Matte
BASIS: 25 x 38
PACKED: Cartons

	1 CTN.	4 CTNS.	8 CTNS.	16 CTNS.
	PRICE PER CWT			
WHITE GLOSS	130.70	114.55	110.90	106.50
WHITE DULL	131.65	115.35	111.65	107.25
WHITE MATTE	128.00	112.20	108.60	104.25

COLOR	BASIS	SIZE	M WT.	SHEETS PER CTN.	1 CTN.	4 CTNS.	8 CTNS.	16 CTNS.
					PRICE PER 1000 SHEETS			
GLOSS FINISH								
White	80	19 x25	80	2000	104.55	91.65	88.70	85.20
White	80	23 x29	112	1100	146.40	128.30	124.20	119.30
White	80	23 x35	136	1100	177.75	155.80	150.80	144.85
White	80	25 x38	160	1000	209.10	183.30	177.45	170.40
White	100	19 x25	100	1600	130.70	114.55	110.90	106.50
White	100	23 x29	140	1000	183.00	160.35	155.25	149.10
White	100	23 x35	169	900	220.90	193.60	187.40	180.00
White	100	25 x38	200	800	261.40	229.10	221.80	213.00
DULL FINISH								
White	80	19 x25	80	2000	105.30	92.30	89.30	85.80
White	80	23 x29	112	1100	147.45	129.20	125.05	120.10
White	80	23 x35	136	1100	179.05	156.90	151.85	145.85
White	80	25 x38	160	1000	210.65	184.55	178.65	171.60
White	100	19 x25	100	1600	131.65	115.35	111.65	107.25
White	100	23 x29	140	1000	184.30	161.50	156.30	150.15
White	100	23 x35	169	900	222.50	194.95	188.70	181.25
White	100	25 x38	200	800	263.30	230.70	223.30	214.50
MATTE FINISH								
White	80	23 x35	136	1000	174.10	152.60	147.70	141.80
White	80	25 x38	160	750	204.80	179.50	173.75	166.80
White	100	23 x29	140	1000	179.20	157.10	152.05	145.95
White	100	23 x35	169	750	216.30	189.60	183.55	176.20
White	100	25 x38	200	750	256.00	224.40	217.20	208.50

Figure 9–11 Still another paper merchant's catalog provides prices "per thousand sheets" as well as "per hundred pounds." Both pricing systems will result in essentially the same cost for the order.

back to Figure 9–10). Because the 500 sheets is fewer than a full carton, the "ream" price is used; but note that $24.20 is not the cost of a ream, but rather the cost of an M if only one ream is ordered. Thus,

$$.5 \text{ Ms} \times \$24.20/M = \$12.10$$

Many paper merchants set up their catalogs to allow for either method of calculation; that is, they show prices by both hundredweight *and* thousand sheets. An example of this combined format is seen in Figure 9–11. If the same order of paper were figured both ways, the two prices should be within five cents of one another. To check this, an order of 2,000 sheets of 25 × 38-inch Sub. 80 Skylark Coated Book Dull will be priced by each method. The first method is by cwt:

1. 2 Ms × 160# (Mwt) = 320#
2. 2 Ms ÷ 1 M (sheets per ctn) = 2 ctns   (use 1 ctn price)
3. 320# × $1.3165/# = $421.28

The next method is by Ms:

1. 2 Ms ÷ 1 M (sheets per ctn) = 2 ctns
2. 2 Ms × $210.65/M = $421.30

Penalty Weights

One aspect of paper marketing that needs to be addressed is the matter of penalty weights. Simply stated, **penalty weights** are substance weights of paper (usually the lightest weight available for a particular sheet) that cost as much to buy as if they were one category heavier. For example, Sub. 13 bond will usually cost as much as if it were Sub. 16 bond in the same size and brand. Referring back to the prices for Northpark Bond (Figure 9–9), imagine buying 5,000 17 × 22-inch sheets of Sub. 13 white bond. Because there are 5,000 sheets in a carton, this order qualifies for the 1 ctn. price of $1.3585 per pound.

$$5 \text{ Ms} \times 26 = 130\#$$
$$130\# \times \$1.3585 = \$176.61$$

Now, calculate the same order except substitute Sub. 16 for the Sub. 13. The 1 ctn price bracket is still appropriate, but notice that the 1 ctn price has gone down from $1.3585 per pound. Therefore:

$$5 \text{ Ms} \times 32 = 160\#$$
$$160\# \times \$1.1045/\# = \$176.72$$

The comparison shows that the costs for Sub. 13 and Sub. 16 bond are essentially the same with a difference of only $.11.

The reasoning behind the practice of pricing lightweight sheets as though they were heavier reflects the mill's costs for producing the paper. The fact that Sub. 16 paper has "more tree" in it than Sub. 13 paper is negligible. The primary cost factor is the time on the paper machine, which runs at the same speed for all weights. Therefore, it takes as long—and therefore costs essentially as much—to produce a sheet of one weight as another.

When the paper merchant's catalog gives different prices for Sub. 13 and Sub. 16 stocks, the buyer's job is made easier because the penalty weight differential is already built into the calculations. This is not always the case, however, and when both weights are given the same price per pound, a buyer of Sub. 13 paper must make the adjustment by using the Mwt for the Sub. 16 sheet in the same size. In this way, the lighter sheet will cost as much as if it were the next higher weight. Figure 9–12 is an excerpt from such a catalog;

```
┌─────────────────────────────────────────────────────────────┐
│  ALUM CREEK BOND                                            │
│                                                             │
│   Size      Weight      Mwt.          Sheets/ctn            │
│   17 x 22   Sub 13      26            5000                  │
│             Sub 16      32            4000                  │
│             Sub. 20     40            3000                  │
│   19 x 24   Sub 13      32            4000                  │
│             Sub 16      39            3000                  │
│             Sub 20      49            3000                  │
│                                                             │
│                     Price  CWT                             │
│               Ream    1 ctns.  4 ctns.   16 ctns.          │
│   Sub 13 and 16  $100.50  $72.80  $62.10   $56.55          │
│   Sub 20          $97.66  $68.45  $57.15   $54.87          │
└─────────────────────────────────────────────────────────────┘
```

Figure 9–12 An example of a paper merchant's catalog that does not provide separate prices for Sub. 13 bond paper.

using it to order 6,000 sheets of 17 × 22-inch Sub. 13 Alum Creek paper provides an example of this method.

1. 6 Ms × 32# = 192# (Mwt for Sub. 16)
2. 6 Ms ÷ 5Ms (sheets in ctn for Sub. 13) = 1.2 ctns (use 1 ctn price bracket)
3. 192# × $.7280/# = $139.78

Note that only the Mwt is borrowed from the next higher substance weight. Because the order is still for Sub. 13 paper, there will be 5,000 sheets in each carton received. Incidentally, 6,000 of the Sub. 16 paper would cost the same, so there is no cost savings to buying the lighter sheet. Only when the finished product is to be mailed would ordering the lighter weight stock ultimately be a cost advantage to the customer.

Another interesting feature of a paper catalog is the ability to move to the *next heavier substance weight* sheet and actually spend less money. This potential bargain can occur when the order's quantity is near a price threshold and is due to differences in the number of sheets in a carton. For instance, there is a situation in which a printer can actually save money by moving from Sub. 13 paper up one substance weight to Sub. 16. For example, imagine that a customer needs 12,000 sheets of 17 × 28-inch Northpark Bond white; the order can be run on either Sub. 13 or Sub. 16 stock. The assumption might be that the lighter sheet would be less expensive or at least the same cost; however, neither is the case when the costs for both sheets are calculated (refer back to Figure 9–9).

Cost for Sub. 13 17 × 28-inch 12,000 sheets:

1. 12 Ms × 33# Mwt = 396# (Note that because the catalog gives different prices for Sub. 13 and 16, the true Mwt is used.)
2. 12 Ms ÷ 4Ms (sheets/ctn) = 3 ctns (use 1 ctn price)
3. 396# × $1.3585/# = $537.97

Cost for Sub. 16 17 × 28-inch 12,000 sheets:

1. 12 Ms × 41# Mwt = 492#
2. 12 Ms ÷ 3 Ms (sheets/ctn) = 4 ctns (use 4 ctn price)
3. 492# × $.969/# = $476.75

Surprisingly, the Sub. 16 is $61.22 less than the Sub. 13 sheet of the same brand because of the difference in the number of sheets in a carton.

Buying a Larger Size for Less Money

Not only can a printer order a heavier sheet at less cost, but sometimes a *larger sheet can also be less expensive*. Once again, a difference between the two orders in the number of sheets in a carton is the key. To see this, compare ordering 3,000 sheets of 17 × 22-inch Sub. 16 Northpark Bond with the same order in the larger size of 19 × 24 inches.

17 × 22-inch Sub. 16

1. 3Ms × 32# (Mwt) = 96#
2. 3Ms ÷ 4Ms (sheets/ctn) = .75 (use pkg price)
3. 96# × $1.8515/# = $177.74

19 × 24-inch Sub. 16

1. 3Ms × 39# (Mwt) = 117#
2. 3Ms ÷ 3Ms (sheets/ctn) = 1 (use 1 ctn price)
3. 117# × $1.1045/# = $129.23

Buying the larger sheet is actually $48.51 less costly. This situation seldom exists outside of the 1 carton threshold. For example, of these two sizes, the larger is more expensive at the 4 carton and 12 carton thresholds.

The purpose of the last couple of pages is to demonstrate the potential for the resourceful paper buyer to pay a lower price than might be possible through a superficial "once only" calculation. Because paper averages between 25 and 40 percent of the overall printing expense, buyers who seek a competitive edge need to be able to use the intricacies of the paper merchant's catalog to their advantage and be on the lookout for opportunities that revolve around quantity, weight, or sheet size.

Paper Measurements in the Metric System

In the middle 1970s the federal government initiated a program of ushering Americans into adoption of the metric system. Twenty years later, virtually nothing has changed for most Americans. Plywood and other lumber products are still measured in feet and inches; distances between cities are still measured in miles instead of kilometers; and bathroom scales still record weight in pounds instead of kilograms. In the paper industry, Americans also still work with pounds and inches, but examination of a paper carton's label will reveal that the sheet size in metric measurements has been included (see Figure 9–13). It is widely held that the 1994 North American Free Trade Agreement (NAFTA) and other milestones in the move toward a global economy will accelerate America's transition to metric—the system used by every other nation in the world.

The paper industry measures paper in linear distance (size and thickness) and weight. At present in America, linear distance is measured by inches, but the metric system uses the *meter* and its increments—the *centimeter* (1/100 of a meter), the *millimeter* (1/1000 of a meter), and the **micrometer** (1/1,000,000 of a meter). To gain perspective, it may help to remember that a meter is 39.37 inches long—slightly longer than a yard stick. A more useful factor in paper calculations is to know that 25.4 millimeters equal one inch. An awareness of this ratio allows the conversion from inches to millimeters. For example, a sheet of 25 × 38-inch paper also measures 635 × 965 millimeters (25 × 25.4 = 635 and 28 × 25.4 = 965). At the same time, the same factor allows a paper caliper of four points (.004-inch) to be converted to 102 micrometers or **microns** (4 pts × 25.4 = 101.6).

Because millimeters are so small (about the thickness of a paper clip wire) and micrometers are even smaller, paper sizes and calipers are rounded to the nearest whole unit. Table 9–1 contains conversion factors for English and metric measurements that are relevant to the paper industry.

Metric Paper Sizes

The metric system for available paper sizes is analogous to the American system of standard sizes, which features increments of the basic size. The reader will recall that 17 × 22 inches is the basic size for bond papers and that 22 × 34, 11 × 17, and 8 ½ × 11-inch papers are, respectively, double, one-half, and one-quarter of that basic size. In the metric system the basic size is 841 × 1189 millimeters. The significance of these dimensions is two-fold. First, when

Figure 9–13 This label provides the sheets' dimensions in both American and metric measurements.

multiplied they equal 1 million square millimeters or 1 *square meter.* Second, the ratio between the two dimensions is equal to 1:1.4142135 and 1.4142135 is the square root of 2; therefore, the ratio for the metric basic size of paper also can be expressed 1:$\sqrt{2}$. The 1:1.4142 ratio creates a proportion that is similar to American basic sizes; for example, 17 × 22 inches is 1:1.294; 20 × 26 inches is 1:1.3; and 25 × 38 inches is 1:1.52. Somewhat like the American system of standard sizes, the available sizes in the metric system are increments of the basic size.

The basic size (referred to as A0) and seven smaller sizes are represented in Figure 9–14. Notice how A1 is one-half the size of A0, but twice the size of A2, which is twice as large as A3, and so on. Table 9–2 lists the available sizes of A-series and B-series paper sizes which are also available and follow the same pattern of each size as an increment of the others.

1 meter	=	39.37 inches
1 millimeter	=	0.03937 inches
2.54 centimeters	=	1 inch
1 gram	=	0.0022 pounds
1 kilogram	=	2.2 pounds
1 pound	=	.4545 kilograms
1 ton (2,000 pounds)	=	909 kilograms
1 metric ton (1,000 kg)	=	2,200 pounds

Table 9–1 This chart allows for commonly used metric conversions.

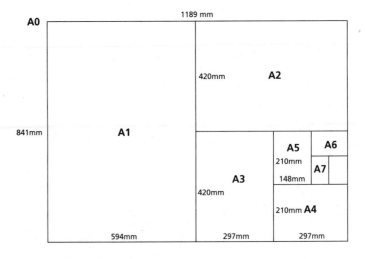

Figure 9–14 Each metric paper size is one-half of the next larger size.

2A	1189 × 1682	B0	1000 × 1414
A0	841 × 1189	B1	707 × 1000
A1	594 × 841	B2	500 × 707
A2	420 × 594	B3	353 × 500
A3	297 × 420	B4	250 × 353
A4	210 × 297	B5	176 × 250
A5	148 × 210	B6	125 × 176
A6	105 × 148	B7	88 × 125
A7	74 × 105	B8	62 × 88
A8	52 × 74	B9	44 × 62
A9	37 × 52	B10	31 × 44
A10	26 × 37		

Table 9–2 Metric paper sizes in A and B series.

Paper Weights in the Metric System

In America, a paper's substance weight is expressed as the weight (in pounds) of one ream in the basic size. The metric system is similar except that *grams,* instead of pounds, are used as the unit of weight measurement and—because the basic size is 1 square meter—the metric equivalent of substance weight represents the weight of a ream of paper cut to equal 1 square meter in area. Instead of substance weight, the metric system uses the term **grammage** (grams per square meter or gsm). Table 9–3 contains factors for converting between substance weight and grammage. To demonstrate the conversion process, consider the grammage (metric equivalent) of Sub. 60 book paper.

60# (Sub. wt) × 1.48 (Factor) = 88.8 (rounded to 89)
(Grammage)

Therefore, under the metric system, Sub. 60 book paper would have a grammage of 89. Phrased another way, an 89 grammage offset paper manufactured overseas is the equivalent to an American Sub. 60 offset sheet. The same kind of conversion between American and metric is possible by using the following formulas.

$$\text{grammage} = \frac{\text{substance wt} \times 1406.5}{\text{basic size area}}$$

substance weight = grammage × basic size area

× .000711

For example, assume that a person wants to know the metric equivalent of Sub. 16 bond. The procedure would be

Classification	Basic Size	Grammage to Ream Weight	Ream Weight to Grammage
Business	17 × 22	0.266	3.760
Book	25 × 38	0.675	1.480
Cover	20 × 26	0.370	2.704
Newsprint	24 × 36	0.614	1.627

Table 9–3 Conversion factors for the four most common paper classifications.

to multiply the substance weight by the conversion factor (16 × 1406.5 = 22,504) and then divide by 374 (17 × 22 inches is the basic size of bond paper, so 374 square inches comprise the area of the basic size). The division of 22,504 by 374 is 60.17, so 60 is the equivalent grammage.

The other formula allows the conversion from grammage to substance weight as demonstrated with the following question. A bond sheet from Scotland has a grammage of 75; what is its equivalent in substance weight? To answer the question, simply multiply the grammage by the basic size area (75 × 374 = 28,050) then multiply by the conversion factor (28,050 × .000711 = 19.94 rounded to 20). The calculation reveals that a 75 grammage bond is equivalent to Sub. 20 bond.

The Metric Equivalent of M Weight

Although M weights are measured by Americans in pounds, the metric system measures the weight of 1,000 sheets in *kilograms* (kg) and a kilogram is approximately 2.2 pounds. To calculate the M weight in kilograms for a particular sheet, simply multiply its area (in millimeters) by its grammage and divide by 1,000,000. For example, consider ordering some 508 × 660 mm cover paper with a grammage of 189 gsm. What would be its M weight in kilograms?

$$\text{Mwt} = \frac{508 \times 660 \times 189}{1,000,000}$$

$$\text{Mwt} = 63.4 \text{ (rounded to 64 kg)}$$

The specifications given in this problem are the metric equivalents for 20 × 26-inch Sub. 70 cover stock, so a quick check of the calculations can be made by recalling that 1 kilogram approximates 2.2 pounds. Multiplying 63.4 (kg/M) by 2.2 gives 139.5 pounds, which is essentially 140—the Mwt for 20 × 26-inch Sub. 70 cover paper.

No one can predict the future with certainty, but it seems safe to predict that America—currently the last nation to resist conversion—will eventually adopt the metric system. When this happens, terms such as millimeters, kilograms, and A3 size paper will become dominant and the procedure presented here will replace the status quo. As has been observed, the formulae are essentially the same as those that are currently in use. For this reason, using metric will not be more difficult than the present system.

Sources of Paper

Most printers order their paper from **local merchants.** These firms serve as "middlemen" between the paper mills that produce the paper and the printers who use it. Naturally, paper merchants (or wholesalers) apply a markup to what they sell, but at the same time provide some key services. First, they maintain a local inventory too large to be practical for most printers to try to warehouse. Second, they will sell orders too small for a mill to bother with. Third, they provide a wide variety of brands because they probably buy from several mills. Fourth, they employ sales representatives to service local accounts. Fifth, they are more likely than a mill to extend a 30-day credit to a new customer.

Nonetheless, a significant amount of paper is sold by the **mills** directly. Most of this volume is bought by book publishers, newspaper and magazine publishers, stationers, and envelope manufacturers; these are firms that can quickly use carloads of the same product. Commercial printers also can take advantage of potential cost savings by ordering direct from a mill; however, this practice is not without its risks. First of all, a printer should check to see that the desired type and quantity of paper is within the mill's inventory. Second, mills usually require that a minimum quantity be ordered. Third, depending upon how far away the mill is located, the order will likely take between one and two weeks to arrive, compared to one to three days if ordered from a local merchant. Last, a printer who buys paper from a mill only to have the printing order cancelled will probably be stuck with the paper.

Ordering directly from a mill is sometimes necessary when a printer needs paper with certain physical characteristics not currently on the market—in short, a custom-made sheet. The inclination of paper mills to produce **custom orders** varies greatly; key factors include how busy they are and how massive their paper machines are. Mills that have retained at least one small machine are better able to make the necessary adjustments required of a custom order. The nature of the desired characteristics will also heavily influence the request's feasibility.

The easiest adjustment to make is size because the function of slitting and sheeting is done in the mill's finishing room after the paper has already been made. Depending upon how evenly the cross-grain dimension cuts out of the large roll of paper that comes off the machine (commonly known as a *log*), the minimum order for a special size can range from four cartons to 5,000 pounds.

Printers that require unique finishes or substance weights also require changes in the actual manufacture of the paper; the first change could involve the coating, calendering, or embossing of the sheet; the second change is in the actual water/fiber ratio of the stock that goes into the paper machine. Both involve more adjustments in production and understandably require a higher minimum order; 5,000 pounds is a common minimum order. The most difficult characteristics to alter are color, opacity, strength, brightness, and dimensional stability because these are determined before the stock reaches the machine. Because a special "recipe" must be used to produce such a sheet, the minimum order can be as much as a carload (36,000 pounds).

Even if a mill is willing to accept special orders and the volume of the order is sufficient, the time frame can run as long as one-half year and, even if the wait is not prohibitive, the added cost may be.

Still another source for the paper buyer is the assortment of **converting firms** that buy large rolls from mills and are then in position to slit and sheet them as needed. Even some large volume paper merchants supply this service. Compared with most mills, these firms—especially the paper merchants—are usually more responsive to small orders and can supply custom-size sheets for less money.

Also to be mentioned is the **paper jobber,** a firm that sells what is termed *job lot paper* and represents the secondary market of the paper industry. Persons who are acquainted with stores that advertise bargains on "buyouts," "special purchase merchandise," or "liquidations" should already have some understanding of what constitutes the secondary market. Clothing manufacturers will often liquidate last year's designs, slowly moving merchandise, or even irregular goods to stores that specialize in these kinds of products. For the consumer, such stores can make available phenomenal savings on perfectly fine merchandise. At the same time, the goods may be flawed and the decision of how important the flaw is must be made. Sometimes a sweater that is marked "20% off" has a missing sleeve.

This same broad range of reasons for goods being available in the secondary market exists within the paper industry. Sometimes a mill's management finds that it has overproduced a particular size, weight, or color of a sheet. The remedy is often to sell off the excess to a jobber and thereby free up valuable warehouse space. In this case, the paper is top quality. There are other situations in which perfectly fine paper can be sold to a paper jobber. When a mill discontinues a product line or merely changes from a certain color, the inventory on hand is often liquidated to be sold as job lot paper. Also, if a mill chooses to upgrade one or more of a sheet's physical characteristics, paper on hand will possibly be liquidated in the same way. Also, a small roll of paper is sometimes left over at the end of the "log" or large roll after a custom order for a special size has been produced. This excess paper occurs when the special size is not an even multiple of the length of the log. These rolls are called **side runs** or *butt rolls* and are often sold to a jobber.

There are also instances in which job lot paper is physically different or even defective in some way. Examples of merely different, but not flawed, stock would include paper that was intended to be Sub. 50 offset but turned out to be Sub. 52 offset, paper with a brightness level not up to specifications, or a color that does not match the swatch books. Even though this stock cannot be sold under the mill's label, it is basically fine paper and should be excellent in most situations. Still other job lot paper is actually defective in one or more ways, but can still be considered acceptable in certain situations. For example, a coated-two-side (C2S) sheet may run fine on one side, but pick badly on the other. Obviously, this paper is inadequate for a brochure, catalog, or any other application involving two-sided printing. However, for a poster or any other product that is printed on only one side, such a sheet would be fine. Similarly, paper with wavy edges may be salvaged by trimming three inches or whatever amount is necessary from the outside and producing a smaller, printable sheet.

As was mentioned in the chapter on problems in the pressroom, reel or wrap curl occurs when paper is sheeted from too close to the middle of the roll. Although this is usually reduced to broke and recycled, sometimes jobbers will offer to buy the remainder of the roll and sheet it themselves. This paper may produce press problems, but, if they can be solved, a printer can certainly gain access to some very inexpensive paper for fliers or other jobs that are not concerned with curl.

The key to buying job lot paper is *good communication* between the jobber and the buyer so that the paper sold is right for what will be asked of it. The jobber should know the history of the paper. On occasion, this is not known and the jobber should at least be willing to stand behind the stock should it prove unsatisfactory. The procedure for the printer to follow if a problem does arise is the same as if first-class stock were purchased from a paper merchant. First, try to preview the stock by running a few sheets in advance of the actual press run. If problems develop, there will be adequate time to replace the stock. If this is not possible, do not open all of the paper at once, but rather just enough paper to find out how it will perform. Alerting the dealer to the problem is the next step. Generally, only that which is unopened, uncut, and unprinted will be bought back.

Clearly, the best rule for buying job lot paper is *caveat emptor*—buyer beware. For those jobs in which the customer wants only the finest materials and top quality work the printer would be well advised to avoid as many unknown variables as possible.

The Influence of Quality Management in Purchasing

Many paper companies have followed in the footsteps of the American automobile industry and adopted the philosophy of *quality management* that allowed Japanese manufacturers to enjoy international success with their products. Mills that have made the transition to quality management have done so to make their products' quality more consistent, thereby reducing customer complaints and improving the chances of maintaining their present customer base and increasing their market share. Mills that have a global market are finding that being able to meet ISO 9000 standards is crucial to remaining competitive.

The implementation of quality management within the paper industry is growing. A 1995 *Pulp & Paper* survey of North American mills found that 93 percent of respondents reported a formal quality system in place and 71.8 percent reported using ISO 9000 as the basis for their system, a 400% increase for ISO 9000 use since 1991.[2]

Among American mills that have quality programs, the most commonly used is **total quality management (TQM)**, a management philosophy developed in the early 1950s by Dr. W. Edwards Deming, an American management guru.

TQM is a fresh approach to doing business that features some unique concepts. First, a TQM manufacturer abandons the idea of controlling quality by inspecting out the faulty products after they are made, but before they are shipped. Instead of product control, the emphasis is on process control; that is, the process is monitored closely enough that poor products are not created in the first place. Second, instead of a manufacturer establishing product standards, manufacturers work toward meeting the standards established by the customer; clearly, TQM organizations are customer-driven. Third, TQM organizations replace the top-down direction of communication with a new corporate culture that trains and empowers employees at all levels to make decisions and work in teams to improve the process continuously, thereby improving the product and service that the customer receives. For a paper mill to make the commitment to TQM requires an enormous investment in employee training and, therefore, is not entered into lightly.

Printing firms that have made the shift to TQM realize that they cannot eliminate defects in their printing, unless the materials that they work with—such as paper—are also produced within a controlled process. For this reason, an increasing number of printers have abandoned the practice of buying paper and other materials primarily on the basis of price. These printers understand that low-priced paper can become the most expensive after lost press time due to curling, slow drying, picking, linting, or other ailments is added to the potential cost of disrupted schedules and dissatisfied customers. As a result, these printers seek to form a partnership with paper companies that will commit to TQM; in exchange for making this commitment, the paper company understands that the printer will not jump to another mill that comes in later with a lower price.

Printing companies that have been able to establish relationships based on a common vision and the reliance on data for decision-making are able to reduce miscommunication and production problems.

Helpful Publications

Persons who buy paper have available to them a number of aids to help in finding a suitable sheet to meet a customer's expectations. On occasion, a customer may walk in with an unknown sample sheet and request that his order be run on the same stock. Such a situation requires the printer to play detective and assembling clues, analyzing the evidence, and narrowing down the possibilities constitute the pattern of action. A paper micrometer can reveal the sheet's caliper. Armed with this information, the finish, and a comparative bulk chart (Table 9–4), an idea of the substance weight can be discerned. For example, if the sample brought in is a glossy brochure with a caliper of .004-inch, then it is a good bet that it is Sub. 80 coated book paper. From here, the printer can go through swatch books of coated book paper, trying to find the best match in color, weight, and finish.

Sometimes a printer cannot supply the exact sheet that a customer requests. If the sheet has been discontinued or if no local merchant carries it, the printer must then try to find a substitute to suggest to the customer. Helpful tools here are the comparative weight chart (Table 9–5), *The Competitive Grade Finder, The Paper Buyers' Encyclopedia,* and *Walden's Paper Catalog.* The first is helpful because it allows a printer to match the weight of a sheet in one category (such as bond) with a sheet from a different category (such as offset). Note on the chart that a sheet of Sub. 20 bond compares closely with a sheet of Sub. 50 offset and that Sub. 65 cover is very similar to Sub. 80 Bristol.

Trying to match sheets from different manufacturers and maintain the same level of quality is a gray area at best. Mills produce various grades of their papers (usually No. 4, No. 2, No. 1, and Premium No. 1 moving from low to high), but there is no industry-wide standard or agency to police these labels. In fact, many printers may feel that one mill's number 2 bond is superior to another mill's number 1. An annual publication called the *Competitive Grade Finder* is of some help in this matter. This book groups the sheets currently available by cost and, to some degree, level of performance. Information about a given sheet's grade, recycled content, and brightness level is often provided, as well. The same publisher also makes *The Paper Buyers' Encyclopedia,* which contains the material of *The Competitive Grade Finder* as well as a listing of suppliers and other information. Still another useful publication is the semiannual *Walden's Paper Catalog,* which not only groups brands by price but also provides a listing of brand names and the manufacturer that produces them. In addition, there is a listing of the merchants that represent a given mill. Figures 9–15 and 9–16 are samples from this publication.

Figuring the Number of "Sheets Out"

As the reader has observed from the same paper catalogs in this chapter, paper merchants stock a particular sheet in a limited number of sizes. For example, in Figure 9–11, Sky-

	Basis	Thickness
BOND		
17 × 22	9	.002
	13	.0025
	16	.003
	20	.004
BRISTOL (Plate)		
22½ × 28½	100	.008
	120	.010
	140	.012
	160	.014
	180	.017
BRISTOL (Antique)		
22½ × 28½	100	.012
	120	.014
	140	.016
	160	.018
	180	.020
COATED (Book)		
25 × 38	60	.003
	70	.0035
	80	.004
	100	.0055
	120	.006
COATED COVER		
20 × 26	50	.00475
	60	.006
	65	.0065
	80	.0075
	100	.0095
COVER		
20 × 26	50	.007
	65	.009
	80	.0105
	100	.0135
	130	.018
EGGSHELL		
25 × 38	50	.0045
	60	.005
	70	.006
	80	.0065

	Basis	Thickness
ENGLISH FINISH		
25 × 38	45	.0025
	50	.0035
	60	.004
	70	.0045
INDEX		
20½ × 24¾	58½	.007
	72	.008
	91	.0105
	111	.0135
	143	.0175
LEDGER		
17 × 22	24	.004
	28	.005
	32	.00525
	36	.00575
	40	.0065
	44	.007
OFFSET		
25 × 38	50	.004
	60	.0045
	70	.005
	80	.006
	100	.0075
	120	.009
SUPER		
25 × 38	50	.0025
	60	.003
	70	.004
TAG		
24 × 36	80	.006
	100	.0075
	125	.009
	150	.011
	175	.0125
	200	.015
	250	.018
	300	.0225
TEXT		
25 × 38	60	.0045
	70	.0065
	80	.007
VELLUM		
17 × 22	20	.004
	24	.005
	28	.006
	32	.0065
	36	.0075
	40	.0085

Table 9–4 The Comparative Bulk Table provides the thickness of a sheet for a given category and substance weight.

EQUIVALENT WEIGHTS OF PAPER
Basis weights in bold—
All weights are for 500 sheet reams

Category	BOND 17 × 22	BOOK 25 × 38	COVER 20 × 26	BRISTOL 22½ × 28½	BRISTOL 22½ × 35	INDEX 25½ × 30½	TAG 24 × 36
	13	33	18	22	27	27	30
	16	41	22	27	34	33	37
	20	51	28	34	42	42	46
Bond Ledger Mimeo Duplicator Writings	**24**	61	33	41	51	50	56
	28	71	39	48	59	58	64
	32	81	45	55	67	67	74
	36	91	50	62	76	75	83
	40	102	56	69	84	83	93
	44	112	61	75	93	92	102
	12	**30**	16	20	25	25	27
	16	**40**	21	27	33	32	36
	18	**45**	24	30	37	37	41
	20	**50**	27	34	41	41	45
Book Offset Text	24	**60**	32	40	50	49	55
	28	**70**	38	47	58	57	64
	31	**80**	43	54	66	66	73
	35	**90**	48	60	75	74	82
	39	**100**	54	67	83	81	91
	47	**120**	65	80	99	98	109
	59	**150**	80	100	124	123	136
	29	73	**40**	49	61	60	66
	36	91	**50**	62	76	75	82
	43	110	**60**	74	91	90	100
Cover	47	119	**65**	80	98	97	108
	58	146	**80**	99	121	120	134
	65	164	**90**	111	136	135	149
	72	183	**100**	124	151	150	166
	58	148	81	**100**	123	121	135
	70	176	97	**120**	147	146	162
Mill Bristol	82	207	114	**140**	172	170	189
	93	237	130	**160**	196	194	216
	105	267	146	**180**	221	218	242
	117	296	162	**200**	246	243	269
	52	133	73	90	**110**	109	121
Printing Bristol	59	151	83	102	**125**	123	137
	71	181	99	122	**150**	148	165
	83	211	116	143	**175**	173	192
	95	241	132	163	**200**	198	219
	43	110	60	74	91	**90**	100
Index Bristol	53	135	74	91	112	**110**	122
	67	170	93	115	141	**140**	156
	82	208	114	140	172	**170**	189
	105	267	146	182	223	**220**	244
	43	110	60	74	91	90	**100**
	54	137	75	93	114	113	**125**
	65	165	90	111	137	135	**150**
Tag	76	192	105	130	160	158	**175**
	87	220	120	148	183	180	**200**
	109	275	151	186	228	225	**250**
	130	330	181	222	273	270	**300**

Table 9–5 The Equivalent Weight Table reveals the substance weight of a given category of paper that would compare to the substance weight of a different category. For example, Sub. 50 book paper feels a lot like Sub. 20 bond paper in terms of caliper.

Offset Enamel
Dull C2S—White

Brand	Mfrs.	Bright-ness	Opacity Range	Basis Weight
Premium - Brightness Range 88 and Above				
Consort Royal Silk	U.K. Paper	98	97-98	100-110
Gleneagle Dull	U.K. Paper	94	96-97	80-100
Golden Cask	Kanzaki	88	95-98	70-100
Golden Cask Dull	Kanzaki	91	94-97	70-100
Golden Cask Dull Recycled	Kanzaki	91	94-97	70-100
Hanno' Art Diaden	Hannover	91-93	95-99	80-100
Ikonofix Dull	Zanders	92	95.5-97	78-100
Ikonorex Dull	Zanders	93	95-96	90-100
MasterArt R/C Dull Recycled	Ahlstrom	90	95-99	80-115
Multi-Art Silk	Stora Papyrus	90	89-96	60-100
Phoenix Imperial Dull	Scheufelen	88-91	95-100	80-100
Phoenix Imperial Chlorine Free	Scheufelen	88-91	95-100	80-100
Preeminence	Kanzaki	99	95-98	80-100
Preeminince Dull	Kanzaki	99	94-96	80-100
Trophy	Arjo-Wiggins	94	94-97	90-115
No. 1 - Brightness Range 85-87.9				
Cameo Dull 1	Warren	86-88	95-96	80-100
Celesta Litho Dull	Westvaco	86.5	94.6-97.3	70-100
Celesta Web Dull	Westvaco	86.5	93-97.6	70-100
Centura Dull 1	Consolidated	86.3		70-100
Consort Royal Osprey Silk (R/C)	U.K. Paper	87	96-97	100-110
Excellence Dull 1	Repap	87	95-97.5	70-100
Gleneagle Osprey:GEO Dull(R/C)	U.K. Paper	86	96-97	80-100
Lithofect Plus Dull 2	Repap	86	93-96.5	60-100
Lustro Dull 1	Warren	86-88	94-96	70-100
Phoeno Grand Chlorine Free	Scheufelen	85-88	95-98	80-100
Phoeno-Star Chlorine Free	Scheufelen	85	90-97	70-100
Quintessence Dull 1	Potlatch/Northwest	86-88	96-98	80-100
Shasta Dull	Simpson	87	95-98	70-100
Signature Cover From Mead-Dull	Mead	87	94-97	80-100
Signature From Mead -Dull 1	Mead	87	86+	80-100
Sterling Web Dull	Westvaco	85.5	93.3-97.2	60-100
Sterling Web Dull-PC	Westvaco	85.5	93.3-97.2	60-100
Tahoe Dull	Simpson	86	95-98	70-100
Vintage Velvet 1	Potlatch/Northwest	86-88	94-97	70-80-100
No. 2 - Brightness Range 83-84.9				
Productolith Dull 2	Consolidated	84		70-100
Topkote Dull	Kanzaki	84	95-97	70-100
Topkote Dull Recycled	Kanzaki	84	95-97	70-100
No. 3 - Brightness Range 79-82.9				
Delta Bright®Dull	James River	81	89-96	40-80
No. 4 - Brightness Range 73-78.9				
Ecolocote	P. H. Glatfelter	76	93-95+	45-70
Mission®Web Dull	James River	75	90-96	40-80
Phoenix Imperial Cream Dull	Scheufelen	74	94-99	80-100
Recolocote	P. H. Glatfelter	76	93-95+	40-70
Surfa®Web Dull	James River	78	89-96	40-80
Monterey®Web Dull	James River	72-74	89-96	36-80

Figure 9–15 *Walden's Paper Catalog* lists the brands of a particular category and grade of paper. The groupings are mostly based on cost. *Courtesy of* Walden's Paper Catalog.

K

Kaleidoscope - Riverside
Smooth Antique 24
Laid And Felt 30

Karma Cover - Potlatch/Northwest
Cover—Offset Enamel Matte C2S—White 11

Karma Embossed 1 - Potlatch/Northwest
Offset Enamel Embossed C2S—White 4

Karma Embossed Cover - Potlatch/Northwest
Cover—Offset Enamel Embossed Matte C2S White 13

Karma Natural Cover - Potlatch/Northwest
Cover—Offset Enamel Matte C2S—Cream 11-a

Karma Natural Text 1 - Potlatch/Northwest
Offset Enamel Matte C2S—Colored 3-a

Karma Natural Web - Potlatch/Northwest
Offset Enamel Dull C2S Cream Web 63-a2

Karma Natural Web Cover - Potlatch/Northwest
Offset Enamel Dull C2S—White 63-a

Karma Text 1 - Potlatch/Northwest
Offset Enamel Matte C2S—White 3

Karma Web 1 - Potlatch/Northwest
Offset Enamel Matte C2S—White 63-b

Karma Web Cover - Potlatch/Northwest
Offset Enamel Glossy C2S—White 63

Kashmir - Simpson
Offset Enamel Matte C2S—White 3
Cover—Offset Enamel Matte C2S—White 11
Cover—Offset Enamel Embossed Matte C2S White 13

Kashmir Natural - Simpson
Offset Enamel Matte C2S—Colored 3-a
Cover—Offset Enamel Matte C2S—Cream 11-a

Kashmir Web - Simpson
Offset Enamel Matte C2S—White 63-b
Cover Papers Matte C2S—White 63-e

Katahdin Blade - Bowater
Offset Enamel Glossy C2S—White 1

Katahdin Matte - Bowater
Offset Enamel Matte C2S—White 3

Kevtone - Holliston
Nonpaper Printing Materials 23-f

Kiana - Hopper
Offset Papers Uncoated—White 19
Offset Cover Papers Uncoated—White 20
Offset Papers Uncoated—Colored 22
Smooth Antique 24
Rough Antique 25
Recycled Offset/Opaque 34
Recycled Text/Text Cover 35
Uncoated Text And Vellum Opaque Offset 64
Smooth Opaque Text And Offset 64-a
Cover & Text Business Papers 72

Kilmory Script - Simpson
No. 1 Sulphite 71

Kimberly ®Cover - Neenah
Cover & Text Business Papers 72

Kimberly® - Neenah
Smooth Antique 24

Figure 9–16 *Walden's Paper Catalog* also has a section that lists a mill's products and merchants. *Courtesy of* Walden's Paper Catalog.

lark Offset matte is stocked in only three sizes: 23 × 35, 25 × 38, and 23 × 29 inches, and in each instance the paper will be grain long (note underlining). This kind of restriction on what is available means that the printer must figure out a way to adapt and maximize the yield from whatever size is ordered. To figure how many small sheets can be cut from a larger sheet, write the smaller sheet's dimensions below the larger sheet and, if the grain direction of the small sheet is a factor, stack the long grain dimensions. For example, to find out how many $\underline{5} \times \underline{7}$-inch cards can be cut from an $11 \times \underline{17}$-inch sheet, write these dimensions as follows:

$$11 \times \underline{17}$$
$$7 \times \underline{5}$$

Observe that the dimensions of the smaller sheet are reversed so that the long grain dimensions are aligned vertically. Next, divide the 7 into 11 and write only the number of the answer under the 7. Perform the same operation with the second column of measurements, then multiply the two quotients.

$$11 \times \underline{17}$$
$$\underline{7 \times 5}$$
$$1 \times 3 = 3$$

Only whole numbers are used as answers at this point; although 11 divided by 7 equals 1.571 and 17 divided by 5 equals 3.4, only 1 and 3 are recorded. By multiplying 1 and 3, we find that three $\underline{5} \times 7$-inch sheets can be cut from a sheet of $11 \times \underline{17}$-inch. Figure 9–17(a) illustrates the way the stock will be cut.

The grain direction of the large sheet is indicated with the arrow, which also verifies that the small sheets will be grain short as the customer requested. The shaded area of the press sheet represents trim waste—paper that cannot be used to print the $\underline{5} \times 7$-inch cards. If the customer had wanted the cards to be grain long ($5 \times \underline{7}$-inch) instead of grain short, then the setup and calculations would proceed as follows:

$$11 \times \underline{17}$$
$$\underline{5 \times 7}$$
$$2 \times 2 = 4$$

Figure 9–17(b) illustrates how such a cutting would appear. Clearly this is a more efficient cut because it produces an extra card and reduces the percentage of waste. Unfortunately, it is not relevant to the restrictions imposed by the customer's specifications and the paper merchant's catalog.

On the few occasions when the large sheet is available in either long or short grain, then both configurations can be tried and compared as follows:

$$
\begin{array}{ll}
11 \times 17 & \quad 11 \times 17 \\
\underline{5 \times 7} & \quad \underline{7 \times 5} \\
2 \times 2 = 4 & \quad 1 \times 3 = 3
\end{array}
$$

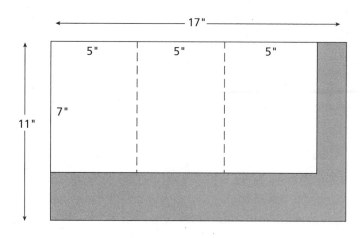

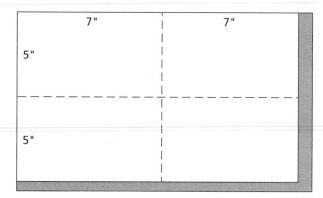

Figure 9–17 Either three or four 5 x 7-inch sheets can be cut from a larger 11 x 17-inch sheet. The shaded areas represent trim waste.

As has already been discussed, the grain direction of either the finished sheet or press sheet is usually an important consideration because it affects stiffness and dimensional stability. If the customer is printing a job as simple as fliers to be placed on windshields, then grain is unimportant and a third alternative is occasionally possible. Consider cutting 8 ½ × 11-inch sheets out of a <u>28</u> × 34-inch stock. Figure 9–18 reveals two approaches. The first will produce eight 8 ½ × 11-inch sheets that are grain short. The second configuration will yield nine 8 ½ × 11-inch sheets that are grain long. The third approach will yield ten 8 ½ × 11-inch sheets, but four will be grain long and six will be grain short (see Figure 9–19). Such an unorthodox configuration is called a *stagger cut* (or *dutch cut*) and, although it minimizes waste, is not often used because (a) most jobs require that the grain direction be consistent throughout the run and (b) stagger cuts produce additional sheets in only occasional instances.

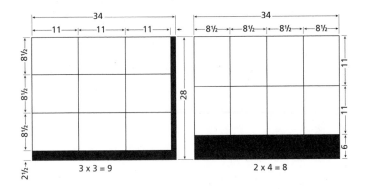

Figure 9–18 Either eight or nine 8-½ x 11-inch sheets can be cut from a larger 28 x 34-inch sheet. In either instance, the grain direction of the resulting 8-½ x 11-inch sheets will be consistent. *From Adams, Faux, and Rieber,* Printing Technology, *4th edition, © 1996 by Delmar Publishers.*

Often, the procedure of figuring the number of "sheets out" must be performed twice for a single printing job: once to find out how many images can be carried on a press sheet and a second time to figure how many press sheets can be cut from the purchase sheet. For example, a printer owns a press that can accommodate a press sheet up to 19 × 25 inches. Suppose that a job for 24,000 6 × 9-inch flyers that must be delivered grain long comes along and the customer has specified a sheet for which 25 × 38 inches is the only available size. To figure how many sheets to order, first see how many press sheets can be cut from the purchase sheet. Next, see how many 6 × 9-inch flyers can be carried on a press sheet. The first step to the solution is as follows:

$$\frac{\underline{25} \times 38}{19 \times 25} \qquad \frac{\underline{25} \times 38}{25 \times 19}$$
$$\overline{1 \times 1 = 1} \qquad \overline{1 \times 2 = 2}$$

From this work, it is apparent that each purchase sheet can yield two 19 × 25-inch press sheets; note that they necessarily will be grain long (see Figure 9–20). The next step is to see how many flyers can fit onto a press sheet.

$$\frac{19 \times \underline{25}}{6 \times 9}$$
$$\overline{3 \times 2 = 6}$$

Observe that the required grain dimensions must be stacked and the printer is not permitted to try the alternative configuration. The job can be printed six flyers at a time. However, if the customer could live with a grain short flyer, or a different paper merchant could supply a 25 × 38-inch purchase sheet, then eight could be run on a press sheet.

Buying Paper 215

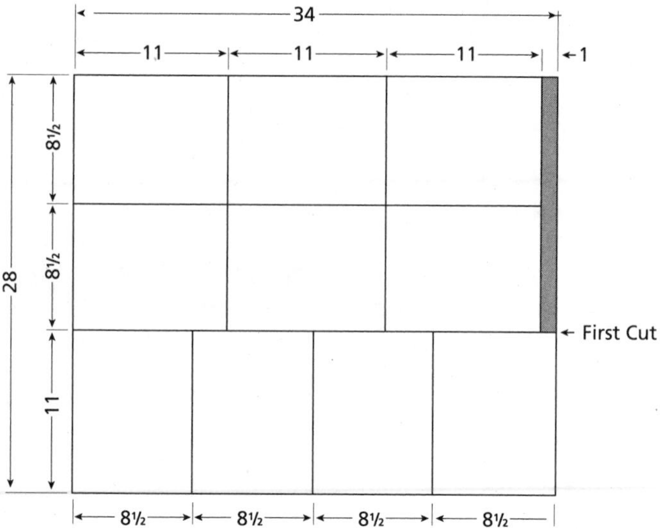

Figure 9–19 A stagger cut will produce ten 8-½ x 11-inch sheets but some of them will be grain long and others will be grain short. *From Adams, Faux, and Rieber,* Printing Technology, *4th edition, © 1996 by Delmar Publishers.*

$$\begin{array}{r} 19 \times 25 \\ 9 \times 6 \\ \hline 2 \times 4 = 8 \end{array}$$

If neither option is possible, the printer will have to live with the higher paper waste and a longer press run of running only six flyers on a press sheet instead of eight, while the customer will have to live with the resultant higher costs.

If a spoilage allowance of 6 percent is ordered to compensate for press sheets that will likely be consumed during the press makeready and the course of the run, then the final calculations follow:

$$\begin{array}{rl} 24,000 & \text{finished units (flyers)} \\ \div\ \ \ 6 & \text{out} \\ \hline 4,000 & \text{net press sheets} \\ +\ 240 & \text{wastage sheets (6\% of 4,000)} \\ \hline 4,240 & \text{total press sheets} \\ \div\ \ \ 2 & \text{out} \\ \hline 2,120 & \text{purchase sheets} \end{array}$$

Because each 25 × 38-inch press sheet will ultimately yield 12 flyers, only 2,120 of them will be needed for this 24,000 flyer order.

In this example, the purchase sheet was available in only one size, thereby simplifying the calculations. When multiple sizes are available, a printer will usually reverse the se-

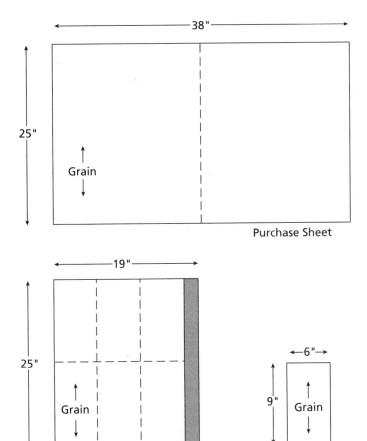

Figure 9–20 A 25 x 38-inch purchase sheet will yield two 19 x 25-inch press sheets with no waste. Unfortunately, only six 6 x 9-inch flyers can result from a single press sheet because of restrictions imposed by grain direction.

quence and figure the number of images that will be placed onto a press sheet. Once the press sheet's dimensions are established, they are applied to the dimensions of the available purchase sheets in an effort to find the one that will yield the least waste.

Buying Paper in Rolls

This chapter so far has dealt with buying in sheets. Printers with web presses will obviously purchase paper in rolls and will need to calculate the weight of these rolls, as well as figure the number of linear feet that a roll will yield.

To calculate the **roll weight** of paper, a formula can be applied that uses the roll diameter, the core diameter, the roll width, and a factor number found on Table 9–6:

Antique finish	.018	Machine finish	.026
Bond	.021	Supercalendered	.028
Ledger	.023	Coated 2 sides	.033
Uncoated offset	.022	Coated 1 side	.030

Table 9–6 These factors are used to calculate the weight of a roll of paper.

$$\text{roll wt} = \left(\frac{\text{roll diameter}^2}{\text{core diameter}^2}\right) \times \text{width} \times \text{factor}$$

To use the formula, first subtract the square of the core diameter (there is no paper in the hollow core) from the roll diameter and then multiply the difference by the roll width and the appropriate factor. A sample problem will illustrate this method. Figure the weight of a 30-inch diameter, 34-inch wide roll of a machine finish stock with a 3-inch core.

1. $\left(\dfrac{30^2}{3^2}\right) \times 34 \times .026$

2. $\left(\dfrac{900}{9}\right) \times .884$

3. $891 \times .884 = 788$ pounds

If only a quick approximation of the weight is needed, Table 9–7 can be used. The procedure is to look under the appropriate roll diameter heading and move down that column to the correct line for paper category. Then multiply the number found at the intersection by the roll width. For the problem just calculated, the solution with the table would be:

$$23.4 \times 34 = 796$$

The 8 pound difference between the two answers is the allowance made for a quicker computation. Note that Table 9–7 is based on a 3-inch core and provides only an approximation of weight because differences in caliper and winding density vary among paper mills. A more accurate chart for a given mill's products is probably available upon request.

Conversions From Rolls to Sheets

Printers who buy rolls of paper also must be able to calculate the number of linear feet of paper that the roll contains. This is the first step in figuring how much of a roll or how many rolls are needed to produce a certain job. The formula for converting rolls to linear feet is expressed as follows:

GRADE	24″	26″	28″	30″	32″	34″	36″	38″	40″
Antique finish	10.4	12.2	14.1	16.2	18.4	20.8	23.3	26.0	28.8
Uncoated offset	12.7	14.9	17.2	19.8	22.5	25.4	28.5	31.8	35.2
Machine finish	15.0	17.6	20.4	23.4	26.6	30.1	33.7	37.5	41.6
C1S paper	17.3	20.3	23.5	27.0	30.7	34.5	38.9	43.3	48.0
C2S paper	19.0	22.3	25.9	29.7	33.8	38.1	42.9	47.7	52.8
Supercalendered	16.1	18.9	22.0	25.2	28.7	32.4	36.3	40.4	44.8
Bond papers	12.0	14.0	16.4	18.9	21.6	24.2	27.0	30.2	33.7
Ledger papers	13.2	15.4	18.0	20.8	23.8	26.8	29.7	33.2	37.1

Table 9–7 These factors provide a quicker, but less accurate, roll weight than the factors in Table 9–3.

$$\frac{\text{Roll wt} \times \text{Total area of one ream in basic size} \times 500}{\text{Substance wt} \times \text{Roll width} \times 12 \text{ (inches in a foot)}}$$

To understand this formula, calculate the footage in a roll of Sub. 60 offset paper that is 34 inches wide and weighs 120 pounds.

$$\frac{120 \times 25 \times 38 \times 500}{34 \times 60 \times 12} = \text{Length in feet}$$

$$\frac{120 \times 950 \times 500}{24,480} = \text{L}$$

$$\frac{57,000,000}{24,480} = \text{L}$$

$$2,328 \text{ feet} = \text{L}$$

Being able to see a roll of paper in light of a certain number of sheets is important to buyers of rolls. To see how this works, imagine that a customer needs a finished unit that consists of a 22 × 34-inch sheet folded three times to produce a 16-page 8 ½ × 11-inch piece. How many finished units can be produced by the 2,328 foot 34-inch wide roll that was just figured?

To solve this, first picture the 22 × 34-inch flat signature before it is folded. The 34-inch dimension matches the roll's width exactly so it will be the 22-inch dimension that will be sheeted from the length of the roll. The 2,328 feet in the roll converts to 27,936 inches and this divided by 22 yields the 1,269 22 × 34-inch sheets that would result if there were no wastage.

It might be noted here that the sheet in question would be grain short because roll paper always has the grain running in the direction of the web.

The math problems presented here in no way represent all that printers need to know about paper math. The questions of signature imposition, number of sheets out of a large one, binding allowances, and postal weight restrictions also deal with paper math but are more involved with the use of paper than with its actual purchase.

Notes

1. Alum Creek, Hawkeye, Saginaw, Northpark, Midpoint, Skylark, and Red Wing are fictitious brands of paper.
2. Kelly H. Ferguson, "ISO 9000 Standards Dominate Mill Quality Efforts in North America," *Pulp & Paper* (June 1995): 61–62.

Questions for Study

1. How does the substance weight of a paper relate to the sheet's basic size?
2. What would be the weight of 2,000 sheets of 11 × 17-inch Sub. 16 bond?
3. Calculate the weight of 3,500 sheets of 22 × 34-inch Sub. 24 bond.
4. Use the Equivalent Weight Formula to figure the weight of 500 sheets of 19 × 24-inch Sub. 20 bond.
5. What is the weight of 7,000 sheets of 25 × 38-inch Sub. 50 book paper?
6. Calculate the weight of 11,250 sheets of a paper with a ream weight of 68 pounds.
7. What would be the cost of 2,000 sheets of 11 × 17-inch Sub. 24 bond paper that is selling for $1.20 per pound?
8. Referring to Figure 9–10, what would be the cost per pound of Sub. 16 white Midpoint bond if one carton is purchased?
9. Referring to Figure 9–7, what would be the cost per pound for five cartons of Sub. 20 white Saginaw bond?
10. How many sheets are in a carton of Sub. 24 22 × 34-inch Northpark bond?
11. What is the Mwt of Sub. 16 19 × 24-inch Northpark bond?
12. What is the available grain direction of 17 × 28-inch Sub. 16 Northpark bond?
13. Calculate the cost of 4,000 sheets of Sub. 20 34 × 44-inch white Northpark bond.
14. Calculate the cost of 12,500 sheets of <u>22</u> × 34-inch Sub. 16 white Midpoint bond (see Figure 9–10).

15. Referring to Figure 9–12, calculate the cost of 18,000 sheets of 17 × 22-inch Sub. 13 Alum Creek bond. Note that Sub. 13 is a penalty weight.

16. What services are provided by paper merchants to printers buying paper that mills do not offer?

17. Identify three reasons for a particular "lot" of paper to make its way into the inventory of a paper jobber.

18. How many 7 × 12-inch sheets can be cut from a 23 × 29-inch sheet?

19. How many 12 × 18-inch sheets can be cut from a <u>25</u> × 38-inch sheet? Note the grain directions.

20. How many 8 ½ × 11-inch sheets can be cut from a 25½ × 30½-inch sheet? The customer has no concerns about consistency of grain direction.

21. If a printer needs 31,216 11 × 17-inch press sheets, how many 22 × 34-inch purchase sheets must be ordered?

22. Calculate the cost to purchase 7,804 sheets of <u>24</u> × 38-inch Midpoint Bond Sub. 16 white.

23. Calculate the weight of a roll of bond paper with a 32-inch diameter, a 36-inch width, and a 3-inch core.

24. How many linear feet are in a 200 pound roll of Sub. 60 book paper that is 38 inches wide?

25. Relative to Figure 9–13, how can a paper have a substance weight and an M weight that are both 20 pounds? Is this a typo on the label? Could the ream weight also be 20 pounds?

26. To what does the term *grammage* refer?

27. What is the area relationship between A1 and A2 metric paper sizes?

28. Calculate the grammage of Sub. 50 litho paper.

Key Words

substance weight	penalty weights	grammage
M weight	micrometer	converting firm
cwt	paper jobber	roll weight
custom order	TQM	
side runs	Equivalent Weight Formula	
basic size	micron	

10 Ink Ingredients

AFTER STUDYING THIS CHAPTER, THE STUDENT SHOULD BE ABLE TO:

■ Distinguish between flushed and dry pigments.

■ List the types of ink resins and their most common applications.

■ Explain the role played by an ink's solvent.

■ Distinguish among non-drying, drying, and semi-drying oils.

■ Describe how manganese, cobalt, plasticizers, waxes, and wetting agents affect an ink's performance.

■ Describe how viscosity and tack affect ink and how they can be modified.

With few exceptions, the composition of modern day printing inks is remarkably complex. In an effort to print well onto a particular substrate, such as paper, by way of a given process, such as lithography, inks must have the proper body, adhesion, drying properties, and chemical compatibilities. If any one of a dozen ink properties is incorrect, the ink could: (a) fail to dry on the substrate; (b) set up or actually dry while still on-press; (c) attack the plate material; (d) fail to adhere to the substrate; or (e) create a printed image that is unacceptable in appearance in a number of ways. Far too often, a press operator encounters problems like these because of one of two mistakes: (1) using the wrong ink in a particular situation; or (2) adding one or more components to an ink without an understanding of the chemistry involved. In either case, the problem lies in not knowing enough about an ink's components and their functions.

The Basic Components Throughout History

At its most basic level, ink is composed of *pigments* and a *vehicle* that carries them. Because the pigments provide most of the optical qualities of an ink, their importance is obvious. However, because pigments are finely ground solid particles—like chalk dust—they are worthless without a liquid that can carry them to the surface to be printed (see Figure

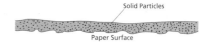

Figure 10–1 Because pigment material is solid, it must be ground and later mixed with a liquid vehicle. *Courtesy of Continental Inks.*

Figure 10–2 The pigment-vehicle relationship is a colloidal suspension of solid matter within a liquid.

10–1). For this reason, modern inks, like the crude inks used 2,000 years ago, are fundamentally pigments carried in a vehicle. For several centuries, all inks were black, usually lampblack (soot) dispersed throughout a vehicle derived from the boiling of linseed oil. Just as today's ink makers are able to provide hundreds of colors through a variety of pigments, high-tech printing is only possible because research has provided a broad range of vehicles, each one being well-suited to a particular printing situation. Although today's inks contain a variety of additional ingredients that distinguish one ink from another, the need for a pigment and a vehicle is common to them all.

Ink Pigments

An ink's solid **pigments** do not dissolve in the vehicle that carries them. A rough analogy can be drawn between the pigment-vehicle relationship and a container of fruit gelatin in which the pieces of fruit are dispersed through the liquid and are held in place after the gelatin sets up (see Figure 10–2). The insolubility of pigments in their vehicles distinguishes them from *dyes,* which do dissolve in their vehicles.

Ink manufacturers obtain their pigments in one of two forms—*dry* and *flushed.* **Dry pigments** are obtained by mixing, in a water solution, two relatively colorless chemicals that combine to form pigment crystals. These crystals become part of the solution, which is then rinsed and filtered to remove impurities as well as excess water. The resulting compound (called *press cake*) is then baked in an oven until the water or other liquid has evaporated out. The dried pigment material is then ground into a powder and mixed with the selected vehicle to form a printing ink. However, dry pigments sometimes produce particles that cluster together and are difficult to grind and disperse evenly through an ink's vehicle. For this reason, an alternative to removing the water by evaporation is to flush it out and replace it with another liquid. While the press cake still has a moisture content of 30 to 80 percent, it is mixed with a different fluid, often an oil. This fluid is selected because of its ability to *draw the pigment particles to it and away from the water;* the water then separates away from the oil-pigment mixture and can easily be poured off. In this way, the pigments have been "flushed" from the water and are termed **flushed pigments.** They are now ready to be combined with the proper vehicle and other ingredients by the ink manufacturer. Because flushed pigments are easier to work with later on in the ink manufacturing process, the extra time involved in producing flushed pigments is usually a wise investment.

Ink Vehicles

The ink **vehicle** is the liquid portion of the ink. Some printing inks are rather simplistic in the composition of their vehicles, while others are an amazingly complex collection of chemicals necessary to impart some desired characteristic to the ink in question. The composition of the vehicle will determine the ink's stiffness, drying rate, ability to adhere to a particular substrate, degree of gloss, rub resistance, and appropriateness to lithography, gravure, or any other printing process. When discussing the components in an ink's vehicle, several publications refer to it as **varnish,** a broad term sometimes said to be the portion of the vehicle that includes an ink's solvent, resin, and/or drying oil—but excludes any drier or wax. In other writings, the term varnish *does* include the drier or wax and then varnish and vehicle become synonymous. As if this lack of precision was not confusing enough, the term varnish also is used in an entirely different context: the clear coating that is often applied on-press to paper to affect the sheet's gloss or protect the printed image. To minimize confusion, the term varnish will herein refer only to the clear coating and not to a portion of the ink.

It is important to reiterate that, with the exception of its color, all of an ink's characteristics are determined by the vehicle. For this reason, a brief analysis of what may be found in a given ink's vehicle is appropriate.

Ink Resins/Binders

Because ink pigments are solid particles that are carried to the substrate by a liquid vehicle, provision must be made to bond these pigments to one another, as well as to the substrate after most of the vehicle has evaporated as part of the drying process. This needed ingredient is a binding agent that ink makers refer to as the ink's **resin** (also known as *binder*). Without a resin, the dried ink film would rub or flake from the surface of the substrate. Several resins are available to the ink maker. The correct one is determined by: the intended substrate (paper, metal, vinyl, etc.); the printing process to be used (lithography, screen, flexography, etc.); the desired gloss level; the required rub resistance; and other end-use requirements such as resistance to heat or chemicals.

There are two basic classifications for resins: *natural* and *synthetic*. A popular **natural resin** found in lithographic and letterpress printing is *rosin*, a natural compound found in pine trees. *Rosin-based resins* are used in both paste inks, such as lithographic and letterpress inks, as well as liquid inks, such as gravure and flexographic. Another group of natural resins contains the *cellulosics*, resins derived from cellulose fibers. In addition to their binding properties, cellulosics can add excellent scratch and rub resistance to an ink's film. Still another naturally occurring resin is *cyclized rubber*, rubber that has been treated with acids that cause its molecular chains to form a circle, thereby causing the rubber to lose its natural elasticity and become an ink resin with excellent adhesion and rub resistance. Also, the presence of cyclized rubber resins is what prevents the rubber-based inks that are commonly used in offset duplicators from setting up on the press overnight. Such inks are said to "stay open" on the press.

In addition to natural resins, ink manufacturers also use a number of **synthetic resins** either to impart particular characteristics to an ink or to be more compatible with other compounds in the ink. For example, because acrylic resins adhere well to most foils and packaging films and are fade-resistant, they are commonly found in flexographic, gravure, and screen inks that are used in the packaging industry.

Other common synthetic ink resins are *vinyls* used in screen inks; *maleics* (muh-LAY-iks), used in lithographic and flexographic inks; *polyamides* (poly-A-midz), used in flexographic and gravure inks; *acrylics,* used in flexographic, gravure, and screen inks; and *epoxies,* used in offset metal decorating inks. Often, two or more resins will be combined to capture the desired characteristics of each.

In the water-based inks that are used in gravure and flexographic printing, resins are also categorized by how they interact with the water. Two types of resins are used—*solution resins* and *emulsion resins.* Solution resins are molecules that have been dissolved by the water and are in a true solution. Emulsion resins cannot dissolve in water but do form an emulsion in water. The characteristics that these two types of resins impart to the ink differ, as well. While solution resins enhance the ability of the water to wet the pigment particles and help to make the ink film glossy, they do not dry well. Emulsion resins have excellent drying properties, but are not useful at enhancing wetting and or achieving high gloss. Therefore, in order for a water-based ink to have good pigment wetting and to dry quickly with high gloss, both solution and emulsion resins must be present in the formula.

Ink Solvents

All ink resins are solid compounds that must be dissolved or nearly dissolved by a liquid before they are useful in printing inks. The liquid that dissolves these resins and maintains them in solution is the ink's **solvent.** The selection of the correct solvent is crucial in ink formulation and involves several considerations. First, the solvent must be a liquid that dissolves the resin, but does not cause the pigment's color to bleed. Second, it must evaporate at an acceptable rate. Third, it must be compatible with the printing plate material and it must impart the flow and adhesion properties that are desired of the ink film.

Ink makers can select from a number of solvents. For example, flexographic and gravure inks rely mainly on alcohols and oil derivatives such as toluene, heptane, or acetate for their solvents; these *solvent inks* dominate the publication and packaging industries. However, many flexographic and gravure inks also use water as their solvent. These formulations are known as *water-based inks* and they are becoming increasingly popular due to growing concerns over the emission of *volatile organic compounds* (**VOCs**), which result from the evaporation of many solvent inks. Chapter 16 examines this issue more closely.

A crucial factor in selecting the proper solvent is its **volatility,** the speed at which it evaporates. An ink with a solvent

a

Partial transfer of ink from sheet below

b

Figure 10–3 Setoff occurs when a freshly printed ink film partially transfers to the sheet above it (a). The original image (b) becomes wrong-reading when it appears on the bottom side of the next sheet.

that lacks volatility will fail to dry quickly enough on the printed sheet, perhaps leading to smearing or setoff (see Figure 10–3). However, an extremely volatile solvent can begin to evaporate while still on the ink train of the press. When ink that is still on the press loses its solvent, it begins to dry prematurely, becomes too tacky, and can pick or tear the press sheet. As can be imagined, getting ink to set or dry immediately after being printed—but not before—is a tricky task. Selecting the proper solvent is a key factor in walking this tight rope.

Ink Oils

The three ink components discussed to this point—pigment, resin, and solvent—are present in all printing inks. However, some inks also contain one or more oils that are necessary for the ink to dry as intended. **Nondrying oils** are at the heart of inks that dry by **absorption** (also known as *penetration*). Newspapers, comic books, and many paperback novels commonly are printed with nondrying *mineral oils* that depend upon the high absorbency levels of the paper. The key

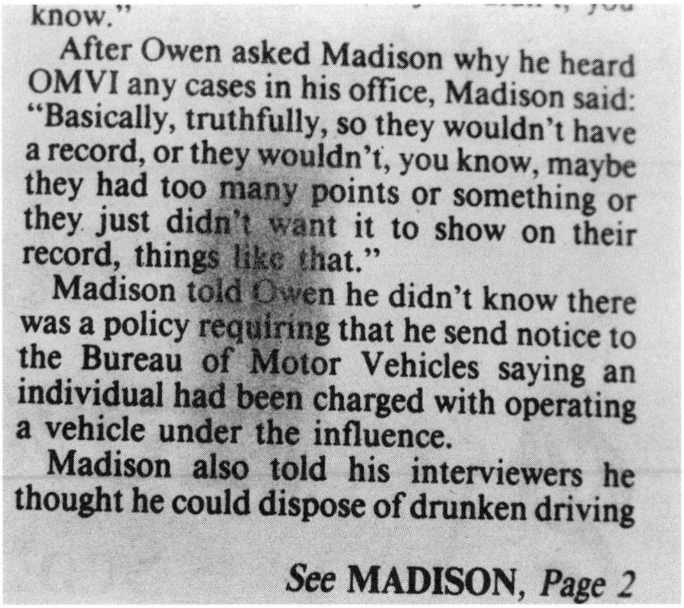
know.”

After Owen asked Madison why he heard OMVI any cases in his office, Madison said: “Basically, truthfully, so they wouldn’t have a record, or they wouldn’t, you know, maybe they had too many points or something or they just didn’t want it to show on their record, things like that.”

Madison told Owen he didn’t know there was a policy requiring that he send notice to the Bureau of Motor Vehicles saying an individual had been charged with operating a vehicle under the influence.

Madison also told his interviewers he thought he could dispose of drunken driving

See **MADISON,** *Page 2*

Figure 10–4 Even though this newspaper was two weeks old, the ink smeared when rubbed.

to nondrying oil inks is that their mineral oil solvent will penetrate the paper and, thereby, leave behind a resin-pigment film that stiffens in the absence of the oil. Clearly, absorption-dependent inks with nondrying oil vehicles are simplistic; the name nondrying is well-chosen because these inks never truly dry. For example, it is common knowledge that the ink on even a two-week-old newspaper will smear when rubbed (see Figure 10–4). When nondrying oil inks are chosen, the primary consideration is low cost rather than high quality.

A very different kind of function is performed by **drying oils,** which are found in offset and letterpress inks that must actually dry to a smear-proof film. Different from mineral oil, which is a petroleum product, drying oils are vegetable oils, usually linseed or tung oil. The valuable property of drying oils is their ability to stiffen and harden when exposed to the oxygen that is present in the air of the pressroom, an interaction that is called *oxidation.* When drying oils oxidize, the oxygen that is absorbed causes the oil’s molecules to join and form molecular chains; the chemical term for the chaining of like molecules is *polymerization* (see Appendix B for an explanation of polymerization). As polymerization continues and the chains continue to grow, they begin to impart a structure to the still-liquid ink film that makes it less mobile and therefore increasingly rigid (see

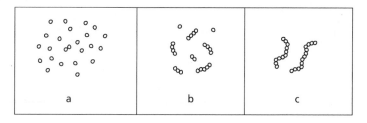

Figure 10–5 Because monomers (single molecules) have a high degree of mobility, the substance itself is liquidy. During polymerization, monomers begin to link up and form chains, thereby losing most of their mobility with the effect that the substance itself begins to stiffen and become more solid.

Figure 10–5). When the entire ink film has become rigid, it is dry. Because most lithographic and letterpress inks dry through oxidation, they require drying oils in their formulas. As will be explained later, the oxidation of drying oils is accelerated through the presence of catalysts called **driers.** Therefore, inks with drying oils will always contain driers, as well.

Press schedules that require the press sheet to be "backed-up" or printed on the second side very soon after the first printing need an ink film that can receive modest handling without smudging *before* it has actually dried. In other words, the ink must set very quickly while it dries at a much slower pace. These inks are called **quick-set inks** and they are very commonly used in lithographic and letterpress printing.

Quick-set ink technology is an extension of the inks previously discussed that dry through the oxidation of drying oils. However, in quick-set inks, the drying oil/resin solution is very carefully matched with a solvent with which it is only marginally compatible. The solvent also must be significantly thinner (more liquid) than the oil. Because the resin/drying oil solution and the solvent did not completely mix, they retain their individual properties, including their different rates of flow. Therefore, when the ink is applied to paper, the solvent immediately penetrates the paper's surface and "drains out" of the ink film (see Figure 10–6). The loss of the solvent leaves behind a very viscous and immobile resin/drying oil ink film. This immobility keeps the ink from transferring to the sheet above it in the delivery pile and the ink is considered to have *set.* Although the press sheets also may be printed on the other side without smearing, they may not be ready for folding until the oxidation process has actually dried the ink. It must be understood, however, that setting and drying are quite different from one another. An ink film can set in one or two seconds, but still require several hours before it is truly dry.

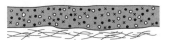

Figure 10–6 Quick-set inks rely upon a solvent's prompt absorption into the substrate. The loss of this solvent by the ink film leaves behind a much heavier drying oil and resin film that is solid enough to resist setoff.

Paper Fibers

∘ Very thin solvent
× Heavier resin/drying oil solution
• Pigment particles

PRINTED WITH
SOY INK™

Figure 10–7 The SoySeal indicates that an ink contains the required percentage of soybean-derived oil as established by the American Soybean Association. Because the SoySeal is a trademark of the American Soybean Association, users must first obtain permission to use it. *Courtesy of the National Soy Ink Information Center.*

In addition to the nondrying oils and drying oils discussed so far, many inks also include **semidrying oils**—oils that do oxidize and stiffen, but at a much slower rate than drying oils. *Soybean-derived oil* (also known as *soya oil*) is the most commonly used semidrying oil because it brings some very desirable properties to an ink. First, soya oil is comparatively clear in color as compared with many other oils. For this reason, news inks in which the petroleum-based oils have been replaced with soya oils are able to print brighter colors. Some printers also report improved ink transfer, reduced dot gain, and improved press stability. Still another force that increases the interest in soya and other vegetable oil inks is concern for the environment with respect to volatile organic compound (VOC) emission levels during printing, ease of de-inking, and the ability to biodegrade when buried in a landfill after its end use. Soy inks contain practically no VOCs, are highly biodegradable, and paper printed with them is reported to be more easily de-inked. In 1994, U.S. Department of Agriculture researchers announced a soya-based news ink formula with a reduced amount of pigment that is five times more degradable than petroleum-based inks.[1]

The American Soybean Association (ASA) was very effective in the late 1980s at promoting the use of soya oils in printing inks and established minimum percentages of soya oil content in order for the SoySeal to be displayed on an ink can or printed product (see Figure 10–7). As of 1994, these thresholds were 30 percent for colored news inks, 40 percent for black news inks, 20 percent for business forms inks, 20 percent for sheet-fed inks, and 7 percent for heat-set inks. The reason that the percentage is highest for news inks is the fact that news inks dry primarily through absorption and vegetable inks can be absorbed as well as petroleum inks. However, inks that dry through oxidation would be adversely affected with so much soya oil in them because its comparatively slow oxidation rate translates into longer drying times.

Because of the very high boiling point of soya oil, it must be severely limited in inks made for high-speed heatset printing in order to avoid drying problems that will occur

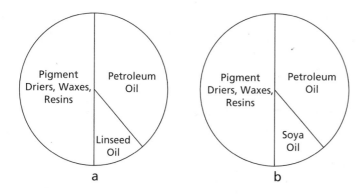

Figure 10–8 These two inks have the same percentages of petroleum oils, vegetable oil, and other ingredients. The only difference between them is whether the vegetable oil is linseed or soya. The inks are equal in environmental friendliness.

when the oil fails to evaporate quickly enough. Even in applications where drying time is not a factor, a potential drawback of replacing petroleum oils with soya oil involves the cost factor. As of 1992, black soy-based news inks cost between 25 to 40 percent more than their petroleum-based counterparts. In addition, it should be noted that the precise environmental benefits of using soya oils have yet to be fully measured. To confuse the matter further, some ink makers earn the SoySeal for their inks by incorporating soya oil in their formulations at the expense of other vegetable oils, like linseed, instead of replacing the ink's petroleum oil component. As a result, this type of soybean-derived ink is no more environmentally friendly than before soya oil was added (see Figure 10–8).

The growth of soy-based inks has been rapid and can be expected to continue. In 1987, six American newspapers used soy ink, but five years later the number had exceeded 3,000. In 1993, over 25 percent of America's commercial printers reported using soy ink on a regular basis.

Ink Additives

Although the inks used by Benjamin Franklin consisted of nothing more than a suspension of a pigment in a vehicle, the demands on today's printed products require a number of other ingredients called *additives*. The first of these additives to be examined is the **drier.** As was explained earlier, most inks form a skin across the top of the ink film when the ink's drying oils combine with the air's oxygen. This oxidation process, however, is greatly accelerated when a *catalyst* (the drier) is present in the ink. The most commonly used drying catalysts are *cobalt* and *manganese* compounds. True to the definition of a catalyst, these metals do not actually dry the ink; instead they merely promote the reaction be-

Ink Ingredients   231

tween the drying oils and the oxygen. **Manganese** promotes drying throughout the ink film and is, therefore, referred to as a *through drier*. One drawback of manganese is that its inherently dark color can contaminate light inks such as white, yellow, or tan. The drier most responsible for getting the ink film to set so quickly is cobalt, a much more active catalyst than manganese. **Cobalt** is most effective at accelerating the oxidation process where the ink film is in contact with the most oxygen, at the surface of the film. Therefore, the presence of cobalt greatly speeds the hardening of the top portion of the ink film, which results in its almost immediate setting, while the remainder of the ink begins its drying process. Cobalt is therefore referred to as a *top drier*. Because it is so active, however, excessive cobalt can be the cause of ink crystallization or, on occasion, gloss ghosting. Both crystallization and gloss ghosting are explained in Chapter 14.

Regardless of the metal used as the drier, an excessive amount of drier can cause the ink to oxidize and skim over while still on the press and also result in filled-in halftones. Excessive drier in a lithographic ink can result in scumming, which usually prompts the operator to increase the amount of dampening solution. Such an adjustment can load both the paper and the ink with extra water and actually retard, rather than speed up, the ink's drying rate. Bear in mind that because driers are nothing more than catalysts for a chemical reaction between drying oils and oxygen, they are found only in inks that dry through the oxidation of drying oils.

Because resins are stiff and brittle by nature, there is the tendency for this property to be imparted to the ink film after it has dried. A rigid ink film on a flexible substrate, however, will clearly tend to crack and perhaps flake off when the surface is bent. Imagine, for example, an old leather boot with dried mud on it. When the boot's leather is flexed, the brittle mud coating is unable to accommodate the bending and soon separates and falls away. A potentially similar situation occurs when a tube of toothpaste is squeezed or when a plastic bread wrapper is handled. In these and other instances requiring flexibility, **plasticizers** are added to the ink formula basically to interfere with the close alignment of the ink's polymer molecules that make the ink so rigid. As a result, toothpaste tubes and other flexible surfaces retain their printed images even when crinkled. (See Appendix E for further explanation.)

Plasticizers can enhance the printed ink film in several other ways. These include imparting (a) increased gloss; (b) improved adhesion to problematic surfaces; and (c) protection from becoming too brittle at very low temperatures. This last property is crucial when printing on frozen food packaging.

Because of their overall usefulness, plasticizers are commonly found in the paste inks of lithographic and letterpress printing as well as the liquid inks of gravure, screen, and flexographic printing. What does vary for different inks is the type of plasticizer that is used, as well as how much is added.

A dried ink's resistance to being rubbed or scratched off is improved by the inclusion of **waxes** in its formula. Obviously, **rub resistance** is especially important in certain printed products such as book jackets, playing cards, packaging, or any other product that will be handled. Other qualities that waxes can enhance in an ink are improved *slip* and *resistance to water*. However, the presence of wax in an ink is also a potential problem because, if the wax particles move out of their emulsion with the vehicle during the drying process, they will migrate to the surface and create a film that will not readily accept a subsequent application of ink or coating (see Figure 10–9). The significance of this phenomenon, called **crystallization,** is readily apparent when a printer tries unsuccessfully to print or "trap" one color atop a previously printed ink film that experienced wax migration when it dried. Crystallization only occurs when one ink film is applied over an ink film that has fully dried, that is, during *dry trapping*. Also, excessive wax can reduce the ink's gloss and increase drying time. The Graphic Arts Technical Foundation reports that maximum benefits occur when around 3 percent of the ink's total weight is wax.

As was discussed earlier, most sheet-fed offset and letterpress inks dry through oxidation. Because the most oxygen is found near the top of the ink film, the first stage in drying is the formation of a skin across the top. To prevent an ink from skinning over in the can or on the press during down time, **antiskinning agents** (also called *antioxidants*) are incorporated into these inks. Sheet-fed press operators sometimes keep their own supply of one of these agents to spray onto the plate and blanket before going to lunch to augment the amount already present in the ink. However, after having been sprayed, the ink then contains so much antiskinning agent that it will not dry properly on the paper, so the first several sheets must be discarded when printing begins again.

Because pigment particles tend to cluster together, air pockets commonly form within and around these group-

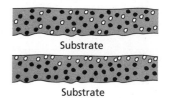

Figure 10–9 At top is an ink film in which the colloidal dispersion of both pigments (dark) and wax (light) particles is maintained. At bottom, the wax particles have migrated to the top of the ink film where they will resist the trapping of a second ink film.

Figure 10–10 The agglomerate of pigment particles on the left has air pockets that prevent the pigments' entire surface area from being "wet" by the solvent. On the right is a pigment agglomerate that has been more thoroughly penetrated by the solvent because of the addition of a wetting agent.

ings. As a result, the pigments do not disperse uniformly through the vehicle. To reduce the vehicle's surface tension and help the vehicle penetrate these microscopic air pockets, the ink manufacturer adds **wetting agents** to the vehicle (see Figure 10–10). In effect, the liquid has been made "wetter." A similar process occurs when a photographer dips recently developed film into a wetting solution before hanging the film to dry. In this instance, the wetting agent will allow the water to spread out across the film more evenly instead of forming beads that will cause water spots. For a more thorough explanation, see Appendix B. Wetting agents are primarily needed with the water-based inks used in gravure and flexography.

As might be expected in ink making, the type of wetting agent and the amount that is used are crucial to an ink's performance. In the case of lithographic inks, any deviation from the proper level could cause excessive emulsification of dampening solution into the ink, which could lead to ink transfer or drying problems. Appendix C explains emulsification.

Because the proper *flow* is so crucial to an ink's ability to move through the press and transfer to the substrate properly, additives are available to adjust an ink's *body. Viscosity* is a term that is usually used when discussing an ink's body. Viscosity refers to a liquid's resistance to flowing. A liquid that pours easily has low viscosity, while one that must be practically pushed along has high viscosity. Chocolate syrup, for example, becomes more viscous after sitting in the refrigerator.

An ink's viscosity impacts on a number of crucial aspects of printing. Viscosity is the most obvious difference between the liquid inks used in gravure and flexographic printing, and the paste inks of letterpress, screen, and lithographic printing. When an ink's viscosity is too high or too low, its performance will suffer. Of all the printing processes, lithography's ink is the most viscous; if an open can of litho ink were turned upside down and held for 60 seconds, none of the ink would move. To achieve this level of viscosity, an ink manufacturer adds **stiffening agents** such as *body gum* or a *binding varnish* to "pull the ink together." On occasion, a press operator will add some stiffening agent to an ink to correct press problems such as scumming, slow drying, tinting, chalking, or excessive emulsification.

A property of ink that is somewhat related to viscosity is its *tack*—the ability to resist splitting when being pulled apart. In lithographic inks, high tack is a desirable quality because it prints sharp dots and fine lines and resists excessive emulsification of the dampening solution. Unfortunately, an ink with extremely high tack will resist splitting to such a degree that the sheet may stay on the blanket too long and develop back-edge curl, as was explained in Chapter 7, or, in extreme cases, wrap around the blanket and tear away from the grippers. However, a much more common problem resulting from excessive ink tack is picking, which was also explained in Chapter 7. To ensure that an ink does not have excessive tack, the manufacturer can add an **ink reducer,** which is usually an oil, grease, or other petroleum solvent such as petroleum jelly. Boiled linseed oil and light varnishes are also used as reducers. It is common practice among lithographic press operators to add a tack reducer to an ink when picking occurs. Although an offset ink's tack can be slightly adjusted upwards with a tack **increaser,** the fact that it already comes with such a high tack level means that it is much easier to reduce tack than to increase it.

Summary

Clearly, there are several components to any printing ink and they are all necessary for that ink to function at an optimum level on a given substrate when applied by a particular printing process. When a change is made in the substrate or the method of printing, a given ink can no longer be expected to work as well. A key point of this chapter is that printing inks can be formulated to achieve nearly any printing goal, but only after thorough communication between the printer and the ink manufacturer. The second key point is that some press operators qualify for the adage "they know just enough to be dangerous" when it comes to tinkering with inks. Because of the delicate balances and interactions involved in modern ink formulas, to alter an ink's makeup in the pressroom is a very risky business, made even more risky if it is done without an understanding of the side effects. Too often, an operator's attempt to cure a press problem by tampering with the ink formula only creates bigger problems.

Because the ink business is basically a service industry, printers can turn to their ink makers for expertise and guidance—instead of merely a product.

Notes

1. Linda Cooke, "Soy Ink's Superior Degradability," *Agricultural Research* (January 1995): 19.

Questions for Study

1. What are the two most common printer-caused reasons for ink-related printing problems?
2. How are pigments different from dyes?
3. Distinguish between flushed and dry pigments relative to their preparation.
4. What is the function of an ink's resin?
5. What compound prevents rubber-based lithographic inks from setting up on the press during the lunch break?
6. Distinguish between solution and emulsion resins— both by their chemical differences and the benefits they bring to an ink.
7. List the three main requirements of an ink's solvent.
8. What problems result when an ink solvent's volatility is either excessive or inadequate?
9. Explain how drying oils interact with oxygen to dry ink.
10. Distinguish between an ink that has set and one that has dried.
11. What is a trade-off of increasing the percentage of soya or other vegetable oil in an ink's formula?
12. What role does cobalt play in an ink's formula?
13. What problem could develop if a lithographic ink were made with no antioxidant?
14. Describe the function of a wetting agent.
15. Assess honey and milk in terms of their viscosity levels.
16. Why is maintaining the proper tack level important in lithographic printing?

Key Words

pigment

flushed pigment

solvent

drying oils

drier

varnish

ink wax

wetting agent

absorption

vehicle

plasticizer

volatility

semidrying oils

manganese

rub resistance

antiskinning agent

stiffening agent

quick-set inks

dry pigment

resin

nondrying oils

VOC

cobalt

crystallization

ink reducer

11 Ink Manufacture

AFTER STUDYING THIS CHAPTER, THE STUDENT SHOULD BE ABLE TO:

■ Trace the history of ink manufacture.

■ Describe the process of ink mixing.

■ Explain why it is necessary for an ink pigment to be adequately wetted.

■ Explain the pigment dispersion process.

■ Describe the function of the three-roll mill.

■ Compare and contrast the ball and shots mills.

■ Describe the role of the filtration process.

History of Ink Manufacture

Although several books have been written to trace the development of paper, very little exists on the history of ink—an omission that is probably attributable to two factors. First, early progress in ink development was very slow, with several centuries passing between improvements; in short, there is not a whole lot to say.

Second, most printers and ink makers appear to have kept their formulations a secret to avoid aiding their competitors. Despite these circumstances, scholars have gleaned enough facts to compile a short history of printing ink.

The first inks to be developed were *writing inks,* developed around 2500 B.C. by both the Egyptian and Chinese cultures. The first pigment was *carbon residue* (a soot called lampblack) from burning an oil. The carbon was then stirred into a solution of *water* and *gum* (a sticky, water-soluble substance exuded from certain plants). The carbon particles did not dissolve in the water, but instead formed a colloidal suspension. The most notable feature of these primitive inks is their endurance through more than 30 centuries, which is attributed to carbon's resistance to the effects of light and moisture. In third century China, tree sap was mixed with ground-up insects to create a reddish dye. Although these substances were not used for printing, they did form the basis for what would evolve into inks used for printing from wooden blocks.

In either the fourth or fifth century, a Chinese craftsman named Wei Tang is credited with refining the combination of lampblack, water, and various gums to create an ink superior to its predecessors. Tang collected the lampblack by burning oil wicks in a vessel that was under a funnel-shaped cover. The cover soon became coated with the lampblack, which was then brushed off onto paper and mixed with gum to form a paste that could be applied to the wood blocks used in printing. Sometimes the paste was poured into molds to form crayon-like sticks that could be used for writing. This type of ink became widely used in the Orient for the next 1,000 years and eventually made its way to Europe by way of India, thereby acquiring the name India ink, which it holds today. Presently, the use of India ink is limited to drawing, but for nearly ten centuries, it was the medium used in conjunction with printing form wood cuts and blocks in both Asia and Europe.

In fourteenth century Europe, it was found that India ink did not work well with the metal plates that were being developed. Although it worked fine when placed on wooden blocks, it did not distribute itself uniformly on metal surfaces; instead, it tended to sit in lumpy globs. To remedy this problem, the European printers took a lesson from the art community that used oil-based rather than water-based paints. The printer's ink base of choice became **linseed oil**— oil extracted from the flax plants that also provided linen fiber.

In the fifteenth century the quality of printing ink improved greatly because Johann Gutenberg did not settle on merely inventing cast movable type. Gutenberg found that boiling linseed oil and a natural resin prior to adding the pigment rendered the ink insoluble in water and enhanced the ink film's bonding to the page. Gutenberg's new ink also printed well and created an intensely black, nonglossy image that did not fade. Prior to Gutenberg's breakthrough, printing inks of that era were inconsistent in quality. A 1457 book of Psalms contains papers with glossy black letters as well as pages with letters that faded enough to require their being touched up with a pen. Other inks did not always adhere to the paper adequately and could be rubbed off.

Gutenberg's ink formula of cooking linseed oil and resin to create a *varnish* to hold the lampblack became the prototype for modern inks and his formulation was altered only modestly for the next 300 years. For example, John Baskerville, the eighteenth century English printer who designed the still-popular typeface that bears his name, also developed an ink that is reported to have been blacker and more velvety than most inks used at that time. Also in the eighteenth century, an English patent was first granted for making colored inks—thereby lifting printing from the black-only era.

Until the nineteenth century printers usually manufactured their own inks, an approach that probably hindered rapid technological advance. However, by the nineteenth century, ink making had become a separate enterprise and printers were generally happy to leave the dirty task of working with the pigment material and the danger of fire from linseed oil boiling to others. Therefore, in the nineteenth century, improvement in ink chemistry started to accelerate. Agents that could speed ink drying rates were discovered; the stiffness of varnishes became controllable for different papers and printing processes; and **vegetable oils** replaced varnish on high-speed newspaper presses. During the twentieth century, ink making entered the realm of sophisticated chemical endeavor with discoveries such as synthetic pigments and resins that today make it possible to formulate inks that can satisfy the demand of our modern Information Age.

Modern Ink Manufacture

Ink manufacturers produce large quantities of **stock inks** that are purchased off-the-shelf by printers who are involved with standard job requirements. Very often, however, the demands of a particular job such as an unusual substrate or end-use will require that an ink be specially formulated to meet the printer's needs. To arrive at this precise formulation, the printer must supply the ink maker with all the pertinent information: not only color, the type of press, and the substrate material, but also any end-use requirements such as light-fastness, absence of odor, or resistance to heat or water. Information concerning the nature of the press run can also be pertinent, such as "Will this ink be laid down first, second, third, or fourth?" or "Will a clear coating be applied over this ink?"

Printers need to supply this kind of information to an ink maker when a *custom* ink formulation is required to handle an unusual printing situation. For more standard printing jobs, the *stock* or off-the-shelf inks that are manufactured from standard formulas to ensure consistency from batch to

Typical Water-Based Ink Ingredients	Typical Solvent-Based (Paste) Ink Ingredients
Pigment(s)	Pigment(s)
Pigment extenders	Pigment extenders
Water	Solvent(s)
Solution resin(s)	Resin(s)
Emulsion resin(s)	Wax
Wax	Plasticizers*
Amines or ammonia	Slip compound*
Defoamer	
Surfactants	
Antifungal and antimicrobial additives	* - used when needed
Plasticizers*	
Coalescing aid*	
Slip compound*	

Table 11–1 The recipe for a pound of ink is determined by the printing process, substrate, and end-use. As a result, a paste ink used in lithography and a liquid ink made for flexography have little in common.

batch are appropriate. An analogy can be drawn here to someone who wishes to repaint a bedroom. If one of the dozen or so ready-mixed colors is deemed acceptable, the customer can merely take one or two gallons from the shelf, pay for them, and drive home. However, if the walls' color must exactly match a color in the bedspread, then the customer must wait around while the custom color is achieved. In ink making, however, the customizing of an ink to a particular application can involve many more issues than color, as the previous chapter discussed.

For either stock or custom inks, the biggest challenge of ink making lies with figuring out the formula that best meets the printer's needs. For example, Table 11–1 illustrates the range of ingredients in the formulas for two inks, solvent-based and water-based. However, once the correct formula has been determined, the actual production process consists of only three basic steps: mixing, milling, and filtration.

Mixing

Mixing is performed in stages, with the first stage introducing the pigment material into the vehicle and then distributing these solid particles throughout the liquid. Mixing is performed in batches on machines that function somewhat

Figure 11–1 An ink mixer's impellers are raised after a mixing operation has been completed. *Courtesy of Braden Sutphin.*

like a cake mixer—except that the impellers are shaped differently and usually move much more slowly (see Figure 11–1). Mixers vary in both design and size with the tubs holding as few as 5 or as many as 1,000 gallon batches. The mixing speed is usually determined by the viscosity or body of the vehicle. When mixing heavy paste inks, the speed begins with a slow stirring and gradually advances to a faster speed. In contrast, the mixing of a liquidy gravure ink can mean speeds of thousands of revolutions per minute.

A key goal of pigment-vehicle mixing is the **wetting** of the pigment, which means to contact as much of the pigment matter with the vehicle as is possible by driving out the air pockets that form around and within the *agglomerates,* or clusters of pigments (refer back to Figure 10–11).

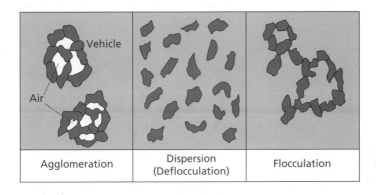

Figure 11–2 Proper dispersion of pigment particles breaks up agglomerates and prevents the pigments from flocculating.

| Agglomeration | Dispersion (Deflocculation) | Flocculation |

The amount of time required for mixing is determined by two factors: the viscosity of the vehicle and the amount of predispersion that may have already occurred. As was explained in the preceding chapter, it is common for the ink manufacturer to use flushed pigments, that is, pigments already dispersed in an oil. When the pigment material has been flushed or predispersed by other means, the mixing process can be greatly shortened; in fact, most often, a thorough mixing is sufficient and the milling step can be omitted entirely. An ink with inadequate mixing is likely to be inconsistent in its perfomance on the press, drying rate, and final appearance because the ingredients have not been uniformly distributed throughout the mixture.

Milling

After the mixing stage has distributed the pigment material as thoroughly as it can within the vehicle, the next step involves the **milling** of the ink—that is, the breaking down and further wetting of the pigment agglomerates. The amount of time required for milling depends on the nature of the pigment being used as well as how effectively the mixing step broke down and wetted the pigment agglomerates. As the mixing and milling steps become more technologically advanced, the distinction between their functions becomes somewhat blurred and the dispersion and wetting functions must be viewed as a process that begins during mixing and continues through milling.

Dispersion is the breaking down of pigment agglomerates into individual particles or, at least, very small agglomerates (see Figure 11–2). Obviously, as agglomerates are broken up, air pockets are flushed out and pigment wetting is improved. However, because pigment particles tend to **flocculate** or reassemble into groups, the dispersion must proceed

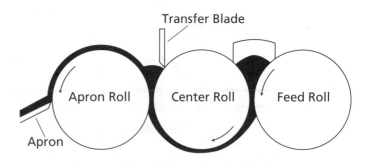

Figure 11–3 A paste ink is loaded into a three-roll mill between the feed roll and the center roll. The pressure and friction at the nips of the rolls grind and disperse the ink's pigment particles before the ink is carried off the apron roll by the apron.

Figure 11–4 The remnants of an ink that has been milled slide down the apron of a three roll mill. *Courtesy of Braden Sutphin.*

to a point at which the pigment-vehicle mixture becomes stable.

A wide range of equipment is available for milling ink. The most common is the **three-roll mill,** which usually consists of three steel rollers that revolve at different speeds, thereby generating friction at the nip (see Figures 11–3 and 11–4). Ink is fed between the feed roll and the center roll; as it moves between them and onto the apron roll, the shearing force that is encountered at both nips disperses the pigment particles. Dispersion efficiency is determined by the mill's speed of operation and the pressure between the rollers. Especially hard pigments can require a second pass through the mill to achieve adequate dispersion.

Not all inks are appropriate for a three-roll mill. Liquid inks lack the viscosity to effectively pass through it. Also, because the workings of three-roll mill are not enclosed, inks with a highly volatile solvent would lose much of it to evaporation during the milling operation. Such inks require that the milling take place in an enclosed operation—hence the **ball mill.** Anyone who has ever seen a rock polisher or a clothes dryer in action will be able to understand the technique of a ball mill. A large cylindrical canister that contains thousands of steel or ceramic media (usually balls) as well as the ink to be milled simply rotates (see Figure 11–5). As the combination of media and ink inside rolls up and cascades down, that friction between the media disperses the pigment particles. Factors that influence the degree of pigment dispersion achieved in a ball mill are the amount of time spent being milled, the size and shape of the media (see Figure 11–6), and the rotation speed.

When light-colored inks that might suffer from metallic contamination are involved, **pebble mills** are often used. These mills work in the same fashion as ball mills except that the mill wall is lined with a nonmetallic material and the steel shot has been replaced with porcelain or ceramic media.

In operations that require continuous instead of batch milling, the **shot mill** is used (see Figure 11–7). Whereas the ball mill is a rotating cylinder that lies on its side, the shot mill is a stationary cylinder that stands on end. The shot mill differs from the ball mill even more greatly in its operation. Imagine a rod with several evenly-spaced propellers on it rotating at high speed inside a cylinder that is filled with steel shot. As ink is pumped into the bottom of the cylinder, it makes its way upward, but not without becoming caught up in the steel shot that is rapidly swirling between the rotating blades. The whirlpool-like flow pattern that the ink encounters between each set of blades (or impellers) breaks down and disperses the pigment particles through the shearing forces that are created, as well as the grinding effect of the shot. As ink reaches the top of the cylinder, it encounters screens that allow the ink to pass, but hold back the shot. Although shot mills are efficient at pigment dispersion, they

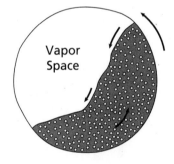

Figure 11–5 As the ball mill's canister revolves, the highest media begin to cascade downward. This movement of the media disperses the pigment particles.

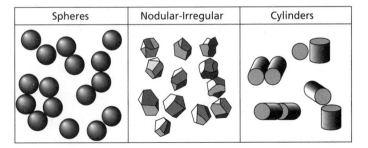

Spheres	Nodular-Irregular	Cylinders

Figure 11–6 Of the three media shapes shown above, cylinder-shaped media produces the least variation in pigment size and, therefore, can produce ink with a high gloss level.

245

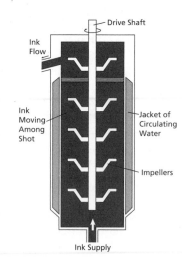

Figure 11–7 Ink enters at the bottom of a cylinder shot mill's cylinder and exits from the top. Shot mills use multilevel impellers to keep the shot in motion and help move the ink upward. *Courtesy of Braden Sutphin.*

are difficult to clean for color changes and have a tendency to clog, which limits their popularity.

A recent development in reducing the tendency of pigment particles to form agglomerates involves coating each particle with a polymer. The *micro-encapsulation* process involves coating finely-milled pigment particles with a polymer that acts as an insulator to eliminate the chemical attraction of one particle for another.

Lithographic inks that are inadequately milled will have excessively large agglomerates that can wear away the plate's image areas. Weak color, streaking, and the piling-up of ink on the plate, blanket, and ink rollers are other potential problems with poorly milled ink.

Filtration

The liquid inks used in flexographic and gravure printing commonly receive an additional processing step called **filtration** to remove grit and dirt by pumping the ink through a bag that traps these contaminants. However, for the vast majority of printing inks, filtration does not contribute to the primary goals of the ink manufacturing process—the uniform dispersion of pigment particles and other ink components throughout the vehicle. These functions are accomplished through mixing and milling.

After the Sale

Ideally, an ink has successfully passed a battery of tests before it is allowed to be sold, but problems can still occur in the field for different reasons. The ink could have been formulated wrong or received insufficient milling. But the main reason is that the ink is being used in a situation for which it was not intended. Too often a printer will try to use a standard off-the-shelf ink on an unusual substrate, under unusual conditions, or for an unusual end-use. Also, a printer may ask an ink supplier for advice on which ink would be compatible for a certain job, but fail to supply all the pertinent details.

After a press run has begun and a printing problem develops, the true culprit may be any one of several factors such as the dampening solution, pressroom humidity, or the surface of the paper. However, the components of the job or the environment are usually difficult or impossible to alter quickly. Therefore, the attention often turns immediately to the ink. In this situation, however, the problem is sometimes made worse when an ink is modified in the pressroom before any testing is performed to identify not only the true cause, but also the best solution. Quality-conscious ink manufacturers understand that their responsibilities do not end when an ink leaves their premises. For this reason, they are prepared to work with a printer who is experiencing problems with one of their products. On occasion, this support is in the form of testing the ink to find a remedy. The next chapter examines the methods available for testing printing inks.

Questions for Study

1. What is meant by "wetting the pigment"?
2. Which pigment material mixes more readily—dry or flushed? Why?
3. What would happen if an ink's pigment dispersion did not become stable?
4. What advantage(s) does a ball mill's design have over that of the three-roll mill?
5. Why is the filtration stage necessary?
6. If an ink receives inadequate milling, what effect might this error cause on-press?
7. What press problem could result from an ink with inadequate dispersion?

Key Words

vegetable oils	linseed oil	stock inks
mixing	wetting	milling
three-roll mill	ball mill	pebble mill
filtration	shot mill	flocculate

12 Ink Properties and Testing

AFTER STUDYING THIS CHAPTER, THE STUDENT SHOULD BE ABLE TO:

■ Discuss how an ink's viscosity, pseudoplasticity, thixotropy, and temperature are related.

■ Explain how an ink's tack and length can affect the print quality.

■ Explain the importance of an ink's fineness of grind and how it is measured.

■ Explain how an ink's specific gravity can influence that ink's press performance and appearance.

■ Define color, color strength, undertone, masstone, opacity, and gloss and explain how these can be measured.

■ Cite products that require one or more of ink's nine end-use properties.

For most printers, the quality of an ink's manufacturing process is measured by how well that ink performs—both on the press and in the hands of the customer and/or end user. An ink's performance, in turn, is generally determined by the levels attained by its various properties during manufacture. For example, a lithographic ink that is too viscous may not transfer well through the ink system of the press and, as a result, print poorly. Fortunately, the properties of a batch of ink can be tested by the manufacturer as well as by the printer to avoid problems at press time.

An ink's properties can be categorized as *working properties* that deal with pressroom performance, *optical properties* that determine how the ink will interact with light, and *end-use properties* that come into play after the printed job has been delivered and starts to be used. The following is an examination of ink properties in each of these three categories, as well as how they are tested.

Working Properties

The first properties to be examined are the **working properties** that largely determine how well the ink will transfer through the ink train of the press and produce a sharp, clean image.

Viscosity and Related Properties

The *flow* of an ink is crucial to its ability to perform well. In ink making, flow is measured in terms of **viscosity,** an ink's ability to resist flow. In fact, the two broad categories of ink—*paste* and *liquid*— are determined by their viscosities. As was explained in the chapter on ink ingredients, the more difficult a liquid is to stir, the higher is its viscosity.

Viscosity is a crucial ink property for several reasons and printers need to remember that an ink's viscosity is reduced when acted on by *heat* and/or *agitation*—two forces that are at work when an ink is on the press. Because lithographic ink is so viscous, it may sit in the fountain and not move down to replace the ink carried away—thereby losing contact with the fountain roller altogether (see Figure 12–1). Although not an accurate description of what is happening, the term "backing away" is applied to this phenomenon. To counteract an ink's "backing away," the operator of a small press may periodically work or stir the ink in the fountain with an ink knife, while larger presses have a conical ink agitator that revolves while moving sideways across the fountain (see Figure 12–2). Whether manual or automated, agitation reduces the ink's viscosity and encourages mobility.

The tendency of paste inks to lose viscosity when stirred or agitated is called **pseudoplasticity** or *shear thinning* and an awareness of it is important for press operators. If a press operator combines any additives with an ink that has not had its viscosity already sufficiently lowered by agitation, then the ink's stiffness will likely prevent the additive from becoming adequately dispersed throughout the ink, with press problems being a possible result. The decrease in viscosity is reversible, however. When the agitation ceases, the ink will regain some of its viscosity in time. When a viscous substance such as lithographic ink is stirred, it moves in lay-

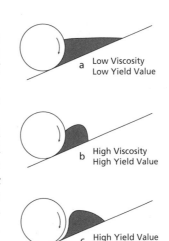

Figure 12–1 "Backing away" refers to a paste ink's being so viscous that it does not flow forward to replace ink carried away by the fountain roller.

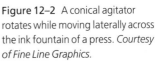

Figure 12–2 A conical agitator rotates while moving laterally across the ink fountain of a press. *Courtesy of Fine Line Graphics.*

249

ers, with the molecules nearest the physical force moving first; this stirring force is called *shear*. As these layers slide against each other, the resultant friction produces heat which, in turn, reduces the viscosity; hence, shear (agitation) lowers viscosity.

Some inks respond more quickly to shear than do others. The property of a liquid's pseudoplasticity being time-dependent is called **thixotropy** (pronounced thiks-AH-tro-pee).[1] Many ink people use the term thixotropy in place of pseudoplasticity, but the two properties are not the same. To understand the distinction, consider the difference between velocity and acceleration; while nearly any car can reach a speed (or velocity) of 60 miles per hour, some cars require more *time* to get to that speed than others. Whereas nearly all inks are pseudoplastic in nature, some thin more slowly when worked, either during mixing or on the ink train. The importance of an ink's thixotropy to the printer concerns how fast the ink sets on the substrate. The time required for a printed ink to set is influenced by the speed with which it regains the viscosity that was lost on the ink train. Highly thixotropic inks require longer periods of time to thin when agitated at a constant force and are also slower to regain viscosity when left alone; hence, they dry more slowly.

The second force that acts on viscosity is *temperature* and its effect is profound; a temperature change of 1 degree fahrenheit (F) will alter an ink's viscosity by 3 to 4 percent, depending on the type of ink. So, a viscosity reduction of over 50 percent could result from no more than a 13° rise in the ink's temperature. Such a rise in an ink's temperature during printing is not uncommon due to the friction that occurs between the ink train rollers. A friction-induced increase in ink temperature is generally beneficial because its reduction of viscosity helps the ink to flow from roller to roller and onto the plate. At the same time, the act of moving onto the cooler plate and blanket causes the ink to reverse course and become more solid, a tendency that continues when it hits the even cooler paper. The resultant increased viscosity, or firming up the ink, gives it a head start toward the setting and drying action that is expected of an ink film when it contacts the paper.

However, if the ink is too viscous when it gets to the blanket, it may not transfer cleanly to the paper but instead may pick—that is, lift pieces from it. In contrast, ink that lacks viscosity may very well be too slow to set and dry. Also, variations in ink viscosity affect the thickness of the printed ink film which, in turn, affects its color and mileage.

Fortunately, an ink's viscosity can be measured by one of three methods. Paste inks are tested with the Laray viscometer (see Figure 12–3), while liquid inks are tested with the Brookfield viscometer. Liquid ink viscosity can be measured inexpensively with an **efflux cup** and a stopwatch. The commonly used Zahn cup and the more accurate Shell cup (see Figure 12–4) simply allow a certain volume of ink to flow out a hole in the bottom, while the required time is measured. Because of ink viscosity's sensitivity to temperature change, a commonality among these three procedures is the need to monitor the ink's temperature and report it along with the viscosity level. In fact, it is recommended that viscosity measurements take place with the ink temperature as close to 77° F (or 25° C) as possible.

Tack

An ink property that is related to viscosity is **tack**, the resistance of the ink film to splitting. Tack becomes especially important when the press transfers its ink film to the paper. At the point of contact (the nip), the ink is sandwiched between the press cylinder and the paper, but, as the cylinder and the paper roll away from one another, the ink should split as roughly half of it stays on the paper and half of it remains on the cylinder. At question is the ink's tendency to split, as opposed to its tendency to resist splitting by holding together. If the tack is too high, the resistance to splitting may be so great that the surface of the paper actually comes apart instead of the ink. This phenomenon is called **picking** (see Figure 12–5) and is discussed in Chapter 7. Tack levels must be controlled when printing on multiple-color presses in order for an ink film to successfully transfer (or trap) to the still-wet ink film that preceded it. This procedure is called *wet trapping* and it involves making sure that each ink color is less tacky than its preceding color. To assist in this effort, ink is manufactured and sold in several *tack levels* identified numerically; that is, 17 is a higher tack level than 9. For example, a printer may use the following color sequence and tack values: black, 16; cyan, 14; magenta, 12; yellow, 10.

Because tack is influenced by an ink's viscosity, which in turn can be altered by agitation, an ink's tack can change during the first ten minutes of being on the press. The ability to resist such variation is called *tack stability*. Inks with good tack stability create fewer headaches for press operators and are, therefore, generally desirable. Both an ink's tack and its tack stability can be measured on a device called an *inkometer*

Figure 12–3 A Laray falling rod viscometer measures the amount of time required for a rod and connected piston to drop a given distance, which reflects the resistance of the ink sample being tested. *Courtesy of Braden Sutphin.*

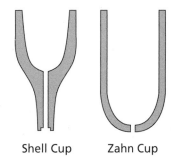

Shell Cup   Zahn Cup

Figure 12–4 Two types of efflux cups are shown in cross section.

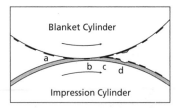

Figure 12–5 The phenomenon called picking is illustrated here. As the impression and blanket cylinders rotate (a) they carry the paper and ink to the nip where the paper is compressed (b). At point c, the ink film splits as the two cylinders move apart. However, at point d, the splitting occurred on the paper's surface instead, creating a noninked crater in what should be an image area. The fiber clump that was lifted by the blanket will accept dampening solution, repel ink, and print as a nonimage area until removed from the blanket.

that is manufactured by the Thwing-Albert Instrument Company in association with the Graphic Arts Technical Foundation (see Figure 12–6). The relationship between ink tack and viscosity is demonstrated by Stefan's equation, which can be found in the Glossary.

Length

The final property related to flow is an **ink's length,** its ability to be drawn or stretched before snapping (see Figure 12–7). For an ink to flow well and transfer from one press roller to another as must occur in lithographic and letterpress printing, it needs to be be relatively *long;* however, if it is too long, *misting* can result from strings of ink that form and break into droplets (see Figure 12–8). Just the opposite is true of screen printing inks, which need to be very *short* to prevent strings of ink from forming after the squeegee has moved forward and the screen begins to lift from the ink film.

There is no precise method of testing an ink's length. The traditional technique of comparing the length of two inks is simply to touch each sample with a finger or knife, lift simultaneously, and observe how far each stretches before snapping. If an ink appears excessively long, causing concern that it will mist on the press, any tendency to mist can be discovered by running the ink on an inkometer in advance.

In addition to its flow characteristics, an ink's *wet film thickness* can cause misting if it is excessive; that is, the ink is

Figure 12–6 An inkometer measures an ink's tack and tack stability by simulating press conditions. *Courtesy of Braden Sutphin.*

Figure 12–7 A long ink will stretch much farther than will a short ink before snapping. *Courtesy of Braden Sutphin.*

printed too heavily. In addition, an ink film that is too thick will be slow to set and dry as well as simply require more ink over the course of the job.

Pigment Dispersion

The chapter on ink manufacture emphasized the concern given to a *uniform dispersion of pigment particles* throughout the vehicle. Ink that has not received adequate milling will contain **agglomerates** (clusters) of pigment particles and is described as having poor *dispersion, poor fineness of grind,* or just *poor grind*. The presence of particle agglomerates can affect the printing operation as well as job quality in several ways. In gravure printing, a poorly ground ink can plug up the cylinder's cells and accumulate against the doctor blade where they can abrade the cylinder and allow ink to migrate into nonimage areas (scumming). On lithographic presses, poor dispersion can result in the accumulation of excessive ink on the rollers (piling), streaking, and a wearing away of plates and rollers. As if these problems were not bad enough, poor dispersion also produces weak or mottled color.

Fortunately, an ink's fineness-of-grind or dispersion can be tested without sophisticated and costly equipment. A device called a **grindometer** (or *grind gauge*) is needed, but it is nothing more than a metal block with troughs that progress from 1 mil to 9 mils in depth (see Figure 12–9). Ink is placed in the deep end and slowly dragged toward the shallow end with a metal spatula that rests on the surface of the block. Ink agglomerates that are too big to slide under the blade will hang up in front of it and leave obvious scratches; the sooner and more often the scratches occur, the poorer is the fineness-of-grind.

Figure 12–8 Misting occurs when an ink is too long. This results in strings forming between the cylinder carrying the ink and the paper. When these strings finally snap, ink particles are sent into the air.

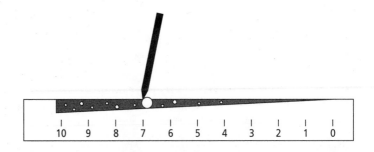

Figure 12–9 A grindometer measures an ink's fineness-of-grind by pulling a blade forward over an ink that is in a trough of decreasing depth. When scratches begin to appear, the size of the largest pigment particles becomes evident.

Figure 12–10 A proof press allows a thin film of a small batch of ink to be quickly laid down onto paper. *Courtesy of Braden Sutphin.*

Ink Drying Rate

One of the most important properties of an ink is its **drying rate.** Once again, the test for this property can be very low-tech. Usually an ink film is either laid down with a *proof press* (see Figure 12–10), a **drawdown** (see Figure 12–11), or merely *tapped out* with a finger. Then, the sample may be covered to simulate its being in a delivery pile and checked every half hour for smudging. For a more high-tech approach, six different instruments are made to test ink drying (see Figure 12–12). However, because drying rate is influenced by many external factors such as the absorbency of the substrate, the area of coverage, and the relative humidity of the pressroom, any accurate prediction is difficult at best.

Problems Related to Drying Rate

An ink's drying rate is crucial because problems can occur from its drying either too fast or too slowly. If an ink dries too slowly, a freshly-printed image can be *set off* onto the back of the next sheet that falls onto it in the delivery pile. In extreme cases, the image can bond to the back of the top sheet so firmly that portions of the top sheet's surface are

Figure 12–11 Two inks being tested for drying rate and a standard (an ink with an already-known drying rate) have been drawn down on a piece of glass. Drawdown tests are also made on paper.

Figure 12–12 The Drying Time Recorder tests ink drying time by measuring the number of time intervals required before a printed ink ceases to set off onto a second piece of paper. *Courtesy of Braden Sutphin.*

actually pulled away by the ink; this phenomenon is called *blocking-in-the-pile* and it is explained further in Chapter 7.

Still another problem can occur when one ink film is applied over a previously printed ink film—a function referred to as *trapping*. For example, a two-color job using yellow and blue inks may call for a trapping of one ink over the other to allow a portion of the design to be green. Another example of trapping is the overlapping of halftone dots that occurs during four-color (or process) printing. If a one-color press is used, the first color already will have dried before the other color is applied—a technique known as *dry trapping*. If a multicolor press is used, the first ink will still be wet when the subsequent color(s) is applied; this technique is called *wet trapping*.

Unsuccessful wet trapping results in first-down inks being lifted by a subsequent ink film. For example, imagine that the yellow and blue job just described were run on a two-color press with the yellow ink being printed first. If the just-printed yellow is lifted from the substrate by the blue printing unit, the job is said to *back trap* and the uneven color that results across the image area is called *back-trap mottle*.

Interesting enough, whereas blocking-in-the-pile and set-off are examples of what can result from ink's drying too slowly, back-trap mottle is a product of an ink film's having dried too fast. When an ink begins to set up while still on the press, it can transfer to the paper, but not adequately penetrate the surface. As a result, the top of the film that has skinned over can be peeled away when it contacts the blanket of the next unit of a multicolor offset press. In this instance, the ink "traps back" onto a blanket rather than staying on the paper and allowing the next color to trap over it. As was explained earlier in this chapter, wet trapping is often controlled by ensuring that each ink has a higher tack level than does the ink that follows it. When this occurs, each ink film being applied is more likely to split than is the previously printed ink film. The ink is not always at fault when back trapping occurs, however. If ink absorption is uneven across a sheet's surface, then ink setting can be uneven, thereby producing the uneven mottled appearance (see Chapter 14 for more on trapping.)

Lithographic printers need an ink that will accept small amounts of dampening solution, that is to say, ink that will **emulsify** with the dampening solution. However, an ink that emulsifies too much becomes waterlogged, resulting in dot gain, slow drying, scumming, and washed out color strength. Waterlogged inks cannot be remedied and must be discarded. An ink's emulsification tendency can be measured with a Duke Water Pickup Tester, a machine that controls the amount of water that makes contact with the ink. At one minute intervals, the ink can be tested for water content. Although an ink's tendency to pick up water is useful information, the Duke tester cannot be expected to duplicate the conditions on a given offset press, so precise predictions about emulsification during printing are not likely.

Pigment-to-Vehicle Ratio

Imagine being at a paint store where two grades of white latex wall paint are for sale; one grade costs $8.00 a gallon while the other one sells for $24.00 a gallon. Excluding a close-out sale or some other unusual condition, the expensive paint probably has a much higher **pigment-to-vehicle ratio**. In fact, although it costs three times as much as the $8.00 paint, if it has four times more pigment, it is the better buy. Remember that most of the vehicle will evaporate during drying and the pigment and resins are what must cover the old paint underneath.

When an ink film is applied, it is the pigment that supplies the color strength and keeps a black from appearing gray or red from appearing pink. Therefore, the pigment portion of an ink is crucial to that ink's value to the printer. A percent solids test will reveal how much of an ink is composed of pigment, resins, and material by merely weighing a sample of the ink, baking it to evaporate off all but the pigment material, and then weighing it again.

Although not an accurate determiner of the pigment-to-vehicle ratio, an ink's *specific gravity* reveals its density or weight-to-volume relationship. Because different colors of ink require different amounts of pigment to attain sufficient color strength, lighter colored inks such as yellow will contain more pigment than a black ink that can rely on its darkness for covering power. Therefore, a gallon of yellow ink may very well weigh three times as much as a gallon of black ink, all other factors being equal. Of course, from the perspective of weight-to-volume, a pound of yellow ink would then be only one-third as large as a pound of black ink. This latter observation has profound implications for how many pounds of ink will be needed to print a job. An ink's **mileage** is expressed by how many square inches a pound of a particular ink will cover when applied to a given substrate—in short, how far a pound of ink "will go." Because of differences in specific gravity, a pound of black may cover 435,000 square inches, while a pound of yellow would only cover 250,000 square inches. Therefore, if 10 pounds of black ink are sufficient for a given printing job, 17.4 pounds would be needed if the ink color were changed to yellow. See Appendix F for an explanation of specific gravity.

Optical Properties

In addition to the numerous working properties just examined, the press operator must be concerned with an ink's **optical properties** because they are usually the most apparent to the customer. *Color, opacity,* and *gloss* are the primary issues at stake when an ink film is examined optically or compared with a customer's standard.

Color

Although it may appear to the average person that color matching is a rather straightforward task, the American Society for Testing Materials has 13 separate tests for measuring *color.* The phenomenon of color perception is complex because it brings together very different sciences—physics, physiology, and psychology. Color perception occurs when light waves from the sun or other source strike an object. Pigments in the object absorb certain components of the light and reflect the rest; that is, different pigments reflect different components of light. The reflected light reaches the eye and stimulates color receptors, which send electrical signals to the brain. Mysteriously, the brain then interprets these signals as visual images. Chapter 13 provides a much more thorough explanation of color theory, but suffice it to say that physics, physiology, and psychology are all involved in color perception. Therefore, any attempt to measure color perception with scientific instruments exclusively would be insufficient. However, instruments can measure how a given pigment will interact with light that strikes it; that is, which parts of the light will be absorbed and which parts will be reflected. By knowing what portions will be reflected, the color that will be perceived by a person who looks at that pigment under the same light source and with normal vision can be predicted. In short, an instrument can measure the nature of the light that an object reflects.

Generally, printers rely on a *densitometer* to analyze **light reflectance** from printed ink. By using colored filters, a densitometer can report how much red light, green light, and blue light are being reflected by the ink. The three readings are measuring the optical density of the pigment. Therefore, if a printer were told to match a printed color to a customer-provided sample, a densitometer could be used to measure how much blue, green, and red light is being reflected, and then compare these values to readings taken from press sheets at various points during the run. Because cyan absorbs green and blue light but not red light, a densitometer with a red filter is used to measure the amount of cyan light that is being reflected (see Figure 13–4 and other related material in Chapter 13). Pantone[2] Purple has the following optical densities: cyan—.35; yellow—.43; magenta—1.14. An ink with significantly different numbers would not look like the same purple. For example, Pantone 513, which is another purple, has corresponding densities of .66, .72, and 1.41.

Instead of analyzing a color with three numerical readings, as does the densitometer, a *spectrophotometer* (see Figure 12–13) plots a curve that profiles an ink's reflectance at every wavelength across the spectrum (see Figure 12–14). Still another instrument used in color measurement is

Figure 12–13 A printed ink film is analyzed for its color with a spectrophotometer. *Courtesy of Braden Sutphin.*

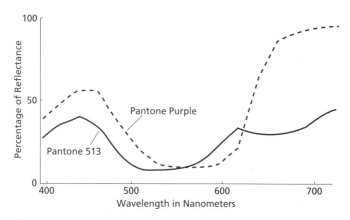

Figure 12–14 A spectrophotometer's printout profiles the amount of light reflected across the visible spectrum. In this figure, two similar-looking inks have been analyzed and plotted.

the *colorimeter,* which numerically analyzes a sample for three characteristics: **hue** (its composite wavelength or apparent color), **saturation** (its color strength or purity), and **lightness** (its black-white value) (see Figure 12–15). Because new versions of densitometers and spectrophotometers are also capable of supplying these data, colorimeters are not as common as was once the case. Both the spectrophotometer and colorimeter are expensive and not commonly found in printing facilities, but ink makers may supply color data that was generated on these instruments to printers, so an awareness of their approach to color measurement is useful.

Once it has been established that an ink has the desired color, there is the question of its **color strength** (also known

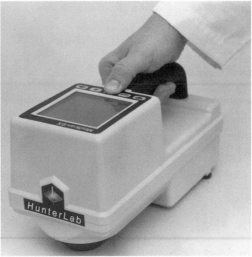

Figure 12–15 A colorimeter (left) generates a numerical analysis of a color in three dimensions. The mini-scan (right) is a portable handheld color measurement system that can function as either a colorimeter or spectrophotometer. *Courtesy of HunterLab.*

as *tinctorial strength*). What is being measured is the relative amount of pigment per unit weight or volume of the ink; in other words, "How concentrated or diluted is the pigment in the ink?" A low-grade ink would likely contain less pigment per pound than a more expensive ink. Using an ink with poor color strength requires a heavier ink film to achieve the desired density, which would necessitate using more ink over the course of the job and increasing the time required for drying. In short, using an ink with poor color strength is counterproductive because ink seldom exceeds five percent of the total printing cost and the potential for savings is too small to risk the much larger costs associated with poor performance on the press and/or problems with drying.

To compare the color strengths of a new ink and a familiar ink, a **bleach test** is performed. A small sample of each is mixed with a much larger amount of white base. The ratio used in the test can be 99:1, 95:5, or 90:10. After each ink sample has been mixed with the tint base and had its color diluted, a small sample of each is placed side-by-side on a sheet of white paper and a spatula is used to draw down both samples and thereby produce a thin film of each. Visual observation or densitometer readings will reveal which of the two inks is standing up better to the dilution of its pigment. If a densitometer is used to measure an ink film, the reading will be the ink's color density. Color strength is a property of an ink while in the can and color density is a property of a film of that ink. Although they are clearly related, color strength and color density are not the same property. The expression bleach test is a misnomer because bleach is not used. The term no doubt originated from the diluted ink's appearing to be "bleached out."

Two aspects of an ink's color are its *masstone* and its *undertone*. **Masstone** is the appearance of an ink in a volume too thick to transmit light and is, therefore, the color of light that is reflected from the ink alone. In other words, masstone is the color of an ink when it is in the can. A property that is of more interest to the printer, though, is an ink's undertone, which is the appearance of a thin ink film on a white surface. An ink's **undertone** results from light passing through the thin film, being reflected from the substrate, and passing back through the ink. Because of the influence of the white substrate, an ink's undertone is much lighter than its masstone, except in the case of opaque inks.

Opacity and Gloss

When an ink drawdown test is performed to check undertone, the paper that is used generally has a black band running across it horizontally so that the ink's **opacity** can be observed, as well.

The drawdown test is also useful when comparing two inks for **gloss,** but only after the ink film samples have been allowed to dry. If a quantitative measurement of gloss is needed, an instrument called a gloss meter can be used (refer back to Figure 6–3).

End-Use Properties

The final category of an ink's properties concerns its **end-use.** Obviously, the need to incorporate one or more of these properties is determined by how the printed product will be expected to perform. When an ink requires a particular end-use property, the ink maker can formulate accordingly and perform the appropriate test to confirm its desired level. The following is a list of end-use properties and one or more examples of a product or situation for which each property would be needed:

Figure 12–16 An abrasion tester controls the amount of pressure and duration of rubbing received by a printed ink film. *Courtesy of Braden Sutphin.*

Abrasion resistance—general packaging (see Figure 12–16)

Wrinkle-crinkle adaptability—bags containing frozen vegetables

Lightfastness (fade resistance)—billboards, bumper stickers (see Figure 12–17)

Absence of odor and/or taste—food packaging

Water resistance—anything exposed to rain or labels for containers such as soda pop that receive condensation

Figure 12–17 A weatherometer tests for lightfastness and resistance to water by exposing a printed sample to both light and moisture under carefully controlled conditions. *Courtesy of Braden Sutphin.*

Proper coefficient of friction(skid)—dog food bags or any other product that is stacked, and so should not slip too easily.

Chemical resistance—labels on food containers; for example, soap wrappers must be alkali-resistant

Heat resistance—packaging pouches that are heat-sealed

Adhesion to vinyl or other unusual surfaces—printed shower curtains, rain coats, plastic bread wrappers

Summary

Ideally, an ink will have successfully passed a battery of tests before it is allowed to be sold. When a problem does occur in the field, there are three approaches to testing the ink. One is for the printer to enlist the *ink maker* in identifying the problem. The second is for the printer to employ an *outside testing facility*. The third approach is for the printer to be in a position to perform some modest—but often revealing—tests *on-site*. Obviously, the last option can provide the answer more quickly, a definite plus when time is money.

Notes

1. Peter V. Robinson, "Consultants' Corner," *Paint and Coatings Industry* (February 1994): 120.
2. Pantone Matching System® and Pantone® are registered trademarks of Pantone, Inc.

Questions for Study

1. Define viscosity and give an example of a substance with high viscosity and a substance with low viscosity.
2. What are two influences that alter an ink's viscosity while on the press?
3. Why is it important that an ink's tack be controlled?
4. Why must lithographic and letterpress inks be comparatively long, while screen inks must be short?
5. How is fineness-of-grind tested?
6. List three factors that influence an ink's drying rate that are not part of the ink formula.
7. What is the significance of an ink's pigment-to-vehicle ratio relative to print quality and mileage?
8. What does an ink's specific gravity suggest about the pigment-to-vehicle ratio?
9. Distinguish between a densitometer and a spectrophotometer relative to their approach to measuring an ink's color.

10. Contrast masstone and undertone.
11. Besides color, what ink characteristics can be evaluated with a drawdown test?
12. List the end-use properties that would likely be required by the following products:
    a. bumper stickers
    b. labels for a vinegar jar
    c. baseball cards
    d. wrappers for bathroom tissue
    e. drinking glasses
    f. soap wrappers

Key Words

working properties
thixotropy
wet trapping
hue
emulsify
color saturation
color strength
undertone
drying rate
optical properties

light reflectance
opacity
proof press
viscosity
efflux cup
picking
grindometer
ink mileage
ink agglomerates
bleach test

end-use
pseudoplasticity
tack
ink length
drawdown
pigment-to-vehicle ratio
lightness
masstone

13 Ink and Color

AFTER STUDYING THIS CHAPTER, THE STUDENT SHOULD BE ABLE TO:

■ Describe the creation of color through the interaction of light and pigments.

■ Describe the visible spectrum and its neighboring bands in terms of wavelength values.

■ Distinguish between the additive and subtractive color systems and list the primary and secondary colors of each.

■ Define each of the three basic properties of color.

■ Use a Pantone Color Formula Guide to numerically identify a color and list the proportion of ingredients needed to create it.

■ Discuss the implications for printers of the various color gamuts being somewhat concentric.

■ Explain metamerism and other potential problems in color matching, as well as a printer's best response.

Few aspects of life are as common and as commonly taken for granted by young and old, rich and poor as is color. The bright colors of a baby's toys bring visual stimulation and delight. As a kindergarten student, the same child will use the color of crayons, paints, or clay to be expressive. As an adult, decisions about color in matters of wardrobe and home furnishings are likely to weigh heavily and—if the person's career choice involves graphic communications—working with color will possibly be an everyday function. And yet, if, during kindergarten, the child were to ask an adult questions such as "What makes color?" or "How do we see color?," few adults—even some working in graphic communications—would be able to respond in a very helpful manner. We know what color is, but we do not know much about it. Part of the dilemma is that the phenomenon of seeing color involves three very different areas of science—physics, physiology, and psychology. And yet, even if a physicist, a physician, and a psychologist were assembled to respond to the child's question, part of the answer would have to be expressed as mere theory because a portion of color perception is still a mystery. Fortunately, this mysterious part does not prohibit an adequate working understanding for people who are involved with color selection, mixing, matching, or printing.

The Requisites for Color

Before color perception can occur, four requisites must be in place. First, there must be a source of *light*. Without light, there is no color. The significance of this statement is that a yellow ear of corn, after being closed inside a refrigerator, is no longer yellow. By the same token, broccoli is no longer green, and cranberry juice is no longer red. This is not to say that their color can no longer be seen; on the contrary, *in the absence of light, there is no color to be seen.*

What the refrigerator's contents do still have is the *potential* for color; that is, they still contain *pigments* that could interact with light to produce color if light were made available. In fact, broccoli and corn are different colors because they have different pigments that interact differently with the light that strikes them. These distinct interactions produce distinct colors. Hence, *color results from the interaction of light and pigments;* if neither light nor pigment is present, there is no interaction and, therefore, no color. When light is missing, the result is black. When pigment is missing, the result is white. For this reason, black and white are not true colors.

Once a color exists through the interaction of light and pigments, the next stage in color perception concerns how it will be viewed. Here, the color receptors of the *eye* accept the light that is reflected by the corn or broccoli and transform it into electrical impulses that stimulate the *brain* to experience the sensation of color.

Therefore, before the sensation of a given color can occur, light must interact with pigments to produce reflected light with certain qualities that are then processed by a normally functioning eye, which finally stimulates the brain. If any of the four components of this system is missing or not functioning normally, color perception either occurs abnormally or not at all. Clearly, this process is very involved and before the printer can produce a colored image that pleases the customer, all four components of the system must be controlled. The remainder of this chapter will examine each of these four components in more detail.

Color and Light

Light from the sun appears to be white, but it is actually composed of several colors of light that, when represented in equal amounts, comprise white. To understand this concept, think about a black-and-white photograph that has been printed in a magazine. Although it appears to be a blend of light, middle and dark tones of gray, examination under a magnifying glass reveals that the photograph is actually composed of black halftone dots. When the large dots of the shadow area, the small dots of the highlight areas, and the intermediate-sized dots of the midtones are all present, they can form any scene. In the case of white light, its components are various colors of light, which, in turn, lose their individuality when blended in equal proportions. The classic method of demonstrating white light's components is to aim a beam of it at a prism and observe how the white light is refracted into its various parts (see Figure 13–1). In fact, every rainbow is nothing more than sunlight being refracted by water droplets in the air.

An understanding of color begins with an understanding of light, which is radiant energy that travels in waves. An examination of Figure 13–2 reveals that light is only one form of energy that travels in waves and that the difference between light waves, heat waves, and radio waves is nothing more than their *wavelengths* (the distance from peak to peak). Within the band of wavelengths that produce light, there are two basic divisions—visible light and invisible light. Just as not all sounds are heard by the human ear (a dog whistle, for example), not all light is seen by the human eye. Infrared and ultraviolet light are forms of invisible light. Infrared light consists of wavelengths that are too long for the human eye to respond to, while the wavelengths of ultraviolet light are too short.

However, the human eye is responsive to electromagnetic wavelengths that measure between 400 and 700 millimicrons; this band is known as the **visible spectrum.** A millimicron (or nanometer) is only a millionth of a meter stick in length and is nearly too small to be imagined. In order to place a 500 millimicron wavelength of light into perspective, consider that there is room for 192 of them within the thickness of a sheet of Sub. 20 bond paper. Within this microscopic world, the wavelength of light determines its color. The difference between red light and green light is simply that green light has a 15 percent shorter wavelength.

Light from the sun appears white because it is composed of equal amounts of red, green, and blue light. However, when the proportions of these three basic components are altered, light begins to assume the property of color. For example, if the wavelengths of a beam of light were limited to between 430 and 450 millimicrons, that light would be a

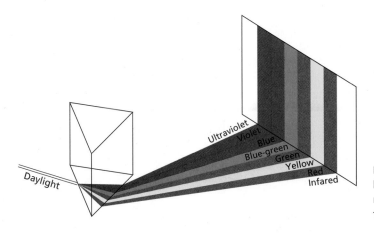

Daylight

Ultraviolet
Violet
Blue
Blue-green
Green
Yellow
Red
Infared

Figure 13–1 A prism refracts white light and reveals its visible components. These colors constitute what is termed the visible spectrum.

bright blue. At the same time, a beam with longer wavelengths between 520 and 540 clearly would be green, and a third beam with even longer wavelengths, between 630 and 650, would be fire engine red. If any two of these beams of light were projected onto the same area of a white surface, their combinations would create new colors.

An examination of Figure 13–3 reveals that whenever two beams of light overlap, the resultant color is lighter than either of the two original colors. The reason for this phenomenon is the simple fact that light is being added to light. Observe also that the overlapping of all three beams produces white. The logic here is understandable. A prism can refract white light into its primary components—red, green, and blue light. When these three are reunited equally, they once again make white. However, if the proportions are varied, then the property of color results. As Figure 13–3 reveals, when the source of blue light is removed, the result is yellow; when only the green light is missing, the result is magenta (a pinkish red) and an absence of red produces cyan. If each of these three light sources were controlled by its own dimmer (like a three-band equalizer), the white area could be made to appear in any color imaginable—merely by experimenting with various combinations of light intensities. The fact that any color can be produced from the correct combination of red, green, and blue light makes these colors the **additive primary colors.** Phrased differently, by adding these colors of light, any other color can be created. Bear in mind that the additive system and its primary colors apply only to the mixing of light.

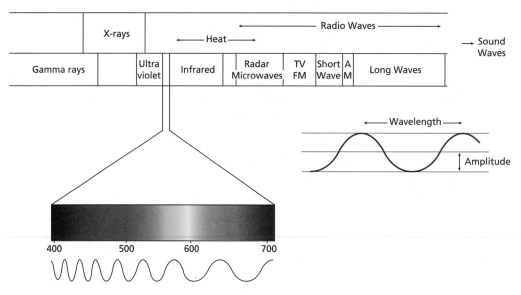

Figure 13–2 The difference among x-rays, ultraviolet light, microwaves, and radio waves is their wavelengths. Only a small slice of the existing wavelengths comprise the visible spectrum. At either end of the visible spectrum are ultraviolet and infrared light, which are invisible.

Pigments and Light

Of greater interest to the printing industry is the **subtractive** system because it describes the interaction between light and the pigments in ink. As was explained earlier in this chapter, the chemical composition of pigments causes them to absorb only certain wavelengths of light. Whenever this selective absorption occurs, the balance that is required for white light is disturbed and color is the product. Figure 13–4(a) illustrates what happens when light strikes a white surface. Because no pigments exist to absorb a significant amount of light, all three additive primaries are reflected to the eye intact and no color is generated. When the same light falls onto pigments that absorb all three of the additive primaries, practically no light is reflected and the visual sensation is black. When a pigment that absorbs only blue and red light is used, something very different happens. The unabsorbed green light is reflected and the observer experiences the sensation of seeing green. Note in Figure 13–4 that the ink is transparent and the paper is doing the actual reflecting. Because printers usually use transparent inks, two or more inks can be applied over one another to (a) combine their absorbing properties; (b) subtract the proper wavelengths of light; and (c) produce the desired color.

Referring back to Figure 13–3, the three primary colors of light are red, green, and blue. Any color that results from equal amounts of two primaries is called a *secondary color*, so the *secondary colors* in the additive system are cyan, magenta,

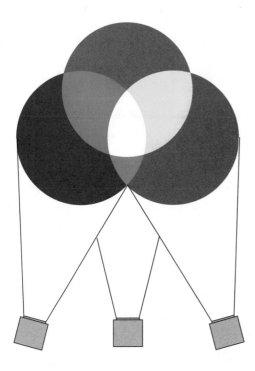

and yellow. If a printer has these same three colors in the form of inks, nearly any color can be created by mixing them in various proportions. The ability to combine light-absorbing (or light-subtracting) pigments to create a broad spectrum of colors is the basis for the subtractive system of color. The secondary colors of the subtractive system are the primary colors of the additive system. This relationship makes possible the printing of postcards, magazines, and all other full-color images from only three colors of ink—cyan, magenta, and yellow—plus black (which is not a true color). These four inks are known as the **process colors** and printing that results from their use is called process, four-color, or full-color printing.

Cyan, magenta, and yellow each absorbs roughly one-third of the visible light that strikes it and reflects the remaining two-thirds (see Figure 13–5). When more than one of these process colors are superimposed or trapped by the printer, they combine their abilities to absorb or subtract from the light that is ultimately reflected by the white substrate beneath. The process inks used in commercial printing are not pure enough to absorb 100 percent of the light that they should (see Figure 13–6). If they were, then each ink color would absorb one-third of the visible spectrum and

a

Visual Sensation is "White"

White Light

White Paper

b

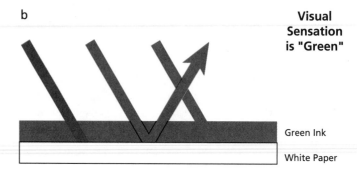

Visual Sensation is "Green"

Green Ink

White Paper

c

Figure 13–4 White surfaces reflect blue, green, and red light (a); therefore, no color sensation results. When pigments absorb the red and blue light, only the green light is reflected (b). When all light is absorbed (c), none is reflected and the object appears black. *From J. Karsnitz, Graphic Arts Technology, © 1984 by Delmar Publishers.*

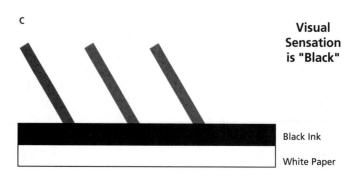

Visual Sensation is "Black"

Black Ink

White Paper

where the three process inks are overlapped the result would be total light absorption and the appearance of black. Because pure process inks would not be commercially affordable, printers achieve black by using a fourth plate that carries black ink.

When a printer wishes to create a **tint** of an ink, screening it into dots has the same effect as mixing the stock ink with white ink (see Figure 13–7). In a similar fashion, an ink can be darkened to a *shade* by printing it with a screen of black (see Figure 13–8). The real versatility of the process colors becomes apparent when they are printed atop one another in varying screen values. An analogy can be drawn here. Just as the texture of music can be influenced by experimenting with the sliding controls of a stereo's equalizer, a printer can control the absorption and reflection properties of a multilayered ink film by controlling the ratio among the three process inks that are printed on the same area. Examine Figure 13–9 and trace how increasing the percentage of coverage for one ink, while decreasing the percentage for a second has a profound effect on the resultant color. When all three of the process colors are involved, the results are even more varied. Examination of any full-color photo in a magazine will reveal how a printer can reproduce a photograph of a July corn field without using any green ink in the press by merely overlaying the correct percentages of transparent cyan and yellow inks over white paper.

The Properties of Color

As has already been demonstrated, a color can be altered by mixing it with a second color, lightening it with white, or darkening it with black. When any of these changes is made, one or more of the original color's *properties* are altered, resulting in a change in appearance. The properties of color are **hue, brightness,** and **saturation.**

To understand the property called **hue,** simply look at the color wheel in Figure 13–10 while remembering that the difference between red light and blue light is their wavelengths. When you are looking at the red area, your eye is being stimulated by reflected light averaging around 660 millimicrons in wavelength, while the blue area is reflecting light with an average of around 440. The 220 millimicron difference that separates these two colors determines their hue and place on the color wheel.

When two nearby colors on the wheel are combined, the result is a new hue. For example, when red ink is equally mixed with yellow ink, the new hue that would result is orange. This new color results from a shift in the predominant wavelength of reflected light. Combining red and yellow pigments creates a mixture with unique light absorbing characteristics—and, predictably, unique light reflecting characteristics. Therefore, a color's hue is determined by its predominant wavelength.

The **brightness** or *value* of a color is just what it sounds like—how light or dark it appears. The vertical series of

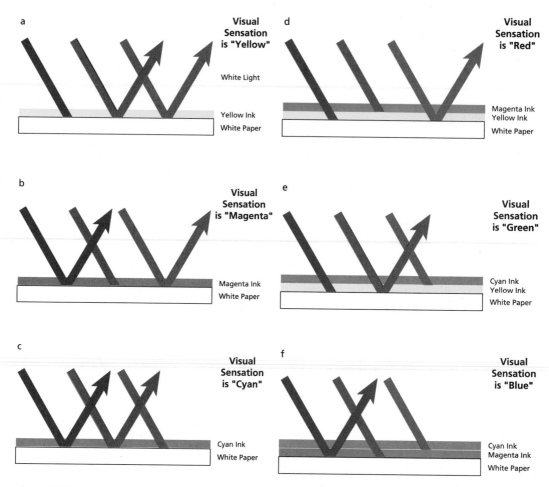

Figure 13–5 When pigments absorb only the blue light (a), red and green light are reflected and the object appears yellow. When only the green light is absorbed (b), blue and red light are reflected, resulting in the appearance of magenta. Cyan appears when pigments absorb the red light and reflect the blue and green light (c). When magenta and yellow are printed on the same area (d), the magenta pigments absorb the green light, the yellow absorbs the blue light, and the red is reflected. Green results when yellow and cyan are overlapped (e), and cyan and magenta inks combine to produce the sensation of blue (f).

squares in Figure 13–11 represents several values or degrees of lightness of the basic color, which is seen in the middle square. As was explained earlier in the chapter, by adding white to a color and lightening it, a tint is created. When black is added, the result is termed a shade. While the wavelength of light determines its hue, the *amplitude* or total height of the wave determines its brightness.

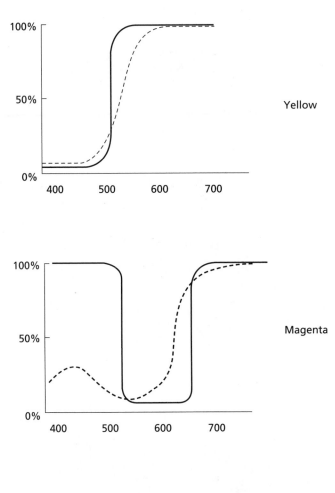

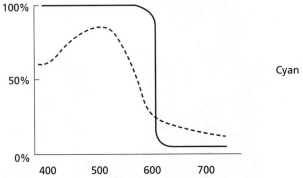

Figure 13–6 The percentages of visible light reflection for commercial process inks are shown as they should be (solid lines) and as they actually are (broken lines). Of the three inks, yellow most closely absorbs and reflects as expected.

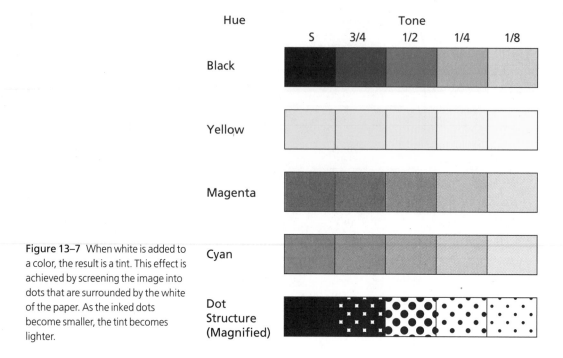

Hue	Tone				
	S	3/4	1/2	1/4	1/8
Black					
Yellow					
Magenta					
Cyan					
Dot Structure (Magnified)					

Figure 13–7 When white is added to a color, the result is a tint. This effect is achieved by screening the image into dots that are surrounded by the white of the paper. As the inked dots become smaller, the tint becomes lighter.

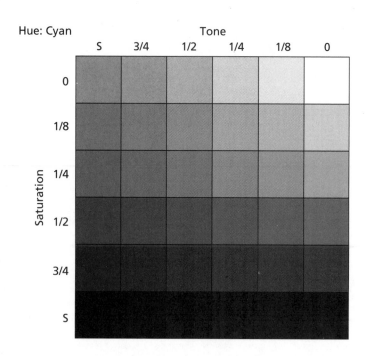

Hue: Cyan

Figure 13–8 When black is added to a color, a shade is the result. In printing, the amount of black is controlled by the size of the black dots.

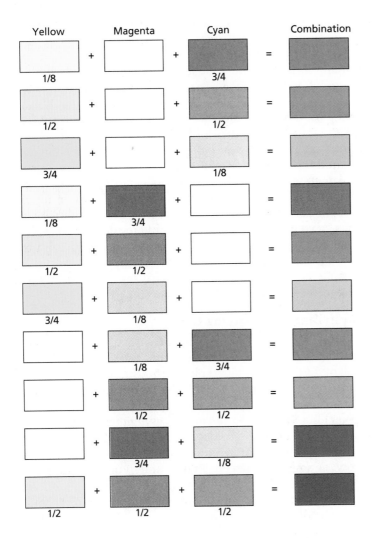

Figure 13–9 By controlling the dot sizes of the process inks, the printer can create a spectrum of colors wide enough to print the world's magazines and postcards.

The final property of color is its **saturation** or *chroma*. Saturation is a measure of a color's purity. Referring back to Figure 13–11, the blue in square number five has a high level of saturation. Any contamination—by black, white, or another hue—results in a dilution of the color and a loss of saturation. Observe that when the blue mixes with a hue from the other side of the color wheel, they tend to neutralize one another and both lose saturation. Hence, saturation results from light that is basically homogenous in wavelength. Colors that are composed of different wavelengths lack saturation, are contaminated, and often look dirty.

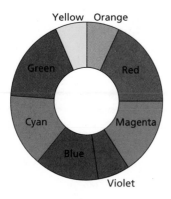

Figure 13–10 When one color is mixed with another, the new color is called a hue. For example, orange is a hue that results from combining yellow with red.

Clearly, the simple color wheel is insufficient at representing the properties of color and the color choices that they can produce. It is much more useful to think of color in *three* dimensions. Imagine a globe of the world with its lines of latitude and longitude. If you can picture the color wheel running around the globe at the equator and an axis, running from pole to pole, that is a series of gradations of gray with white being at the North Pole, black being at the South Pole, and middle gray being at the mid-point, then you would have a more useful model of how color's properties relate to one another. With the model in mind, imagine being on a given location on the color wheel/equator, for example, red. Any shift east or west would create a new hue, any movement north or south would result in a new value, and any movement toward the center (resulting from the red's being mixed with a color on the opposite side) would tend to neutralize the red and make it appear more gray (see Figure 13–12).

Systems for Classifying Colors

Several systems have been devised to classify colors; the Ostwald and Munsell systems are among the most widely used. All of these systems view color in three dimensions (see Figure 13–13). When faced with the task of mixing an ink to achieve a specified color, most printers use the Pantone Matching System®.[1] The Pantone system provides formulas for more than 1,000 colors and requires that only 14 Pantone inks be kept in inventory: Process Blue, Red 032, Yellow, Orange 021, Green, Reflex Blue, Purple, Rhodamine Red, Warm Red, Rubine Red, Blue 072, Violet, Black, and Transparent White. The ubiquitous Pantone Color Formula Guide (formerly known as the PMS book) contains narrow pages with up to seven ink colors on each page (see Figure 13–14). Located near the middle of each page is a hue and moving away from that hue in one direction are increasingly

Figure 13–11 Movement up and down the vertical axis demonstrates how color saturation is reduced by adding either black or white. Movement across the horizontal axis reveals how saturation is lowered by combining colors from opposite ends of the color wheel.

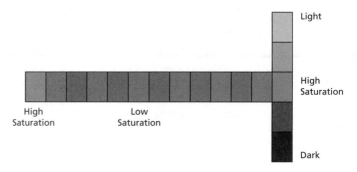

276

Figure 13–12 Most systems for viewing color create a three-dimensional model to show the relationships between hues, tints, and shades. Hues that would result from combining colors that are opposite one another on the color wheel (complementary colors) would appear *inside* a model like this.

light tints and moving in the other direction are increasingly dark shades. Therefore, all of the colors on a Pantone page are values of the same hue, which makes each page **monochromatic.** Because of differences in ink holdout and color appearance between coated and uncoated paper, the Pantone Formula Guide prints all colors on both types of stock.

The primary value of the Pantone Matching System for the printing industry is the means of systematically cataloging a wide range of colors and their formulas. With Pantone Color Formula Guides, a graphic designer in Iowa can telephone a printer in Maine, specify either Pantone 513 C or Pantone 513 U, and nothing more needs to be said. Due to uncontrollable fading of ink pigments, vehicle discoloration, and paper aging, Pantone recommends that its color matching guides be replaced regularly. A well-known graphic arts consultant tells the story of the young production manager who was required to personally verify the match between a press proof and the desired color in a Pantone Color Formula Guide before signing off on the makeready. For every job that specified a Pantone color ink, he would be called to the pressroom where the color in a Pantone guide would be held against the just-printed press sheet for his approval. What he

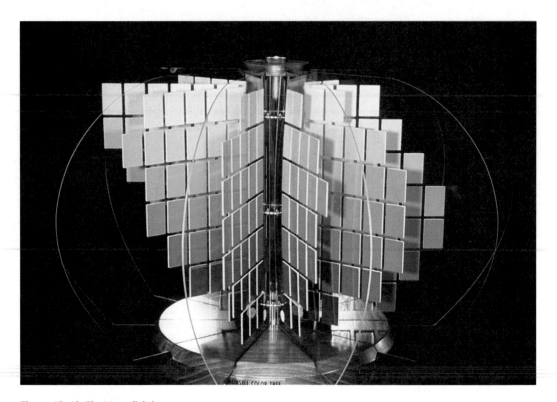

Figure 13–13 The Munsell Color System is represented by this three-dimensional model. *Courtesy of Macbeth, division of Kollmorgen Instruments.*

did not know was that the press operators, who kept several Pantone guides of various ages, would merely go through them until they found one that matched the color on their press sheet, and use it to show the production manager.

The Limits of Process Colors

The range of possible colors that can be produced by a given system is called a **color gamut.** Although the range of colors that can be printed with the process inks may seem boundless, the fact is that the gamut that can be created on a computer's *color monitor* is greater; the gamut that can be captured on *color film* is even greater; and the gamut that can be seen by the *human eye* is greater still (see Figure 13–15). Because of technical limitations, the process inks are unable to faithfully reproduce all colors that can be seen. One technique to expand the printed-ink gamut is to augment the process inks with one or more additional inks such as a pink, blue, violet or red. For this level of color fidelity, presses with five, six, or even eight units are often used.

Generally, the limited possibilities of the process inks are considered adequate by print buyers. Nonetheless, there will

likely be the occasional customer who will not understand why the printed magazine cover does not look precisely like his original color transparency. Printers wishing to define the color gamut for their own inks and equipment can purchase, from the Rochester Institute of Technology, a Process Ink Gamut (PIG) Chart kit, which consists of a set of films that a printer can use to make plates and print a hexagonal pattern of color patches (see Figure 13–16). **PIG charts** not only demonstate the capabilities of a given printing facility, but also allow a printer to experiment with ink brands, paper stocks, color sequences, and print densities in an effort to expand the color gamut that can be achieved. They can also be useful in enlightening customers on the limitations of printing process work with ink on paper.

Color Perception

When an image is focused onto the rear half of the eye, the projected light waves strike the retina and stimulate two types of light-sensitive nerve cells—*rods* and *cones*. **Rods,** which outnumber the cones 20 to 1, contain a pigment that is sensitive to small amounts of light energy. In low-light situations—such as a photographic darkroom—the rods are able to absorb enough light energy to transform it into electrical signals that are sent to the brain. These signals enable the brain to distinguish shades of gray. Because rods are un-

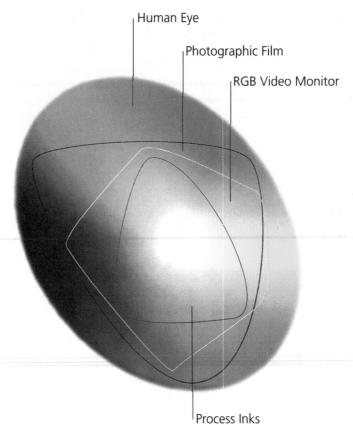

Human Eye

Photographic Film

RGB Video Monitor

Figure 13–15 The gamut or range of achievable color is smaller with printed process colors than with a color monitor. The gamut for light is larger still; therefore, a printer cannot faithfully reproduce every hue that may appear on a color slide.

Process Inks

able to distinguish colors, objects in a low-light situation usually lose much of their color and take on more of a grayish cast. To test this phenomenon, look at a colored object under medium light while slowly closing your eyelids. Although the outline of the object will still be apparent in the dimming light, it will begin to lose its color.

Color perception depends on **cones,** which, in turn, depend on much greater light levels than rods require. A second way that cones are different from rods is that cones will contain one of three different pigments—one that absorbs red light, green light, or blue light. When a cone's pigment absorbs light, the light energy is transformed into electrical signals that are carried to the brain by the optic nerve.

Once these electrical impulses are sent out from the cones, they must be decoded and integrated by the brain. Exactly what occurs along the way is unknown and the serious student of color perception can choose among three theories. The *three-components theory,* first proposed in 1801, maintains

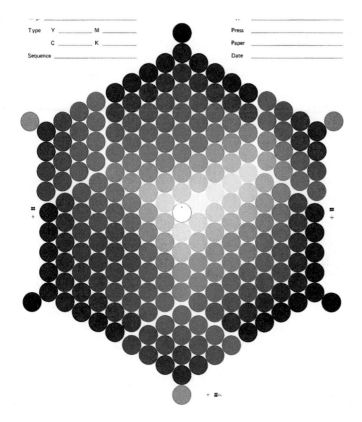

Type Y _____ M _____   Press _____
      C _____ K _____   Paper _____
Sequence _____   Date _____

Figure 13–16 The RIT PIG chart allows a printing firm to access its own gamut. *Courtesy of Rochester Institute of Technology.*

that the electrical impulses are sent directly to the brain where they are processed into color sensations. In 1874, the *opponent theory* proposed that only two of the cones' receptors discern color; one type reacts to either yellow or blue light and another type reacts to either red or green light. The third type of cone measures brightness and darkness. The opponent theory helps to explain why there is no such color as yellowish-blue; the yellow/blue-sensitive cones are either sending yellow *or* blue signals to the brain.

Because neither of these theories adequately explains the after-image phenomenon (see Figure 13–17), other theories have been proposed; the most popular is the *opponent-process* (or zone) *theory,* which combines elements of both earlier theories. The opponent-process theory agrees with the three-component theory in that there are three, not two, types of color-sensitive cones, but proposes that, between the cones and the brain, intermediate nerves transform the three signals into two signals that go on to the brain—yellow or blue and red or green. The concept that the brain re-

Figure 13–17 To experience the after-image phenomenon, stare at the black square inside the green rectangle for 30 seconds. Then immediately shift your attention to the black dot in the white rectangle. You should see the white square assume a rosy hue because red is the complementary color of green.

Figure 13–18 Are the two green triangles shown at right the same? Oddly enough, they are. To find out for yourself, cut two holes in a sheet of white paper so that only the green areas can be seen. Clearly, color perception is influenced by neighboring colors.

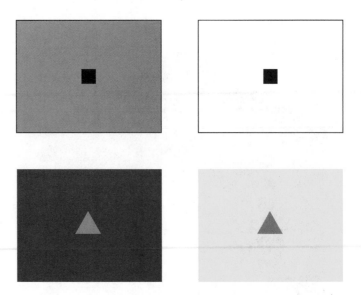

ceives signals from only two sources reinforces the opponent theory. Nonetheless, because these are all theories, it must be said that exactly what follows retinal stimulation to create the sensation of color in the brain is still a mystery.

Color Matching

As the previous discussion has pointed out, four groups of factors are involved in ink color perception: the pigment itself, the observer, the light source, and the surface. The implications of this statement are profound for a printer who is asked to supply a certain color. Even if the printer maintains consistency with the pigment of the ink film, a change in any one of the remaining three factors can result in the ink color's appearing different. In short, the same ink color can appear differently if viewed on two different surfaces, under two different lights, or by two different people.

The most complex component of color perception is the *observer*. In a previous portion of this chapter, it was explained that the pigmentation of cones in the retina of the eye determines which color of light will be absorbed by an individual cone and thereby trigger an electrical impulse that is sent to the brain. When people do not have the necessary pigmentation in all of their cones, then certain wavelengths of light that enter the eye are not absorbed and are, therefore, not perceived by the viewer; this condition is called *daltonism* or, more commonly, color blindness. Whereas only 1 percent of the female population is color blind, 8 percent of males have trouble with color perception to some degree. Most color blind people do not see red. Only a very few do not experience the sensation of color at all.

Even if the retina's cones are sending the correct impulses, strange things can happen when the brain decodes these impulses to produce the sensation of color. First, color cannot be accurately remembered without a basis of comparison. Even if a color is visually studied, it would probably not be correctly identified later if surrounded by several similar hues. Second, color perception is influenced by surrounding color. Figure 13–18 allows the reader to experiment with this phenomenon.

Moreover, most adults have had the experience of the same color appearing differently under more than one *light source*. For example, someone shopping for clothes may carry a garment to the window "to see how it looks in the daylight." In the same vein, if a photographer used the same color film to shoot a night football game and a day football game, the colors on the prints would be very different. In both instances, the natural and artificial light sources produce light with differing proportions of red, green, and blue light. Light from the sun is comprised of roughly equal amounts of the primaries, but fluorescent lighting is heavy with green light, and your basic 100-watt incandescent light is heavy with red light. Under natural sunlight, a yellow ink receives equal amounts of red, green, and blue light; its pigments then absorb the blue light and reflect equal amounts of the red and green light. However, if the same ink is viewed under a light source that contains a smaller amount of red light, less red is available to be reflected and the resulting color will appear more green. Color photographs made under fluorescent lighting without a corrective filter will have this greenish cast. The exact same ink viewed under a light source with a high proportion of red light, such as a basic 100-watt tungsten lightbulb, would appear more orange than yellow.

The color temperature of a light source correlates with its temperature. As the temperature of a piece of iron is increased, its color will move from red to white. The Kelvin (K) scale results from recording the temperature of a light source in degrees Celsius and adding 273°. With the Kelvin scale, the color balance of a light source can be measured. When a change in illumination causes the same pigment to generate a different color sensation, the phenomenon is called **metamerism** and a color that changes its hue under different illumination is a *metameric color*. Because of meta-

Figure 13–19 To standardize the assessment of colors, a printer can purchase a viewing booth that allows proofs to be viewed under controlled conditions. *Courtesy of Foster Manufacturing Company.*

merism, a press sheet's color may appear to match the Pantone Color Formula Guide 1000 or other standard for comparison supplied by the customer. Under the lights in the customer's office, however, the two colors may appear significantly different. To avoid problems associated with metamerism, ink colors must be viewed under standardized light temperatures and 5,000° K is recommended. To control the viewing conditions of a printed ink, a standardized viewing booth can be purchased to ensure standard light temperature and level of intensity, as well as provide a neutral gray background (see Figure 13–19).

Finally, the *surface* that receives an ink film is also a factor in determining the ink's color. As was examined in Chapter 4, a substrate's smoothness level determines its gloss and the highest gloss levels occur with coated paper. Coated papers also are less absorbent than uncoated papers and the combined influence of gloss and absorbency levels is so great that the Pantone Color Formula Guide displays each of the sample hues on both coated and uncoated paper. The first half of the Pantone guide is printed on coated stock, while the remainder is on uncoated stock. A quick comparison of the same color, such as Pantone 312, as printed on coated (312C) stock and uncoated (312U) stock will dramatize the influence that the substrate has on color perception.

Using Color in Printing

With so many variables involved in the complex process of color perception, it is little wonder that so many problems between the printer and the customer revolve around color matching. Clearly, the best approach to avoiding these problems is two-fold: *standardization* of the process and accurate *communication* between the customer and the printer. Many confrontations grow out of an unawareness of the variables examined in this chapter. People who are unaware of the intricacies of color often think that they are being more precise than is actually the case. Either the printer or the print buyer should take the initiative and ask the right questions to ensure that all involved parties are thinking alike.

Mixing Inks to Achieve a Color

When printers are asked to create an ink color that matches a customer-supplied sample, they will most likely need to arrive at the particular color by mixing standard inks that are kept in stock. If the customer specifies a color that is accompanied by a formula, such as a Pantone color, the printer's task is less challenging. A larger task is to match the color on a swatch with no clue to its formula other than its appearance.

Mixing ink to attain a color match is a skill that comes from substantial experience and an understanding of the color wheel. The necessary equipment includes a scale for weighing the component parts, a slab palette, two knives for performing the actual mixing, and a record on which to note the weight of each ingredient. The color circle seen back on Figure 13–10 serves as a guide. If the desired color is a derivative of red, then a red stock ink will serve as the starting point. Notice on the color wheel that mixing red with an adjacent color, such as orange, will generate a new hue that will not severely reduce the color's saturation. In contrast, to add an ink from the other side of the circle, such as cyan, will neutralize the saturation of the red and introduce pigments that absorb much of the light not absorbed by the red. The result will be a dark color that is "dirty." In fact, a dark red can be achieved by starting with a saturated red ink and adding either black or green ink (which is the complementary color to red).

After a small amount of the mixed ink has been created, its color is appraised by reducing it to a film approximately as thick as will likely occur during printing. A proof press, drawdown, or simple tapping out with a finger can produce the ink film to be tested, but the last technique commonly leaves a film that is too thick.

Any color appraisal needs to be done on a sample of the same stock that will be used on the intended job. Before the final color determination is made, the ink film must have already dried. Whether performed by the ink maker or the printer, all of these efforts are intended to ensure that the ink color that is applied on press will meet the customer's expectations. Even with all these procedures in place, though, it still must be remembered that a customer's color perception is subjective in nature and, for that reason, is outside the absolute control of either the ink maker or the printer.

Notes

1. Pantone Matching System® and Pantone® are registered trademarks of Pantone, Inc.

Questions for Study

1. What are the four requisites of color perception?
2. What distinguishes radar, radio, and light waves?
3. Explain the additive color concept.
4. Why is it that when the three primary colors of light are superimposed, the result is white, but when ink's primary colors are superimposed they make a very dark brown?
5. Distinguish between hue, brightness, and saturation.
6. What is the significance of the process inks' having a smaller color gamut than that of a computer monitor?
7. You have probably seen at least one science fiction film in which a person becomes invisible. Could a

person function normally after becoming invisible? To become invisible, a person's body would have to become completely transparent, which means it could contain no pigment cells whatsoever. If that were the case, would the person be able to see or would the person be blind as well as invisible?

8. What causes color blindness?
9. Explain the phenomenon of metamerism.
10. How does the surface of the substrate affect color?
11. Explain two approaches to start with a highly saturated orange and attain a dark orange.

Key Words

visible spectrum
tint
brightness
color gamut
hue

process colors
shade
saturation
PIG chart
rods

additive primary colors
subtractive primary colors
monochromatic
metamerism
cones

14 Ink and Paper for the Printing Processes

AFTER STUDYING THIS CHAPTER, THE STUDENT SHOULD BE ABLE TO:

■ Distinguish between physical and chemical printing.

■ Explain how letterpress inks dry.

■ Describe the configuration of a flexographic press and explain the properties that an ink would contain in order to be appropriate.

■ Identify printed products that are good candidates for a flexographic press.

■ Identify printing problems that can occur in flexography, as well as their potential solutions.

■ Describe the gravure printing process and appropriate physical characteristics of gravure inks.

■ Identify printing problems that can occur in gravure printing, as well as their potential solutions.

■ Describe the screen printing process and the appropriate physical characteristics of screen inks.

■ Describe lithographic printing and the appropriate physical characteristics of lithographic inks and paper.

Just as there is more than one way to do most everything, there is more than one way to get ink onto paper or any other substrate. Differences among printing jobs determine the most appropriate printing process to use. The length of the run, the substrate to receive the ink, and the nature of the image to be reproduced are all factors in the decision of which printing process to use. Once the best process has been determined, the next step is to select a substrate and ink that are appropriate to that process. Failure to match the right materials with the right printing process has resulted in catastrophic results that were entirely preventable. See Table 14–1 for a list of the processes.

Physical printing processes are those in which the image and nonimage areas of the plate are physically different. As seen on Table 14–1, there are three subcategories: *relief* processes, in which the image is raised; *intaglio* (in-TAL-yo), where the image is recessed; and *screen*, with an image area that is porous. In contrast, is *lithography*—the *chemical process*—in which the image and nonimage areas of a plate are basically on the same plane, but are very different in their chemical composition. The final category contains processes that do not qualify as either physical or chemical processes.

Sometimes referred to as *evolving processes* are electrostatic and ink jet printing, but they will not be examined in this chapter. Also excluded are the mimeograph and spirit duplicator methods commonly used by schools and churches.

Physical Processes

Relief processes

Letterpress	Once dominant, but now generally restricted to unique situations.
Flexography	Very common on packaging, labels, and corrugated products.

Intaglio processes

Engraving	Generally limited to business cards and letterheads.
Gravure	Ideal for long-run jobs: postage stamps, catalogs, wallpaper.
Screen	Versatile enough for signs, metal decorating, T-shirts.

Chemical Processes

Lithography	Dominates the industry—newspapers, magazines, books, promotional materials, business forms.

Table 14–1 Each printing process is well-suited to one or more sets of job requirements including the substrate, press speed, and the nature of the image to be printed.

The Letterpress Printing Process

One form of relief printing is **letterpress,** one of the oldest of the printing processes. When printing was first performed in eighth century China, an image was carved onto one side of a block of wood. A rough comparison can be made between these blocks and the wooden blocks with raised letters used by small children. When the block was either placed against ink or had an inked roller moved across it, only the image area was high enough to contact the ink and receive it. Then, like a rubber stamp, the block could be pressed against a surface and the image transferred. Unfortunately, because of their limited size, these blocks were not useful when a sizable amount of information needed printing. However, in 1456, Gutenberg's invention of cast movable type unleashed the potential of letterpress printing and the world was never the same afterward.

For centuries, letterpress was clearly the dominant printing process, as it produced the books that fueled the European Renaissance as well as the newspapers that carried word of the Declaration of Independence, the San Francisco earthquake, and the attack on Pearl Harbor. Since the early 1960s, there has been a steady erosion of the volume of ma-

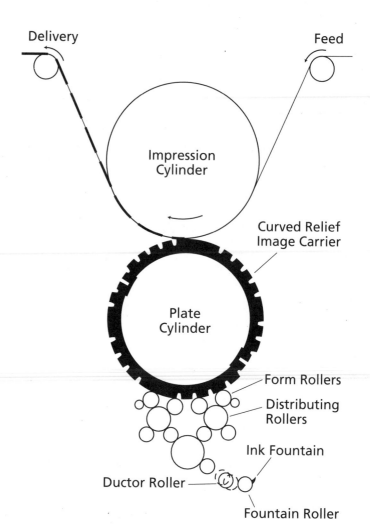

Figure 14–1 A rotary letterpress requires a rather elaborate ink train that begins with the fountain roller that contacts the ink supply and includes the ductor roller, several distributing rollers, and more than one form roller.

terials printed by letterpress. Today, 75 percent of all letterpress inks are confined to the printing of newspapers. Nonetheless, letterpress paper, ink, and presses are still manufactured for commercial printers because it remains the most appropriate of the printing processes for certain situations.

Letterpress Ink

Because letterpress printing applies ink to the raised areas of a plate which then transfers it to the substrate, a *paste ink* (as opposed to a liquid ink) must be used to avoid its running down from the raised surfaces (see Figure 14–1). Letterpress ink formulations vary considerably with regard to the intended substrate and method of drying.

The most basic of letterpress inks is *news ink,* which typically consists of a carbon black pigment and a mineral oil vehicle. Because no resin or other binder is added, news inks are easily identified by their lack of rub resistance. News inks dry through the penetration of their vehicle into the lightly absorbent newsprint. With such a simplistic formula, news inks are very inexpensive and used in such large quantities that they are often not packaged but instead are pumped from delivery trucks into storage tanks located in the pressroom.

Newspapers that contain spot or process color will stock news inks in colors known as ROP (run of press or run of paper). Because ROP colors are standards of the American Newspaper Publishers Association, they are mixed off-press to achieve a certain hue.

In contrast to news inks that dry through penetration, *heat-set* letterpress inks dry after the just-applied ink film is heated to drive off volatile solvents in the vehicle and then immediately chilled. Another letterpress ink that dries quickly is the quick-set variety which works when part of the ink's vehicle is absorbed by the substrate and triggers the speedy oxidation of the remaining vehicle. A third type of letterpress ink that dries quickly is the *moisture-set* variety that dries when the ink film absorbs moisture, which causes the ink's adhesive to fall out of solution and bind the pigment to the substrate. Because they are quick-drying and nearly odor-free, moisture-set inks are commonly used on food packaging. Still another type of letterpress ink is the *water-washable* variety, which allows press washups to be performed with water. These inks are generally used on corrugated boxes.

Job inks constitute the final category and include inks intended for a wide range of presses and substrates. Job inks represent the one-size-fits-all philosophy and are intended for the printer who wants a stock ink that will work on most any job that comes along.

Letterpress Paper

The letterpress process does not place as many demands on the paper as does the more popular lithography. The major concerns when ordering paper for letterpress printing are that it be smooth and not too compressible. The thousands of tiny raised areas that constitute a halftone photograph on

Figure 14–2 The compressible image of this rubber flexographic plate (left) is ideal for high-speed printing on materials like this roll of polypropylene (right). *Courtesy of Columbus Cello-Poly Corporation.*

a letterpress plate require a uniform substrate surface in order to make uniform contact across the photograph. In fact, the development of the letterpress halftone spurred the need for coating paper. When paper is too soft and compressible, the raised images can actually emboss their image into the stock, instead of merely transferring the ink.

The Flexographic Printing Process

A second printing process that carries its image areas above the nonimage area is **flexography.** Except for using relief plates, flexography and letterpress have little in common. Instead of metal plates, flexography uses a resilient image carrier such as a rubber or a photopolymer material. By using a flexible, compressible plate, it is possible to print onto flexible, nonpaper substrates such as the clear films of bread wrappers and thin foils used to package potato chips (see Figure 14–2).

Because of the unusual nature of many flexographic substrates, the presses that print on them bear little resemblance to letterpress machines, and the inks are very different in their viscosity, chemical formulations, and methods of drying, as well. Whereas the letterpress ink train relies on a large number of ink rollers and form rollers to uniformly distribute the paste ink that is used, flexographic presses have a much shorter ink train with no more than one or two rollers

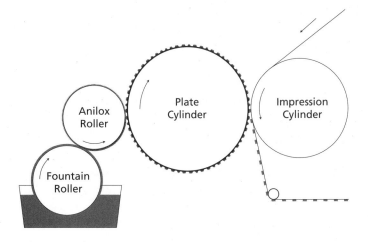

Figure 14–3 In sharp contrast with the complex ink train of the letterpress process, flexography uses only a fountain roller to lift ink onto an anilox roller, which limits the volume of ink carried to the plate. A short ink train is necessary because flexographic ink vehicles are so volatile that they would dry while still on a longer ink train.

which deliver a very fluid (or liquid) ink to the plate (see Figure 14–3).

In flexography, a fountain roller picks up the ink and transfers it to the **anilox roller,** a device that is found exclusively in flexographic printing. The function of the anilox roller is to *meter* the ink or, in other words, control the amount of ink that is transferred to the plate. To accomplish this task, the anilox roller is engraved with thousands of tiny depressions that give its surface the pattern of a waffle and hold the liquidy ink. Because the fountain roller and the anilox roller revolve at different speeds, the resulting rubbing action at the nip removes ink from the surface of the anilox roller but leaves it in the cavities. When the ink on the anilox roller makes contact with the raised surfaces of the plate, the ink is pulled from the cavities and the transfer is made. Once on the plate, the fluid nature of flexographic ink causes the dots to run together immediately and form a solid ink film before making contact with the substrate.

Some flexographic presses utilize no fountain roller and, in the design, the entire ink train consists of nothing more than the anilox roller and plate cylinder. Without the interaction between the anilox and fountain rollers to remove ink from the surface of the anilox, a doctor blade is required to remove excessive ink.

When printing on clear plastic films or other materials that tend to stretch, the material cannot be allowed to travel from the nip of one printing unit to another because the resultant pulling and distortion of the film would cause misregister of the different inks' images. To ensure proper register in multicolor printing on stretchable film, the central

Ink and Paper for Printing   293

impression cylinder was devised. As Figure 14–18 illustrates, once the film contacts the impression cylinder, contact is not released until all printing has been completed, so the film remains dimensionally stable.

Flexographic Inks

The short ink train of flexographic presses is required due to the high *volatility* of the inks used. In short, the inks dry so quickly that they would never make it completely through the extensive roller system found in either letterpress or lithographic presses before drying. While the paste inks of letterpress and lithography set before they actually dry, flexographic inks dry promptly—usually by the rapid evaporation of the solvent. The speed and simplicity of evaporation drying make flexography ideal for printing on plastic films, which are nonabsorbent and often, just moments after being printed, receive further on-machine processing that requires a completely dry ink film. Not only do flexographic inks dry quickly, they also can be nearly odor-free, which makes them very popular in the food packaging sector of printing. (Although a person may not think of a dried ink as having an odor, if he were to simply open up a new book or magazine, put his face close to the pages, and inhale, he might change his mind.)

Still another characteristic of flexography is its ability to use *water-based inks*, which is attractive to printers who wish to avoid problems with ink solvents that are considered environmentally unfriendly. In 1994, water-based inks outsold solvent-based inks by a 2:1 ratio within the North American flexographic industry.

With these features—drying speed, low odor, and substrate versatility—flexography has become the preferred process for so many packaging applications that over 90 percent of all flexographic ink is used in packaging (see Figure 14–4). Although flexographic printing did not originate commercially until the 1940s, by the 1970s, the process had already surpassed letterpress in the volume of printing performed. By 1994, $675 million of flexographic ink was sold in North America alone.

As was mentioned, flexographic inks have very low viscosity; that is, they are very fluid and flow so easily that if they were transported or stored at their working viscosity levels, their pigment material would settle out. To maintain pigment suspension, flexographic inks are shipped in a concentrated form and then **let down** (diluted) by the printer

Figure 14–4 Most grocery carts will contain one or more examples of flexographic printing (left) as they are wheeled to the check-out area. Flexographic printing can be identified by a halo around the image (above).

when needed for a given job. This dilution can be done with either a *reducer,* which is no more than a blend of the ink's solvents, or an *extender,* which is basically the ink without its colorant (pigment material). When the printer wishes to let down the ink without altering its viscosity, an extender is used; if a viscosity change is desired, a reducer is used.

Establishing the correct viscosity through the press run is crucial in flexographic printing. Variations in viscosity can result in color variation, dark halftones, or halos around line art. With water-based inks, maintaining the correct pH is also crucial because if it is too low, the ink can become too viscous and print poorly. Efflux cups are used to monitor viscosity if the press is not equipped with a system that automatically maintains the correct level.

Flexible packaging	43
Corrugated containers	22
Food containers	7
Labels/envelopes/commercial	7
Bags	6
Newspapers	4
Household paper	3
Wall coverings	2
Miscellaneous	6

Table 14–2 This ranking of the applications of North American flexographic printing in 1994 by percentage of ink sold reflects the diversity of flexography as a printing process. *Source: FLEXO.*

Flexographic Substrates

In addition to being printed on plastic films and metal foils, flexographic inks also find their way onto paper and paperboard (see Table 14–2). In the case of printing on corrugated cardboard or any other soft substrate, flexography's resilient plates are ideal because they do not crush or emboss the compressible stock. Water-based flexographic inks work especially well on these porous substrates because they can absorb the ink's water component and require that evaporation remove only the remaining solvents, thereby speeding drying.

Printing with Flexography

When flexography is used in publishing, the nonimage areas between the plate's halftone dots can fill in with ink and print dirty halftones; the cause may involve all three principal players—the ink, paper, and plate. For example, the ink may pick up loose fibers or even aluminum sulfate particles from the paper; particles of dried ink may be present; and/or a plate that is inappropriate for the ink's solvents can absorb them and soften or swell.

Three problems can occur with flexographic printing when the ink's solvents evaporate too soon and the ink dries too quickly. One is **feathering,** which is the term for ragged edges; a second is **fill-in,** the plugging of open areas in small type sizes; and the third is **mechanical pinholing,** which describes the pattern of the anilox cells that can appear on the substrate when the ink dries before it can run together and form a solid area. Mechanical pinholing has the general appearance of a 90 or 95 percent screen tint. All three problems can be corrected by reducing the volatility of the ink's solvent.

In the other extreme, if ink dries too slowly or improperly on a web that is rewound after printing, blocking can result as the ink adheres to the paper above it as well as below it.

The low viscosity of a flexographic ink makes it more susceptible to one or more of its ingredients *falling out* of the mixture and profoundly altering its performance. When a resin falls out of solution, the problem is called *precipitation.* If an ink absorbs too much moisture from the air, then its nitrocellulose (an additive that helps form a thin film) precipitates out—an occurrence known as *souring.*

The simplicity of flexography's ink train reduces the potential for most printing problems to three categories: (1) inappropriate resin for the substrate; (2) incorrect volatility level of the solvent; and (3) faulty viscosity control.

Intaglio Processes

In the relief printing processes just examined, ink is carried above the nonimage portions of the plate. **Intaglio** printing processes carry their ink in recessed areas and, in this sense, are just the opposite of the relief process. There are two basic types of intaglio printing—*engraving* and *gravure.*

The Engraving Printing Process

Engraving is by far the oldest of the intaglio methods, dating back to the sixteenth century. An engraved plate carries ink in recessed troughs that are physically carved into the metal surface. Printing occurs when the surface of the ink plate is wiped clean (leaving ink in the recessed areas) and a sheet of paper is pressed down upon it, causing the ink to adhere to the paper (see Figure 14–5). As might be imagined, engraving is a slow and therefore expensive means of printing. In commercial applications, it is largely limited to letterhead stationery, wedding invitations, and business cards; however, as a point of interest, U.S. currency is still printed from engraved plates. Because of its tiny slice of the overall volume of commercial printing, little will be said here about the engraving process.

Engraving inks are paste inks that dry through *oxidation.* When faster drying is needed, water-based inks are sometimes used because the water can penetrate the paper and leave a smaller amount of oil to evaporate. Because of their pastelike nature, engraving inks can produce a very thick ink film—so thick that ink gloss can be achieved even on

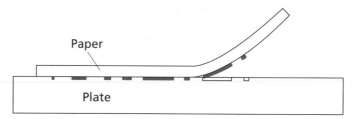

Figure 14–5 (Top) A close-up view of an engraved plate reveals the troughlike image areas. Engraving depends upon the absorbency of the sheet to pull ink out of the low image areas of the plate.

Paper

Plate

uncoated paper. Because most engraving is done on un-coated stocks such as bonds and bristols, an ink with high viscosity is most appropriate. Ironically, engraving prints type on soft, uncoated sheets with sharper definition than can be achieved on most other printing processes.

The Gravure Printing Process

Of the two intaglio processes, the dominant one is definitely **gravure** (also known as *rotogravure*). Instead of carrying ink in carved troughs, a gravure cylinder uses tiny cells—150 to 300 in a linear inch—that are usually *engraved* into the copper cylinder (the exception is the older technique in which cells are etched with acid). These cells carry not only half-tone images, but line images such as type, as well. By carrying its ink in tiny cells, a gravure plate can rotate at high speed with the ink's surface tension keeping the ink from being flung outward and into the air. Gravure printing has no ink train because the plate cylinder actually rotates while partially submerged in the ink reservoir (see Figure 14–6). A steel *doctor blade* then cleanly scrapes ink from the cylinder's surface, leaving ink in only the tiny cells. As the cylinder continues to rotate, the ink contacts the paper web that is pressed against the cells by an impression cylinder.

Unlike rotary letterpress and lithography, the plate cylinder of a gravure press does not hold the plate; it *is* the plate. The plate cylinder is a solid cylinder of copper, with a chrome-plated surface that has been engraved with the tiny cells. Three different types of cell patterns exist (see Figure

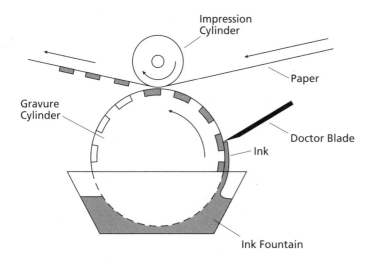

Figure 14–6 The shortest ink train is found in the gravure process as the gravure cylinder actually revolves in a bath of ink. The doctor blade removes excess ink, but leaves it in the thousands of cells. *From Adams, Faux, and Rieber,* Printing Technology, *4th edition, © 1996 by Delmar Publishers.*

14–7). *Conventional* cells are all square-shaped and, unlike halftone dots, all are equal in size. The variation in the darkness of a printed halftone is achieved with deeper cells in the shadow areas. These deeper cells carry more ink to the substrate and print a thicker ink film. A second type is the *variable area* cell pattern in which the width of the cell varies—with large cells in the shadow areas and small cells in the highlight areas—much like the halftone dot pattern in relief and lithographic plates—but the cell depth is uniform. The third type of cell, *variable area and depth,* is just what the name implies; highlight area cells are small and shallow, while shadow area cells are wide and deep. Understandably, this last type of cell is commonly preferred because it offers the printer the greatest control over the tonal range of photographs.

Because gravure has the potential to apply a thicker ink film than can be achieved with letterpress, flexographic, or lithographic printing, it can produce superb quality in black-and-white and four-color halftone printing. For example, *National Geographic* and many other high-quality magazines are printed, at least in part, by gravure, as are numerous newspapers' Sunday magazines and supplements. Part of gravure's success with printing photographs lies with the higher ink gloss that is possible with thicker ink films. In addition, thick ink films allow a lower concentration of pigment material and make gravure well-suited to printing with metallic inks.

All three cell types can be created through etching, but the much more commonly used process of engraving produces

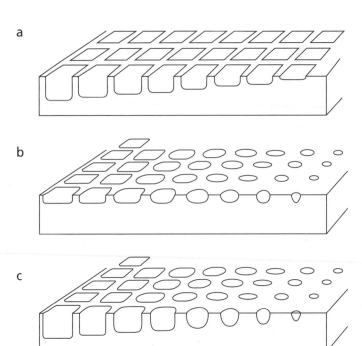

Figure 14–7 Three types of gravure cells are uniform area/variable depth (a); variable area/uniform depth (b); and variable area/variable depth (c), which is the most commonly used.

variable area and depth cells nearly exclusively. For this reason, variable depth and area cells dominate the gravure industry.

Because a gravure press—like its flexographic counterpart—has such a short ink train, it can use highly volatile solvents in its inks, which equates to rapid drying and fast press speeds. In fact, of all the printing processes, gravure is the fastest and, therefore, the most productive. One advantage that gravure has over flexography is the fact that gravure inks can contain solvents that would attack the materials used to make flexography plates. Also, gravure's wide range of acceptable solvents makes it appropriate for numerous substrates. Examples of these substrates include paper, foil, vinyl film, vinyl-coated fabrics, and vinyl flooring.

An advantage that gravure has over all of the major printing processes is the fact that its image is carried below, rather than above, the plate surface. As a result, the image's details are more protected from the wearing effect of long press runs so the expense of replacing a set of plates is usually avoided. While high-quality lithographic plates may be good for 100,000 impressions, a gravure cylinder can typically produce millions of impressions. For this reason, gravure is usually chosen for jobs requiring hundreds of thousands of copies. Postage stamps, imitation wood paneling, cigarette cartons, wallpaper, gift wrap, labels, and long-run magazines are all examples of typical products of gravure printing. For example, before its demise in 1992, as many as 14 million copies of the Sears catalog were printed annually by gravure. One gravure plant is totally dedicated to printing only one product day after day; it produces 65 billion boxes for a single brand of breakfast cereal every year.

Figure 14–8 Gravure printing can be identified by the fact that even type and other line work consists of tiny cells on the plate, thereby producing serrated edges.

Clearly, gravure is well-suited to long-run jobs. In fact, from an economic standpoint, long runs are necessary to compensate for the high cost of plate cylinder preparation and press makeready. When short-run jobs are printed by gravure, the customer is willing to pay high costs in order to get the contrast and gloss that gravure can provide. Examples of short-run gravure jobs are art prints, annual reports, and high-quality magazines with modest circulations.

There are other advantages to the gravure process. Because it does not involve clamping a plate around a cylinder, the gravure process can produce seamless printing, which makes it ideal for wallpaper and similar products. Also, the simplicity of the gravure process can generate high consistency throughout the press run as well as freedom from concerns over plate and blanket compatibility.

To identify a gravure-printed product, look at a sample of text type or other line art under a magnifying glass. If the straight lines actually appear to be a chain of overlapping dots, it was printed by gravure (see Figure 14–8).

Gravure Substrates

When ordering paper for gravure printing, the most important factor in ensuring good printability is a high level of *microsmoothness,* a property of a coated sheet with adequate polishing on the supercalender. When an uncoated sheet with its rough texture of fibers is used in gravure printing, the ink-filled cells do not always make contact with the most recessed areas of the paper's surface, so there is no ink transferred from these cells. The resulting white specs that appear in image areas are called **snowflakes,** skips, or speckles. One method of reducing snowflaking is to use an **electrostatic assist** on the press. This technique involves loading the impression cylinder with an electrical charge that moves through the paper to pull the ink up from the cells and improve the chances of making contact.

A second factor in a paper stock's appropriateness for gravure is *porosity.* Ironically, if the stock is excessively polished to achieve high microsmoothness, then the surface can be-

Publication Applications

A	Inks for newspapers and other publications on uncoated paper
B	Catalogs, magazines, and other publications on coated paper
P	Proofing ink for coated or uncoated paper
W	Water-based inks for coated or uncoated paper

Packaging Applications

C	Inks for nitrocellulose-coated cellophane, aluminum foil, metallized paper
D	Inks for polyethylene, polyester, and polymer-coated cellophane; also top lacquers for paper and paperboard
E	Low-odor inks and lacquers for paper and paperboard; inks for aluminum foils and nitrocellulose-coated cellophane
M	Inexpensive lacquers for cartons and labels
T	Alcohol resistant inks; high-gloss inks for paper and paperboard
V	Inks for vinyl surfaces
W	All-purpose water-based inks
X	Residual category (for inks that do not belong elsewhere, like fluorescent)

Table 14–3 Gravure inks are categorized according to their properties as well as the substrates for which they are formulated.

come so hard (nonabsorbent) that the ink from the deeper cells simply piles up on the coating and impairs printability.

Gravure Inks

Gravure inks are similar to flexographic inks in several ways. First, they are very fluid so that the ink can rapidly penetrate the tiny cells of the plate cylinder. Like flexographic inks, gravure inks are shipped and stored at higher viscosity levels and diluted in the pressroom to prevent a settling out of heavier components.

Second, because of the lack of a significant ink train, gravure inks can use very *volatile solvents* and dry very quickly through evaporation of the solvent. For this reason, they are not limited to absorbent stocks and can be formulated for a wide range of substrates and end-uses. Reflective of this versatility are the 12 classifications for gravure inks, each represented with a capital letter (see Table 14–3). Four of the 12 are publication inks (A, B, P, and W) and the remainder are used in packaging.

In most printing processes, the ink contacts some sort of elastomeric material such as the rubber ink train rollers found in letterpress and lithography or the rubber or photopolymer plates used in flexography. As a result, certain solvents and binders that would harm these surfaces cannot be used. In gravure, however, no such limitations exist, providing the printer with the ability to print on substrates that would not otherwise be practical. At the same time, increasingly stringent standards imposed by OSHA (Occupational Safety and Health Administration) and the EPA (Environmental Protection Agency) have prohibited or discouraged the use of certain highly volatile chemicals so, although gravure is still a very versatile method of printing, it is not as versatile as it once was.

The evaporation speed of gravure solvents requires care that the ink does not dry too soon. If it does, the result can be *screening*, the appearance of a nonimage screenlike pattern in what should be a solid area of ink coverage. Screening occurs when ink from the plate cylinder's cells transfers to the stock, but dries before it has the time to spread out across the paper and form a solid. The result looks something like an 80 percent screen tint.

Pinholing is the term for the small white areas that occur in what should be areas of solid coverage. The problem is not an excessive drying speed, but usually the ink's inadequacy at distributing itself across the substrate's surface or, in other words, *wetting* the surface (see Figure 14–9). Because of gravure ink's low viscosity, it must be agitated on-press to avoid precipitation. As a result, pinholing can occasionally result from a foam bubble getting trapped in a cell and preventing the penetration of ink.

If air bubbles become trapped against the doctor blade, their presence can produce streaks of missing ink within image areas. Because ink makers are aware of the on-press agitation that gravure ink requires, *antifoamers* are included in ink formulas to prevent foaming problems; however, these additives can be overcome if the press agitation or ink pump pressure is excessive.

Finally, the low viscosity of gravure and other liquid inks makes maintaining the suspension of pigment particles within the vehicle more difficult than is the case with paste inks. *Sedimentation* is the term for the settling-out of pigment material when bulk ink is stored for long periods of time or if ink that has been diluted to working viscosity is left unagitated in the fountain. A similar falling-out of suspension by pigment particles can occur suddenly if certain solvents are added too quickly to an ink, a phenomenon called *flocculation*.

Figure 14–9 Three degrees of wetting capability are shown. The liquid at left has a high surface tension that holds the molecules in close; moving to the right you see liquids with increasing abilities to overcome surface tension.

Figure 14–10 Of the uppercase letters, only seven (a) have nonimage islands that require a screen to hold together the separate portions of the screen. The remaining letters (b) could be carried on a stencil that was simply cut out of a piece of paper.

Research is underway to develop water-based gravure inks. To date, the use of these inks has largely been limited to the packaging sector of the printing industry.

The Screen Printing Process

The final physical printing process to be examined is screen printing. Second only to relief in its longevity, screen printing's roots have been traced back to eleventh century Japan. In screen printing, the nonimage area is a stencil that sits on the substrate and protects it. The image areas consist of open areas in the stencil that allow ink to pass through them and adhere to the substrate.

A simple stencil cut from a sheet of paper and used in conjunction with a can of spray paint serves as an elementary means of understanding the distinction between image and nonimage areas in screen printing. Nineteen of the uppercase characters could be printed with a can of spray paint, a sheet of paper, and a knife with which to cut letter-shaped open areas. The remaining seven letters would prove more of a challenge, though, because they contain islands of nonimage areas that are not connected to the rest of the stencil (see Figure 14–10). Because of these islands a fine-mesh screen is required to hold all of the nonimage areas in place without leaving a trace of its presence on the finished product. Very fine-mesh screens can reproduce fine detail.

The image areas of the stencil itself can be cut out by hand, but are nearly always produced by photographing the artwork or type that needs to be printed. Once the stencil has been properly adhered to the screen, printing is performed by depositing ink across the top of the screen and dragging a rubber-tipped squeegee across the screen to force ink through the open areas of the stencil and onto the substrate (see Figure 14–11).

The uniqueness of the screen process makes it the printing process of choice in several situations. First, its equipment demands are modest and it can be performed by hand, thereby making modest financial demands on equipment for image preparation and transfer. As a result, screen printing is ideal for short runs such as 100 yard signs for a local election campaign—although high-speed screen printing

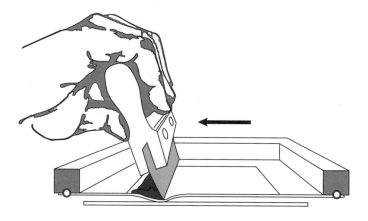

Figure 14–11 A squeegee moves screen ink forward and down through the openings (image areas) of the stencil and onto the substrate.

machines also exist. Second, because it is essentially a stenciling operation, screen printing is adaptable to products that are inappropriate for traveling through a press. Short-run posters, T-shirts, baseball caps, mirrors, highway signs, light bulbs, clock faces, bottles, wallpaper, shower curtains, and drinking glasses that carry a printed image are all representative of the versatility of the screen process.

Third, screen printing can apply a thicker ink film than is possible by the other processes, making it ideal for bumper stickers and posters that require heavy ink coverage and a light-fast image. The mesh of the screen regulates the amount of ink deposited on the substrate, with coarse screens printing the heaviest ink films.

Screen Printing Substrates

The versatility of screen printing creates a long list of potential substrates that includes plastics, glass, ceramics, metals, cloth, wood, and the most commonly used—paper and paperboard. Because screen printers often create yard signs, posters, and other large format work, the expansion of the sheet due to moisture gain can create major problems with misregister on multicolor jobs. Sometimes the screen printer will actually anticipate and accommodate the sheet's growth by enlarging the images of subsequent colors.

Because of the low viscosity of screen inks, the sheet should not be too absorbent; for this reason coated stocks are usually preferred. At the same time, uncoated sheets are less likely to present drying problems and smear.

Sheets with high caliper (bulk) are better able to resist being lifted by the screen when it makes temporary contact with the substrate under pressure from the squeegee. High bulk sheets are also less likely to warp when receiving the thick ink films associated with screen printing.

Screen Inks

This versatility of substrates dictates that screen inks' formulas be just as diverse. Still, there are two traits that all screen inks share. One is that they are all very *short inks;* that is, they cannot be stretched into strings before snapping. Short inks are needed to prevent strings from forming when the forward motion of the squeegee allows the screen to snap back up and away from the newly-printed ink film. The other trait is that most screen inks dry through the evaporation of very volatile solvents.

Despite the high volatility of their solvents, screen inks can be rather slow to dry because the ink films are so thick. The evaporation speed is commonly increased by using microwave, hot air, or flame driers. As an alternative to evaporation drying, some screen inks are formulated to cure nearly instantly through polymerization when bombarded by ultraviolet light. The high cost of *UV-curable inks* limits their use to those products where the investment can be affordable, such as printed electronic circuit boards, plastic bottles, billboards, posters, and pressure-sensitive decals. UV inks are discussed at the end of this chapter.

Plastisol inks contain liquid plasticizers that ensure a flexible ink film that will not fall off substrates that are bent, folded, or otherwise treated roughly. Plastisol inks are typically used on clothing and other textile materials because plastisol inks get increased adhesion from wrapping around fibers.

When ordering inks, screen printers must be especially clear in specifying the exact nature of the substrate to which the ink must bind. To say "I need an ink to print on plastic bags" does not clarify whether the ink will need to adhere to nylon, acrylic, polyester, polyvinyl chloride, polystyrene, polyethylene, or polypropylene. Many a screen printer has applied an ink to plastic only to see the dried ink flake right off.

Lithographic Printing

With physical printing processes, the image areas are physically different from the nonimage areas. In **lithography**, however, the distinction between image and nonimage areas is not physical, but rather is *chemical*. Lithography relies on the chemical principle that water and grease try to avoid mixing with one another. To understand how this principle applies to printing, it may be useful to know that the earliest lithography was performed by writing on a stone slab with a greasy substance (*litho* and *graphy* are derived from the Greek words for stone and writing). When the stone was then wiped with a wet rag, the water did not transfer to the greasy image, so it remained dry, surrounded by wet nonimage areas that did accept the water. At this point, an ink with a greasy chemical composition was rolled over the entire stone. Because of the incompatibility of grease and water, the ink was repelled by the wet nonimage areas that remained free of the water. The stone was then ready to transfer its inked image onto paper. The image areas had to be wetted and inked between each impression or printing.

A modern lithographic press works on the same basic principle applying ink to a plate that has already been wetted. Two key technical developments since the era of the stone are: (1) nearly all presses are rotary; and (2) the plate never touches the substrate, but instead transfers the inked images to another cylinder. Lithography is generally unique among the printing processes because it has a third cylinder that accepts the inked image from the plate cylinder and then, in turn, transfers it to the substrate. Because this extra cylinder, called the *blanket cylinder*, has the image offset onto it by the plate, lithography is formally referred to as *offset lithography* and more commonly as merely *offset* (see Figure 14–12).

The **dampening system** of a lithographic press consists of a fountain roller that revolves in the fountain, a *ductor roller* that picks up the solution and transfers it to a *distributing roller*, and one or more *form rollers* that transfer it to the plate's nonimage areas. The dampening system is designed to meter (control the amount of) the solution that is applied uniformly to the plate. Either inadequate or excessive dampening solution creates major problems with printing quality, as will be discussed later.

Similar to the dampening system in its basic function, the **inking system** is designed to meter a uniform application of ink onto the plate (see Figure 14–13). Many more rollers are required to perform this task, though, than are required in the dampening system. Numerous rollers are needed to work the ink to the correct viscosity for good transfer and act as a reservoir for replenishing the plate after each revolution. The flow of ink onto the *fountain roller* is controlled by a *doctor blade* that can allow extra ink onto parts of the plate

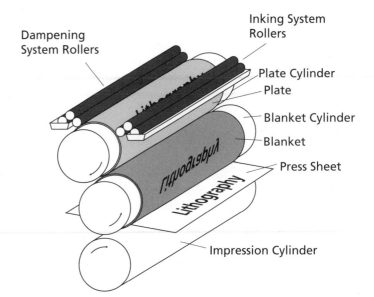

Dampening System Rollers

Inking System Rollers

Plate Cylinder
Plate
Blanket Cylinder
Blanket
Press Sheet
Impression Cylinder

Figure 14–12 In lithography the plate does not touch the substrate, but instead offsets the image onto a blanket. Note also that lithography is unique among the printing processes because it requires a dampening system.

that have the greatest concentration of image areas (see Figure 14–14). The *fountain roller* then transfers its ink to the *ductor roller.* In both the dampening and ink systems, the ductor roller alternates contact between the fountain rollers and a *distributing roller.* This back-and-forth movement is useful in measuring the amount of dampening solution or ink that is transferred onward. To enhance the uniform distribution of ink, the distributing roller that receives the intermittent contact with the ductor roller oscillates side-to-side. The solution that is carried by the dampening system is called both *dampening solution* and *fountain solution.*

The ink that finally makes its way to the form rollers comes in contact with the entire surface of the plate, but refuses to transfer to any portion of the plate that is already wet with a film of dampening solution; in other words, the nonimage area. Like the dampening system, the inking system must control the amount of ink that is carried on the rollers or poor printing is the result. For example, the thickness of the ink film influences the tack level which, in turn, influences one ink's ability to print over a just previously-printed ink (a process called *trapping*). When the dampening system functions separately from the inking system as just described, the dampening system is termed as *conventional* or *direct.* In contrast, *indirect dampening* systems (such as the dahlgren system) link the ink and dampening systems by actually applying the dampening solution onto one of the ink-

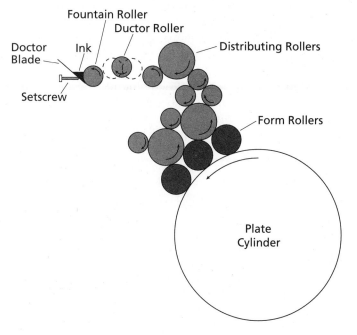

Figure 14–13 The ink train of a lithographic press is similar to that of letterpress machines in its complexity and number of rollers. Such a large ink train requires a highly viscous, pastelike ink. *From J. Karsnitz,* Graphic Arts Technology, *© 1984 by Delmar Publishers.*

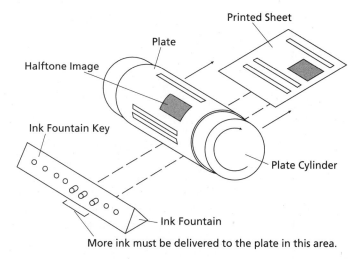

Figure 14–14 Screws (called ink keys) can be adjusted to control the pressure against the doctor blade of an ink fountain and control the amount of ink needed in various areas across the plate. *From Adams, Faux, and Rieber,* Printing Technology, *4th edition, © 1996 by Delmar Publishers.*

ing form rollers where it rides atop the ink film until it is further transferred to the plate's nonimage areas.

Lithographic Paper

While physical printing processes such as letterpress and gravure require paper with high smoothness, the smooth and compressible nature of a lithographic blanket makes it able to flatten at the nip between the blanket and impression cylinder. As a result, lithography is appropriate for textured or comparatively rough paper stocks.

Although lithographic printing does not require a surface with high smoothness, it does require *high surface strength*. Because of the high tack of lithographic ink, the ink film that contacts the paper at the nip does not split easily and paper's surface fibers must withstand the pull of the ink. If the ink-to-paper bond is stronger than the fiber-to-fiber bond the sheet will *pick* (refer back to Figure 12–5). The high tack level of lithographic ink is also the primary culprit in *tail-end hook*, making the *stiffness* of the sheet a more critical issue in lithography. Both picking and tail-end hook are described in more detail in Chapter 7.

Because lithography is the only printing process to employ dampening solution, it is unique in certain demands made on its substrates. First, the paper must have received surface and internal sizing. If letterpress paper is used on a lithographic press, dampening solution penetrates the sheet and reduces the fiber-to-fiber bond. This softening effect becomes more pronounced when the sheet receives multiple passes through the press with multiple exposures to the dampening solution. Second, litho or offset paper cannot contain materials that could mix with the dampening solution and alter its pH or, once again, print quality is impaired. Last, paper's dimensions expand when it picks up moisture, producing potential problems in holding multicolor images in register. Because paper with a low moisture content has the greatest potential for dimensional change when exposed to dampening solution, lithographic paper has a higher moisture content than is found in stocks intended to be printed by other processes. However, when a heatset ink is used, the stock must have a lower moisture content than is normal for litho stock because excessive moisture inside the sheet can *blister* if it is heated so fast that the steam is unable to escape (see Figure 14–15). Because a *perfecting press* (which is described later in the chapter) prints on both sides of the sheet at the same time, there is the potential for twice as much absorption of the dampening solution and an even higher risk of blistering. Many lithographic perfecting presses are web presses that print on a continuous roll of paper instead of individual sheets, so web litho stock is sold with a low moisture content. Unfortunately, if a stock's moisture level is too low, either through manufacture or excessive heatset drying, the stock will be more likely to become brittle and lose fold strength.

Figure 14–15 In this blistering incident, a sheet with excessive moisture content is heated to set the ink, but the ink film itself acts as a barrier to the escape of moisture, which vaporizes inside of the sheet and causes it to rupture.

Substrates appropriate for lithographic printing include films and metals, but paper and paperboard are by far the most commonly used.

Inks for Lithography

Because lithography requires that its ink share the plate area with dampening solution, the ink maker must formulate litho inks to maintain a delicate ink/water balance. As was stated earlier, oil-based litho ink and water do not readily mix. Although this statement is fundamentally true, the two materials can create an **emulsion** (see Appendix C). In fact, litho inks are designed to accept a certain amount of tiny water droplets in order to actually enhance viscosity and print quality. The challenge is for the ink maker and the press operator to meet the challenge of maintaining the optimum emulsification level. Excessive water pickup causes an ink to become irreparably waterlogged and too short to transfer well in the ink train and useless.

Too much water-in-ink emulsification can also lead to the pigment particles coagulating into clumps and piling on the ink train. Occasionally piling is due to a pigment or vehicle that is inappropriate for lithographic printing, as well as paper coating particles that transfer to the blanket. Because coated papers are less absorbent than uncoated sheets, there is less margin for error in **ink-water balance** when printing on a coated stock.

Poor ink-water balance can also result in an *ink-in-water emulsion*, which is undesirable to even a small degree. When ink infiltrates the dampening solution, it is carried to the nonimage areas of the plate and results in dirty printing called **tinting** (uniform contamination) or **scumming** (streaks or specks of contamination).

Because the image area of most lithographic plates is not more than a thin water-repellent coating, these plates are the most susceptible to being worn away either by an abrasive substrate or an ink whose pigment particles were inadequately ground during manufacture. Although special

Ink and Paper for Printing   311

plates, such as deep-etch and bimetal, are more resistant to wear, fineness-of-grind and uniform pigment dispersion are always important features in litho inks.

Lithographic ink formulations vary with (a) the method of drying that will be used; (b) the gloss level that is desired; and (c) whether the press that will be used is sheet-fed or web. There are five basic categories of sheet-fed inks. *Penetrating inks* dry largely by the substrate's absorption of the vehicle's oils; *oxidative inks* depend on their drying oils combining with oxygen in the pressroom; *quick-set inks* dry by a complex chemical reaction triggered by the selective absorption of only part of the vehicle; *radiation curing inks* do not dry, but instead cure, with ultraviolet, infrared, or electron beams; and the last category is comprised of *gloss inks* that employ special resins. Web litho inks are categorized as *penetrating inks* that dry when part of the solvent is absorbed by the substrate; *radiation curing inks* that harden when bombarded with certain ultraviolet light or electron beams; *heatset inks* that dry when their volatile solvents evaporate; and *thermal curing inks* that use heat to initiate a **crosslinking** reaction within the vehicle (see Appendix E).

Although lithographic printing depends on ink and water not mixing, a *water-washable offset ink* was introduced commercially in the middle 1990s. Using a vegetable-based formula, the ink repels water during printing, but can be made water soluble during washup. A major advantage of this ink formula is its solvent-free nature, which allows printers to lower the volume of *volatile organic compounds* (VOCs) given off by drying ink. Standards for VOC emission levels are established and monitored by the Environmental Protection Agency.

Types of Lithographic Presses

Sheet-fed presses print on individual sheets of paper. They can range from a single printing unit, which applies a single color to one side of each sheet, to multicolor presses that utilize more than one printing unit and can print two, four, five, or six colors to one side (see Figure 14–16). An advantage of a multicolor press in lithography is that all the images on one side of the sheet can be printed within a couple of seconds, thereby preventing sheet expansion due to moisture gain between printings. When the sheets' dimensions do not grow between printings, holding different color images in close register is made simpler.

Potential problems with multicolor presses revolve around the need to apply an ink film that will immediately receive

a

Figure 14–16 The wide range of sheet-fed lithographic presses that are available is represented by a one-color duplicator press (a) (*Courtesy of A. B. Dick Company*) and a six-color perfector press (b) (*Courtesy of MAN Roland Inc.*).

b

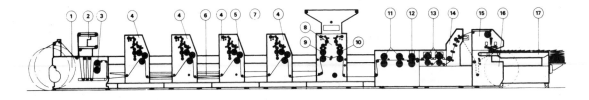

1 50″ unwind
2 Web guide
3 Web infeed system
4 Wet offset tower
5 Wet or dry offset printing insert
6 Turnbars (standard or air loaded)
7 Drive shaft
8 Double numbering (reversible)
9 Imprint insert

10 Numbering insert
11 File punch
12 Marginal punch
13 Cross perforator
14 Vertical slitter/perforator
15 Spiral folder
16 Rewind unit
17 Folder delivery table

Figure 14–17 A web press prints from a continuous roll of paper. This lithographic web press contains special features for printing business forms. *Courtesy of Müller-Martini Corporation.*

one or more other ink films on top of it. Placing one ink film over another is called *trapping,* and *wet trapping* is required when printing with multicolor presses. As explained in Chapter 12, wet trapping depends on using a first-down ink with a tack level that is higher than the subsequent ink's tack level. Oddly enough, none of the inks on a multicolor press should set before the sheet has passed through all of the printing units because a wet ink film actually has more tack than it will after setting.

When an ink film is applied to paper, some of the vehicle penetrates the surface and triggers the *setting* process. If the absorption of the surface is uneven, certain portions of the ink film will set and lose their tack faster than other portions. When a first or second color's ink film contacts the blankets of subsequent printing units, the subsequent colors should split and trap onto the ink that has already been printed. However, if some portions of this preceding ink film have already set, they will likely find greater adhesion with the ink film being applied over it than to the substrate. If this occurs, the already-applied ink film actually will leave the stock and trap back onto the subsequent blanket. This phenomenon is known as **backtrapping.** Backtrapping usually results from irregular absorbency across the surface of the stock, with the result that it produces spotty areas of non-uniform color. Any speckled pattern of varying ink density where uniformity should appear is called **mottle** and when it results from backtrapping, it is called *backtrap mottle.* Back-trap mottle can be caused by ink that sets too quickly, paper that is too absorbent, or paper that is inconsistently absorbent across its surface.

Mottle in both color and gloss can also develop when only one color is printed if there is enough variation in the sheet's absorbency. In this case, the ink film does not leave the

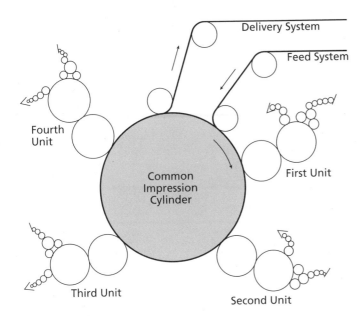

Figure 14–18 A common impression cylinder allows more than one blanket to use the same impression cylinder, thereby holding the substrate in place until it has contacted all of the printing units. When films and other stretchable materials are being printed, reduced substrate distortion and misregister results with the CIC configuration instead of the more commonly found blanket-to-blanket configuration.

stock, it simply sets and dries differently across an image area and these differences are reflected in its final appearance. If a printer is faced with printing on a stock that has a history of mottling, an ink can be formulated with an extra high viscosity to even-out its penetration into the paper.

Web-fed (or *job web*) presses print on a continuous roll of paper or other substrate instead of on individual sheets (see Figure 14–17). The primary advantage of web over sheet-fed presses is speed. By not having to feed, print, and deliver a sheet at a time, web presses can attain speeds in excess of 50,000 impressions an hour. When web presses are used in lithography, several designs are possible. One design is the *in-line* press, which resembles a multicolor sheet-feed press in that it consists of a series of printing units in the normal plate-blanket-impression cylinder configuration. The in-line configuration can be configured for two-sided printing if the just-printed and dried web is sent through a turning bar that rolls the ribbon of paper over before it enters another set of printing units. Presses that are capable of printing on both sides of the sheet or web are called **perfecting** presses.

The *common impression cylinder,* as its name suggests, uses a single extra large impression cylinder in conjunction with multiple plate-blanket units (see Figure 14–18). The common impression cylinder press is able to hold tighter register because the web does not pass from one unit to another, but is held by the same impression cylinder until that side of the web is printed. If two such units are linked, both sides of the

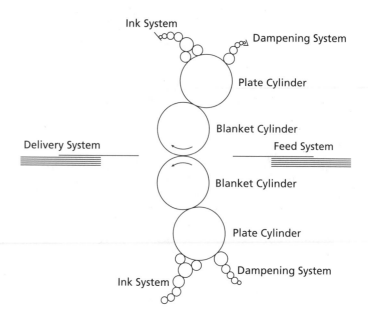

Ink System

Dampening System

Plate Cylinder

Blanket Cylinder

Delivery System

Feed System

Blanket Cylinder

Plate Cylinder

Dampening System

Ink System

Figure 14–19 Perfecting presses print on both sides of the sheet and a blanket-to-blanket. Both sheet-fed and web-fed presses can be configured in this manner.

web can be printed in one pass. As was explained earlier in the chapter, the common impression cylinder configuration is also used in flexography for printing on plastic films that would stretch between printing units and defy close register printing if sent from one printing unit to another. The final design is a perfecting press which has no impression cylinder at all, but uses a *blanket-to-blanket* configuration so that it can print on both sides of the sheet simultaneously (see Figure 14–19). Although blanket-to-blanket presses are very common in the web offset sector of the industry, they have a tendency to cause blistering on some coated stocks. Because they apply an ink film to both sides of the substrate before it enters the driers, the moisture within the stock can get so hot that it vaporizes and, before it can escape through the pores of the coating, explodes within the sheet (refer back to Figure 14–15). The threat of blistering reinforces the need for even a gloss-coated sheet to have adequate porosity.

Waterless lithography or *dryography* is a new but growing method of printing that allows a modified lithographic press to print with no dampening solution. The key to the waterless technique is a plate that consists of two layers. Image areas consist of a photopolymer layer and nonimage areas are represented by a slightly higher silicone layer (see Figure 14–20). Instead of the presence of dampening solution preventing the transfer of ink to the nonimage areas, the chemical nature of the silicone resists the ink and keeps ap-

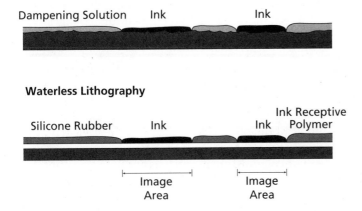

Conventional Lithography

Dampening Solution   Ink   Ink

Waterless Lithography

                                      Ink Receptive
Silicone Rubber   Ink   Ink   Polymer

Image   Image
Area   Area

Figure 14–20 Waterless lithography uses a plate that consists of a silicone (nonimage) layer over a photopolymer layer (image area). The silicone has a low surface energy level that resists the transfer of an ink with an adequately high viscosity.

propriate portions of the plate clean. This principle is called *differential adhesion.*

The absence of a dampening solution immediately eliminates several production problems such as finding the appropriate ink/water balance, optimum levels for emulsification and dampening solution pH, as well as dimensional stability and curl problems resulting from the sheets' absorption of water. As a result, fewer makeready sheets and minutes are needed. Relative to print quality, the absence of dampening solution produces roughly one-third to one-half of the dot gain experienced with conventional lithography, thereby allowing greater resolution and the capability of using much finer screens.

The key to waterless printing is ensuring that the ink film is more attracted to itself than it is to the silicone of the nonimage areas of the plate. To perform well in this process, inks must be formulated with special resins and other additives to produce higher viscosities than are found in conventional lithographic inks. Because an ink's viscosity is affected by temperature, waterless lithographic presses must carefully regulate the temperature of the ink and control the tendency of ink to lose viscosity from friction-generated heat in the ink train. In conventional lithography, the dampening solution provides a cooling effect. When an ink gets too warm and loses viscosity, it will tend to transfer to the silicone of the nonimage areas and produce *dry scumming.* When the ink's temperature is too low, viscosity and tack become excessive, thereby inviting problems with picking.

To regulate the temperature of the ink on the plate, infrared sensors monitor the ink's temperature on the plate or

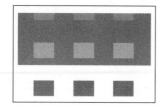

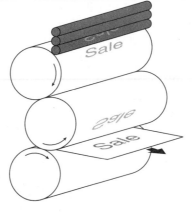

Direction of travel
through the press

Figure 14–21 A mechanical ghost can appear as one or more light images within heavily inked areas. These are sometimes aptly referred to as ink starvation ghosts. The ghost image becomes less distinct farther on the press sheet from the original image.

form roller. Based on this reading, either cool or warm water is pumped through hollow vibrator rollers to adjust the ink's temperature. Waterless lithographic ink, therefore, must be made to respond predictably to temperature changes.

Potential Problems with Lithography

Ghosting is the appearance of a faint and unintended image within a heavily inked or solid printed area. The unintended or ghost image can be found on either another part of the same plate that carried the background area or on a plate used to print the other side of the press sheet, depending on whether the ghosting is mechanical or chemical in nature.

Mechanical ghosting is much easier to both comprehend and remedy, so it will be discussed first. Mechanical ghosting is sometimes called *press ghosting* and occurs in both sheet-fed and web presses, but exclusively in lithographic printing. The ghost image on the left of Figure 14–21 typifies a mechanical ghost. Note that the subtle image of the letter is defined by that part of the rectangle being slightly lighter than the rest of it. The light area of the rectangle was created by an uneven application of ink; that is, the plate simply received less ink on one part of the image area than it did elsewhere. This phenomenon is termed *ink starvation* and it results from the form rollers being unable to sustain the plate's image areas' need for a uniform ink film. Uneven inking across the plate is transferred to the blanket and finally to the paper.

To better understand this type of ghosting, examine Figure 14–22. As the paper moves through the press, the portion of the blanket that carries the word SALE transfers ink onto the paper. On up the ink train, this ink loss from the corresponding area of the plate must be replaced by the form

Figure 14–22 The interplay between the ink train's form roller, plate cylinder, and blanket cylinder area is shown along with the printed sheet and ghosts.

318

rollers. However, recall from Figure 14–13 that the form rollers are much smaller in circumference than is the plate cylinder and if the ink loss on the form rollers is not fully replenished by the upper portion of the ink train, the starved image area will repeat itself across any heavily inked area on the remainder of the plate, blanket, and paper. Each repetition of the word SALE represents a revolution of the form roller. Notice that the ghost images become less obvious as the depleted image areas on the form rollers start to catch up with the remainder of the rollers' surfaces. Unfortunately, when the next impression is made, the process will begin all over again.

Presses with larger ink trains and three or four form rollers are less likely to generate mechanical ghosts. Eliminating ghosts involves several techniques. In the examples shown in Figures 14–21 and 14–22, the problem could be remedied by merely mounting the plate backwards so that the large rectangular image area is printed first. A thoughtful imposition during prepress can prevent much ghosting. When the graphic design thwarts this kind of preventive action, pressroom measures include reducing the dampening solution, increasing the ink film thickness, attaining maximum color strength, and switching the job to a press with a larger ink train.

Ghosts can also be darker than their backgrounds, as is apparent on the press sheet shown in Figure 14–23. By now, the reader can probably assess the problem. The form rollers are not giving up ink to the rectangular image area. For this reason, each revolution of the form rollers replenishes their depleted areas and adds ink to the circle area that never lost any in the first place, thereby creating an excessive ink film that gets transferred to the plate. In the example shown in Figure 14–23, turning the plate around would remedy the problem, but some designs require one or more of the more elaborate "ghostbusting" techniques just mentioned.

Chemical ghosts are a more formidable opponent than are the mechanical variety. The good news is that chemical ghosting does not occur with one-sided jobs or when printing with a web press; the bad news is that the sheet-fed printer is never aware of chemical ghosting until at least four hours after the sheets have been printed. The reason for the delayed appearance is that the effect is produced when vapors from a drying ink film affect the drying of ink on the second side of the sheet. The resulting undesired image is not lighter or darker than the surrounding ink, but appears either more dull or glossy because it reflects light differently.

Figure 14–23 When an image is reversed out of a large, heavily inked area, the corresponding image area on the form roller does not give up ink to the blanket and the resultant buildup can create a dark ghost.

Actual file

w/ "MOM" as black
(for proofing)

Figure 14–24 To get some idea of the result of chemical ghosting, imagine a job in which MOM is printed on one side and a solid rectangle is printed on the other. The faint MOM represents an area in the rectangle that is more glossy than the remainder of the rectangle. This chemical ghost resulted from uneven drying rates across the solid rectangle. The word MOM was printed first on side one and its drying caused a chemical contamination of side two which, in turn, later received the rectangle.

Figure 14–24 provides a simulation of chemical ghosting that resulted when the first side of the sheet was printed with the word MOM. Because the job was large in volume, the printed sheets sat in a tall stack while they dried in preparation for being sent through the press for a second time to print the second side. As the drying ink film oxidized, heat was generated by the chemical reaction, which produced escaping vapors of the ink's drier, but the weight of the high pile of press sheets forced out most of the air that would ordinarily be between them to reduce the concentration of these fumes. With nowhere to go, the vapors built up in concentration and penetrated the bottom side of the sheets above them (see Figure 14–25). Because the vapors were produced by the ink in the word MOM, only areas directly above this image were chemically altered. When the pile of

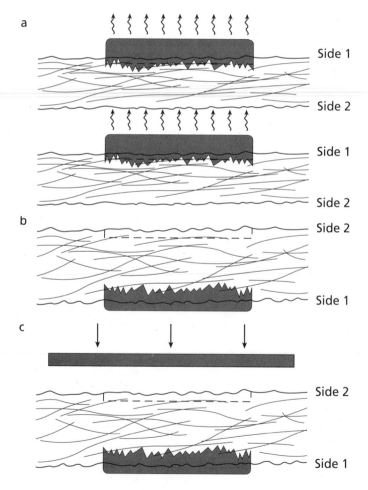

Figure 14–25 After the first side of a two-sided job has been printed (a) vapors from the drying ink evaporate and move up into the surface of the second side. The pile of paper is flipped over to print the second side (b); however, the dotted line represents the area of the paper already chemically contaminated by the first side image. Some of a large solid image laid down on this portion of the second side will fall onto this contaminated area (c) and may dry either faster or slower than the remainder of the image area.

320

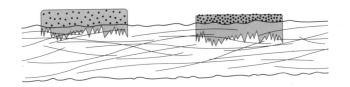

Figure 14–26 The ink film on the left dried quickly enough to leave plenty of vehicle around the pigment particles. The smooth surface that results will reflect light evenly and create gloss. The longer drying time for the other ink film allowed excessive penetration of the vehicle into the paper, leaving some pigment particles to protrude above the film surface, scatter light, and appear more dull.

paper was finally turned over and sent through the press for the second pass, the large solid rectangle was laid down with only a portion of that ink film over the vapor-contaminated areas of the surface. Although a uniform ink film was applied during the second pass, this ink did not dry at a uniform rate. Ink applied to the contaminated surface areas dried more quickly, with the result that less of the ink's vehicle drained into the surface. Instead, it stayed at the top of the ink film to create a smooth, glossy surface.

Like the surface of a sheet of paper, high microsmoothness in an ink film generates specular reflection, which is needed for gloss. The difference in the drying rates caused a portion of the large rectangle's ink film to stay soft longer, lose more vehicle, and dry with a less uniform surface (see Figure 14–26). Because lack of oxygen during the drying of the first-side image is the primary cause of chemical ghosting, the most effective prevention technique is to break up the delivery pile into shorter stacks that can be winded or aired out to replenish the supply of oxygen between the sheets and dissipate the gases. Where appropriate, the printing sequence should permit the large area to be printed first.

Chemical ghosting is more of a problem with coated sheets than uncoated sheets and if a particular job appears to be a likely candidate, the ink maker can formulate a vehicle with less dependency on oxidation for its drying.

Attaining Ink Gloss

The key to obtaining a glossy ink film is to achieve a uniform surface. Gloss inks contain resins that dry without leaving pigment particles uncovered. However, no ink formulation can produce high gloss if the substrate itself is not smooth.

Gloss can be enhanced through the application of an *overprint varnish*. Often, a five-color press will carry the four process colors and a varnish to wet trap over the colors. An oil-based overprint varnish is basically an ink that contains no pigment material.

If a water-based varnish is desired, it is fed onto the plate by the dampening system and the ink train of that printing unit is disconnected. If resistance to scuffing, water, and many solvents is desired, a UV-cured varnish can be wet-trapped over the ink—assuming that a UV drying system is present.

Heat-Set Inks

Because high-speed printing depends upon high-speed ink drying, several approaches have been used to accelerate the process. **Heat-set inks** are used on presses that apply heat to the fresh ink film to drive off the liquid and then chill and solidify the remaining pigment-in-resin component. A potential problem with heat-set printing is that the heat can drive enough moisture from the paper itself to make it more brittle and lower its fold strength.

Infrared-Setting Inks

Even less drying time can be achieved with the use of **infrared inks,** which are used on presses that radiate infrared rays onto the printed image to accelerate the polymerization process and also hasten evaporation. Infrared drying units will speed the setting of any ink that dries through the oxidation/polymerization process, but they are most effective when used with inks formulated for an infrared application. Infrared drying units can be found on both sheet-fed and web-fed presses. As is the case with conventional heat-set drying, the infrared radiation can also heat the paper to the point of driving out too much moisture.

A major concern for heat-set printers is staying within compliance on federal, state, and local guidelines for the emission of volatile organic compounds (VOCs) that are given off when heat-set inks evaporate.

Heat-set printers that generate enough VOCs to warrant regulation must generally install equipment that provides 90–95 percent of VOC removal or destruction. Cooler condenser/filter systems and incineration systems are available but their purchase price can range from $70,000 to $900,000 and can cost $30,000 in annual operating expenses. Due to the high cost of installing control systems that eliminate nearly all of VOC emissions, many printers have chosen to take a different approach and use inks with lower VOC content. Several options are available, including high-solids inks, curing inks, and water-based inks.

High-Solids Inks

One approach to reducing VOC emissions is to use **high-solids inks** with as little as one-third the VOC content of conventional heat-set inks. In many regions, the switch to high-solids inks is insufficient because even greatly reduced VOC levels may still exceed the local environmental limits.

Curing Inks

Printers that traditionally needed highly volatile inks to print at high press speeds or on nonporous substrates have found that they can use inks that have virtually no solvents to evaporate. These inks do not dry in the sense of losing liquids to either the substrate (penetration) or the air (evaporation). Instead, they move from a liquid to a solid state in a fraction of a second when the ink's individual molecules (or *monomers*) respond to a stimulus by immediately locking together or crosslink to form a firm, solid film. Because these inks do not dry in the conventional sense, they are said to cure and are termed **curing inks.**

One method of starting the chemical reaction that cures inks is to expose the freshly-printed ink film to high levels of *ultraviolet* (UV) light. Upon bombardment of UV light, special chemicals in the ink called *photoinitiators* respond to the light energy by triggering the curing. These inks are called *ultraviolet-cured* or UV-cured inks and they are used in lithographic, letterpress, screen, and flexographic printing.

Printing with UV-cured inks eliminates the concept of drying time and allows postpress operations to occur immediately or in-line. Other advantages include the attainment of very high gloss levels, as well as the elimination of (a) the odors resulting from the chemistry of conventional drying; (b) problems of the ink setting up on the press during downtime; and (c) VOC emissions.

In addition to inks, UV-curing varnishes are also available and can be especially well-suited to certain printing and end-use applications because they are resistant to scuffing, water, and many solvents.

Another method of curing inks is to bombard the freshly-printed image with *electron beams* (EB) instead of UV light. EB-curing and UV-curing inks are very similar with the key exception that EB-curing does not require photoinitiators, thereby slightly lowering the cost and avoiding any potential problems that could arise from the presence of unreacted photoinitiators in the finished ink film. Also, because electrons are not absorbed by pigments, as are light particles, EB-curing ensures a more thorough penetration of the entire ink film, which reduces the likelihood of odors caused by uncured chemicals. For this reason, EB-cured inks are commonly used on food packaging. Electron-beam curing inks are used in flexography and lithography and are limited to web presses because the metal grippers of sheet-fed presses are incompatible with electron beams.

Water-Based Inks

Still another way of eliminating VOC emissions is to replace solvent-based (oil-based) inks with **water-based inks.** Ordinarily, ink resins (binders) were originally solids that were dissolved by the solvent. Upon drying, the solvent evaporates and leaves behind the pigment and resin to form the dried film. With water-based inks, the resins are dissolved in water, but they must dry to a nonwater soluble film upon the substrate.

In addition to their lack of VOCs, water-based inks are slower to evaporate than solvent inks, and therefore are less likely to evaporate on the ink train and create a shift in viscosity. Also, they present a greatly reduced fire hazard.

Water-based inks are used in gravure and screen printing but are especially popular in flexographic printing because they are more compatible with the photopolymer plates than are solvent-based inks that can attack the material.

Questions for Study

1. What factors determine the most appropriate printing process for a particular job?
2. Distinguish between moisture-set and water-washable letterpress inks.
3. What is the role of the anilox roller in flexography?
4. What mechanism causes flexographic inks to dry?
5. What causes pinholing in flexographic printing?
6. Distiguish among these terms: intaglio, engraving, and gravure.
7. Describe the three types of cell patterns by which gravure cylinders can control the amount of ink deposited onto a given area of the substrate.
8. Why is gravure the most cost effective printing process for long-run jobs?
9. What is the function of an electrostatic assist during gravure printing?
10. What drying mechanism is used in gravure printing?
11. What is the cause of pinholing in gravure printing?
12. Why is screen printing able to print on a wide array of objects?
13. Why is it important that screen inks are "short"?
14. Why is it important that lithographic and letterpress inks are "long"?

15. Why is lithography not considered a physical method of printing?
16. Describe the function of the dampening system of a lithographic press. Use terms such as distributing roller, form roller, and ductor roller.
17. Describe the function of the ink train of a lithographic press. Use terms such as fountain roller, distributing roller, ductor roller, and doctor blade.
18. Describe backtrapping in multicolor printing.
19. What is an advantage of a common impression cylinder press?
20. What causes mechanical ghosting? What can be done to prevent or correct it?
21. What causes chemical ghosting? What makes it so difficult to remedy? What can be done to reduce the frequency of its occurrence?
22. Distinguish between ink drying and ink curing.
23. Describe the role played in ink drying by infrared rays.
24. Why are many printers concerned about volatile organic compounds in their inks?
25. How can high-solids, UV-curing, and EB-curing inks reduce VOC emissions?
26. What are photoinitiators and in what inks are they found?
27. List three advantages that water-based inks bring to flexography.

Key Words

letterpress
let down
intaglio
doctor blade
sedimentation
emulsion
scumming
backtrapping
waterless lithography
fill-in
inking system
heat-set inks

infrared inks
curing inks
flexography
feathering
engraving
snowflakes
lithography
ink-water balance
mottle
mechanical ghosting
anilox roller
mechanical pinholing

gravure
electrostatic assist
dampening system
tinting
crosslinking
web-fed press
chemical ghosting
sheet-fed press
perfecting press
high-solids inks
water-based inks

15 Estimating and Ordering Ink

AFTER STUDYING THIS CHAPTER, THE STUDENT SHOULD BE ABLE TO:

■ Describe the concept of ink mileage and explain the influence of each factor that contributes to it.

■ Explain why ink wastage cannot be avoided completely.

■ Calculate the weighted average mileage for two or more inks with different ink mileage.

■ Define the safety percentage that is allowed for ink and explain why the safety percentage varies with the nature of the ink and press size.

■ Describe the types of containers in which printing ink can be purchased.

■ Describe the situations for which custom-mixed inks are appropriate.

■ List the information that should accompany an ink order to ensure that the ink manufacturer will select the optimum formulation.

When a potential printing customer requests an estimate for producing a job, the printer will dutifully predict the cost of the paper, plates, and appropriate operations such as camera, press, and bindery work. Too often, the ink cost is not considered because it is thought of as negligible. This attitude can exact a heavy toll when the printer gets stuck with an unexpected $200 ink bill that erodes what was intended to be the job's profit. Although ink is usually a small percentage of the typical printing job, certain jobs—such as those with a high coverage of an expensive ink on an especially absorbent stock—are very much the exception and require that the printer be able to estimate ink costs when appropriate.

Any procedure to predict the amount of ink that a given job will require must include the key variables that affect ink consumption. Many printers use very simplistic formulas that are quick and easy, but not very effective at avoiding large amounts of leftover ink or—worse yet—preventing press downtime due to a prematurely empty ink fountain. Accuracy comes from not overlooking variables that distinguish one printing job from another.

Ink Consumption Factors

Even the most simplistic ink estimating systems recognize that the **area of coverage** is a factor in the amount of ink a job will require. Clearly, a design that calls for ink to cover 80 percent of the area will require around four times as much ink as one with only 20 percent coverage. The trick is to attach a percentage figure to a custom-supplied design. Most beginners overestimate the ink coverage of type. Although a text block may comprise one-half of a design, most of that text block is composed of nonimage areas between and within the letters. Also, the average halftone photograph consists of only 40 percent coverage.

Ink Film Thickness

Less obvious factors that affect ink consumption influence the *thickness of the ink film* that the press operator will have to apply in order to attain a good image. Several factors influence ink film thickness and one of these is the *printing process* itself. The screen process applies the thickest ink film. Even a thin screen-printed ink film of .0012-inch is over twice as thick as is typical in gravure printing, six times thicker than in lithography, and 15 times thicker than a lithographic ink film.

The ink's *covering power* is a second factor in ink film thickness because lighter colored inks will probably require a thicker ink film than will black and other dark colors. In an effort to improve their covering power, lighter ink colors usually carry a higher proportion of pigment material than exists in darker colors with more innate coverage power. Because pigment materials weigh more than the same volume of vehicle, inks with high pigment content have a higher specific gravity or density. Inks with high specific gravities occupy a proportionally small volume and do not "go as far" on the press. Appendix B explains specific gravity in more detail.

Even when the ink color remains constant, the *pigment-to-ink ratio* can vary significantly from one brand of ink to another. Because pigment materials represent comparatively expensive components in a can of ink, reducing the percentage of pigment is one way to lower the cost of producing a given ink. An ink with half the normal proportion of pigment would require twice the normal application in order to attain the needed coverage power.

Even when the ink brand and color are the same, changes in the *paper stock* can mean a surface with different levels of absorbency. Sheets with high absorbency require the application of a heavier ink film to compensate for the ink that will be lost to the sheet's interior (see Figure 15–1). Therefore, the ink's color, specific gravity, and percentage of pig-

Figure 15–1 Two ink films of the same thickness have been applied to sheets with different absorbency levels. The sheet on the left allows only modest penetration and "holds out" most of the ink atop the surface. The more absorbent sheet on the right allows much more penetration and most of the ink is lost to the sheet's interior. Therefore, to attain adequate ink density, heavier ink films must be applied to absorbent substrates to compensate for the volume of ink that leaves the surface.

ment content, along with the substrate's absorbency influence ink film thickness, which, in turn, affects ink *mileage*. Mileage is measured in terms of square inches of surface that can be covered by one pound of an ink.

Ink Wastage

After the last sheet has moved through the press, the normal ink film will remain on the ink train's rollers as well as the plate and blanket. Unless the next scheduled job needs the same ink color, this amount of **ink wastage** will need to be washed from the ink train. The total amount of ink involved will range from one-quarter pound to a whole pound, as determined by the number of rollers and their width. Any prediction of ink requirements will need to include ink wastage to avoid running out of ink before the end of the run.

Still another variable is the fact that one press operator will lay down a heavier ink film than will another, thereby requiring more ink to run the same job. To avoid running out of ink due to subtle ink film variations during the press run or when the makeready becomes troublesome, an appropriate *safety percentage* is added to what would be needed under normal circumstances.

Factors that determine the appropriate **safety percentage** are the *size of the press* and whether the ink color is a *stock* (right out of the can), *mixed,* or *custom-mixed color* (supplied in the desired color by the ink maker). The threat of running out of ink in the middle of a job is greatest with custom-mixed inks because of the potential delay in getting the new ink from an outside source. If the ink runs out during the second or third shift or on the weekend, the lost time could be catastrophic. The press size is a factor because if a safety percentage (for example, 8 percent) that may be appropriate for a short-run job being printed on a small press were used for a long-run job on a large press, an excessive amount of extra ink would probably be generated. After all, 8 percent of 1.2 pounds is less than a tenth of a pound, while 8 percent of 13 pounds is over a pound.

Ink Mileage

An ink's **mileage** is the number of square inches that can be covered by 1 pound and it can be determined by two methods. First, a printer can use data provided on an ink mileage chart that resembles Table 15–1. These charts are available from ink suppliers, but many printers report that those ink mileage figures are as difficult to attain in real life as are the

	Gloss Coated	Dull Coated	Machine Finish	Smooth Offset	Vellum/ Antique	Newsprint
Black	394	322	305	228	182	192
Rubber-based Black	412	338	320	239	191	202
Process Black	392	322	294	224	193	203
Process Cyan	361	323	298	198	173	182
Process Magenta	347	292	251	186	157	165
Process Yellow	337	279	254	188	164	173
Yellow	259	204	198	148	124	130
Purple	358	288	276	205	172	181
Persian Orange	285	264	276	228	191	201
Warm Red	272	242	234	166	150	158
Rhodamine Red	324	284	246	173	145	152
Rubine Red	272	242	234	165	150	159
Reflex Blue	353	282	276	192	161	170
Overprint Varnish	420	344	314	242	204	214

Table 15–1 An ink mileage chart contains the average number of square inches *in thousands* that a pound of a given ink will cover on a given paper surface. For example, a pound of purple ink will cover 358,000 square inches on gloss coated stock, but only 181,000 square inches on newsprint.

gasoline mileage figures supplied by automobile manufacturers. The most accurate technique is for each printer to keep accurate records of stock colors' performance on different types of paper. Although the data in Table 15–1 may be accurate for some printers, differences in presses, operator techniques, and ink formulas preclude a universal mileage schedule.

When a mixed color is being printed, the mileage of the component inks should be averaged in the same proportion as the formula for the desired ink. For example, if an ink consisting of 12 parts reflex blue and 4 parts yellow were being printed on a smooth offset sheet, the *weighted averaging* of the two mileages would proceed in this fashion:

$$
\begin{aligned}
12 \text{ parts} \times 192,000 &= 2,304,000 \\
+\ 4 \text{ parts} \times 148,000 &= +592,000 \\
\hline
16 \text{ total parts} & \quad 2,896,000
\end{aligned}
$$

$$
\frac{2,896,000}{16} = 181,000 \text{ square inches of coverage per pound}
$$

The same procedure would be used to arrive at the mixed ink's cost per pound. Once the mileage and cost are known, the following formula is used:

$$\frac{\substack{\text{square inches} \\ \text{of substrate per copy}} \times \substack{\text{percentage of} \\ \text{ink coverage}} \times \substack{\text{number} \\ \text{of copies}}}{\text{average mileage per pound}} = \text{net pounds (np)}$$

$$(\text{net pounds} + \text{wastage quantity} + \text{safety factor}) \times \text{price per pound} = \text{cost of ink}$$

The **safety factor** is a percentage of the pounds of ink required to cover the substrate during normal conditions or, phrased differently, a percentage of the *net pounds* (np).

The following problem will demonstrate the formula's use. Calculate the cost of ink to print 18,000 copies of a 12 × 19-inch poster with a 30 percent coverage of reflex blue on a dull-coated sheet. The ink train of the lithographic press will require a half pound and a safety factor of 2 percent will be used.

$$\overbrace{}^{\text{net pounds}}$$

$(12 \times 19 \times .30 \times 18,000/282,000) + .5 + (.02\text{np})$
$(1,231,200/282,000) + .5 + (.02 \times \text{np})$
$4.366 + .5 + (.02 \times 6.89)$
$4.366 + .5 + .138$
5.003 or 5 pounds

Ink Packaging

The small-town printer who usually produces several short run jobs in a single day and the giant commercial plant with large multicolor web presses all order ink, but certainly not in the same volumes. Correspondingly, the ink industry has several ways of packaging ink for the small, medium, and large volume customer.

Small-to-medium printers usually purchase ink in **cans,** which can be ordered in the following weights: ¼, ½, 1, 2, 5, and 10 pounds. The 5-pound can is the most popular size. Because various inks have different specific gravities, 1 pound of black ink may occupy nearly twice the volume of 1 pound of a yellow ink. For this reason there are several sizes of cans for each of the available weights just listed. Also, the actual weight of a 1 pound can of ink may be slightly greater or less than 1 pound. In an effort to attain a better seal between the ink and the coated paper that is placed over it before the lid is placed on, ink makers concern themselves more with filling the can than hitting the labeled weight precisely. A good seal between the top of the ink and its covering will reduce oxidation and extend the ink's working life.

For larger volume jobs, a **kit** is the common container. Despite its name, a kit is merely a covered bucket and its capacity is measured in gallons instead of pounds. Available sizes include 2, 3, 3-½, 5, and 6 gallons. The liquid inks used in gravure, flexographic, and screen printing are often ordered in 30 and 55-gallon **drums** when the usage justifies such volume. A hand pump can be attached to the drum so that small or larger amounts can be removed as needed for individual jobs. Some very large volume gravure and flexographic plants actually receive their ink from tanker trucks or railroad tank cars, which is pumped into storage tanks and fed to the presses as needed.

Ink Pricing

Ink is priced by the pound and the *price per pound* of the same ink maker can vary widely due to three factors. First, some inks' ingredients are more expensive to purchase than others. The greatest variance lies with the *pigment material cost* because certain pigments are comparatively rare. As a result, purple and some red inks can cost over twice as much per pound as a black ink.

A second factor is the type of *packaging*. Like paper, an ink's price per pound is lower when packaging costs are reduced. For example, ink will cost less if it comes in one 10-pound can instead of ten 1-pound cans. The final factor is the *size* of the order. Inks' price per pound (also like that of paper) is lower with larger orders. Common price thresholds are 5, 20, 30, 60, and 100-pound quantities. Because ink has a shelf life, the prudent printer needs to avoid the false economy of purchasing a greater volume of ink than can be used in time to avoid its oxidation.

Custom-Mixed Inks

When an ink must be matched precisely to a customer-supplied standard or contain certain characteristics, the printer may order a custom ink instead of a stock ink. In this instance, the ink maker will formulate and mix the required amount and package it for the printer. The cost for handling an individual order ranges between $30 and $80. If the order is large, like 50 pounds, the addition charge for customizing the ink may be only 15 or 20 percent more. However, for an order of only 3 pounds, the custom charge could nearly double the cost, thereby prompting some printers to avoid custom orders wherever possible. In the case of mixing inks to achieve a color match, the actual mixing is often done by the press operator, often resulting in another false economy when one considers the labor costs involved in the actual mixing procedure and press time tied up in the likely modification required before achieving a true match. Although the charge for supplying the printer with a custom-matched ink seems high, it may be a bargain compared to the printer's actual cost of doing it in-house.

Ordering Ink

Continuing with the economics of ink purchasing, it should be remembered that the potential savings that can be had by purchasing an inexpensive ink are tiny when compared to the costs of poor performance in the pressroom. Poor mileage, slow drying, inadequate color strength, picking, and mottling represent some of the ways that a so-called bargain ink can end up costing the printer 100 times more money than the amount that he had hoped to save. Therefore, printers should choose a brand of ink on the basis of performance rather than purchase price. The price-per-pound savings are simply too small when compared to the risk in wasted ink, paper, and press time.

When ordering ink, the printer can avoid the problems just listed by being specific about the nature of the job for which the ink is intended. Information that should be included with any ink order includes the following:

1. the printing process (lithography, letterpress);
2. type of press (sheet-fed, web);
3. drying system (heat-set, UV, etc.);
4. type of substrate (gloss paper, vinyl);
5. color (Pantone number or supplied sample);
6. sequence for printing the color (cyan, magenta, yellow, black);
7. drying time requirements;
8. end-use requirements (light-fastness, rub resistance);
9. converting or post-press processing requirements (laminating, heat-sealing); and
10. desired finish (gloss, matte).

Questions for Study

1. How does an ink's pigment-to-ink ratio affect ink mileage?
2. What is the role of the safety factor in calculating ink usage?
3. What factors influence the proper safety factor for a given ink calculation?

4. Why does a custom ink require a larger safety factor than a stock ink?
5. Why bother to calculate weighted averages for mixed inks?
6. Referring to Table 15–1, calculate the weighted average mileage for a mixed ink with 4 parts warm red and 12 parts yellow printed on gloss-coated stock.
7. What four components comprise the net pounds in an ink's calculation?
8. Calculate the ink cost for printing 20,000 19 × 25-inch posters on gloss enamel stock (one-sided). The color is Pantone 259, which has a formula of 12 parts Pantone Rubine Red and 4 parts Pantone Process Blue. The ink coverage is 35 percent; the ink train of the press requires one-half pound of ink; and a 3 percent safety margin will be used for this mixed ink.
9. Calculate the ink cost for printing 80,000 two-sided cards that measure 5 ½ × 8 ½ inches. They will be printed on smooth finish offset stock. On the front, the cards will have 30 percent coverage of black and 20 percent coverage of green. The back will have a 15 percent coverage of black.
10. Distinguish among cans, kits, and drums as containers for ink.
11. Discuss the cost effectiveness of having an ink supplier custom-mix an ink instead of the printer performing the mixing in-house.

Key Words

area of coverage	ink wastage	kit
can	drum	safety percentage
mileage	safety factor	

16 Paper, Ink, and the Environment

No enterprise as omnipresent as the forest-product, paper, and ink industries can function in the twentieth century without having to strike a balance between meeting consumers' demands for low-cost products that have desirable attributes and being careful not to harm the environment. Consumers want an ample supply of inexpensive timber, paper, and other forest products. Printers' customers want their paper to be bright and their inks to be available in the full range of colors. The printers themselves want inks that print well and dry quickly and they also want to be able to conveniently dispose of ink that is left over from a job. In the past, all of these desires were much more likely to be realized than is presently the case because each of these outcomes is inextricably linked to some environmental concern.

The protection from logging of some of the Pacific Northwest's most valuable timber by the Endangered Species Act clearly impacts on the supply and, therefore, the cost of many forest products. The ability to inexpensively create pulp with high brightness levels is affected by regulations on the chemicals used in bleaching. When the most available ingredient for a given color pigment is a heavy metal that is declared a potential hazard, the ability to create certain hues is rendered more difficult. Printers who end up with 2 pounds of ink leftover from a job may suddenly be faced with laws that govern the handling and disposal of hazardous wastes.

For most of America's history, much of the nation's industries functioned with comparatively little regard for their effects on the environment. Much of this apparent recklessness was actually due to an ignorance of the subtle and complex ecological linkages among various plant and animal species. Today, environmental regulations abound and continue to increase in number. One reason for this surge is that many compounds once thought to be benign are now either known to be or suspected to be harmful. Also, today's evolving technology allows the detection of compounds in traces so slight as to have gone unnoticed ten years ago. The result of these technological breakthroughs and an expanding ecological awareness have created a more difficult world for people who work with paper or ink.

Logging in the National Forests

An ongoing controversy is raging between some environmental groups and the forest industry over the various uses of America's National Forests. At issue is the proportion of National Forest land that should be preserved as wilderness areas.

Not to be confused with the National Parks, the National Forests have been managed since 1905 by the U.S. Forest Service, which was charged by Congress "to furnish a continuing supply of timber for the use and necessities of the United States for the greatest good of the greatest number over the longest time." Toward this goal, over three-quarters of this land's approximately 190 million acres (see Figure 16–1) has been allocated for **multiple-uses,** including recreation and logging.

Recreation is a major use, as hundreds of thousands of Americans visit the National Forest annually to camp, hike, ride horseback, and simply drive through and view the scenery. Also, the National Forests contain most of the nation's ski areas including major resorts such as Jackson Hole, Breckenridge, Snowshoe, Stratton, Mammoth, Waterville, Big Sky, Purgatory, Snow Basin, and Timberland.

However, a primary purpose of the National Forest is also to provide an adequate *supply of timber* for the nation's needs. Every year since the late 1800s, timber rights to portions of the National Forests have been leased to logging operations and in recent history a year's cut has exceeded 12 billion board feet (see Figure 16–2 and Table 16–1). A third

National Forest Lands

Figure 16–1 The 187 million acres of National Forest land are divided among 152 forests throughout the country. One-third of America's softwood timber comes from National Forest land.

allocation of National Forest land goes toward preserving a natural forest setting. To this end, nearly 18 percent of National Forest acreage has been designated as **wilderness** forest areas and, accordingly, protected from logging or the creation of roads, trails, or other permanent "works of man." The existence of wilderness areas allows pristine forests to remain unspoiled by elements of contemporary society. However, a huge argument exists over the proportion of the National Forest that should be designated as wilderness. Some environmental groups such as the Forest Conservation Council and the Sierra Club argue for the protection of all remaining old-growth stands within the National Forests of the Pacific Northwest.

These groups point out that hundreds of years were required to establish these stands of magnificent Douglas fir. They argue that much of the currently protected wilderness areas are not old growth and that only half of the remaining old growth forests are protected. They further argue that because much of the protected old growth is fragmented, animal species that require old growth habitat are restricted to one area, which could produce inbreeding and other ecologically harmful consequences. Therefore, it is the position of the Sierra Club and other groups that all remaining old-

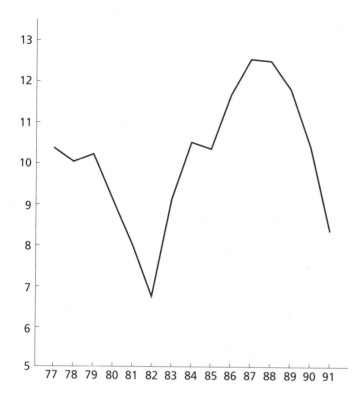

Figure 16–2 This chart shows the volume of timber (in billions of board feet) cut on National Forest land from 1977 to 1991. *Source: U.S. Forest Service.*

growth stands be left alone and that they be connected by protected corridors.

Timber vs. Owls

In an effort to halt logging in unprotected old-growth forests, environmental groups have turned to the Endangered Species Act, a 1973 federal law intended "to provide a means whereby the ecosystems upon which endangered species and threatened species depend may be conserved, to provide a program for the conservation of such endangered species and threatened species." The Act charges the Department of Interior's Fish and Wildlife Service with designating which species of plants and animals are endangered or threatened. Because species that are listed can have their habitats protected by law, environmental groups have claimed that the endangered spotted owl can survive in only the old-growth forests of the Pacific Northwest and they have sought to end timber harvesting within these areas of both public and private land. This argument has been met with research indicating that the spotted owl can easily

	1970	1980	1985	1990
Timber cut				
Volume in million board feet	11,527	9,178	10,941	10,500
Value in million dollars	$308	$730	$721	$1,188
Receipts for timber use				
Value in million dollars	$284	$625	$515	$850
Payments to local government				
Value in million dollars	$73	$240	$229	$368

Table 16–1 The volume and value of timber sold, through logging rights, by the National Forest Service. Local governments receive 25 percent of the revenue from these timber sales. *Source: U.S. Forest Service.*

adapt to secondary forests, but the real issue at hand is not about the spotted owl or any particular species of wildlife. The goal of the environmental groups is to protect the old-growth forests and their ecosystems; therefore, invoking the Endangered Species Act on behalf of a particular plant or animal is a means to that larger end.

At odds with environmental groups is the forest products industry within the Pacific Northwest region. Included in this group are logging companies, lumber mills, and pulp mills. Logging operations and lumber mills have long viewed old-growth forests as the timber of choice with simple logic: large, tall trees contain a greater volume of board feet per tree and provide a higher percentage of knot-free lumber. For these reasons, logging and milling operations in the region have largely been equipped to handle old-growth rather than the smaller trees of secondary-growth forests. For these operations to limit themselves to smaller trees and to replace their equipment imposes a financial burden that many companies claim to be insurmountable, causing them to cease operations. From 1979 to 1991, forest industry employment in the Pacific Northwest fell from 138,817 to 100,001—most of this loss is attributed by the industry and independent research to a reduction in timber availability.

Arguments exist over the magnitude of the *future* economic impact of protecting the remaining old-growth timber. In 1990, some environmental groups projected the loss of timber-related jobs in Washington, Oregon, and northern California to be 2,300, but in 1992 the Fish and Wildlife Service predicted the number to be 21,100—nearly 14 times higher. The forest industry itself has claimed that as many as 100,000 jobs may be lost.

Americans who purchase lumber and other forest products such as paper and paperboard are eventually affected by the availability of timber in the Pacific Northwest. In 1991, enough old-growth timber was protected from harvesting by court injunction to produce a 20 percent average increase in lumber prices nationwide.

Environmental groups argue that the process of re-tooling the lumber industry in the Pacific Northwest is inevitable, claiming that, with current practices, the remaining unprotected old-growth forest will last no more than five to ten years. Other projections run as high as 50 years, but clearly these old-growth stands will not last many more generations.

Environmental groups also maintain that the erosion of timber jobs in the region started long before the spotted owl controversy. They maintain that the timber industry has been phasing out its operations in the Pacific Northwest and transferring them to the Southeast in anticipation of dwindling old-growth stands. They argue that thousands of forest industry jobs were lost in the region between 1980 and 1989, the same time period that saw over 55,000 new timber jobs created nationwide.[1]

Still another factor in the economic decline of the region, as argued by environmental groups, is the increased exporting of whole logs harvested from private forests to Asian nations that, in turn, performed the milling operations. These groups point out that, at its peak, 25 percent of the timber cut in the Pacific Northwest was exported, thereby contributing to the closing of 150 saw mills in the region.[2]

Although the Pacific Northwest still has millions of acres of timber, the premium timber of old-growth forests is needed to offset the high costs of logging in the area. Therefore, when old-growth logging is reduced, either by legal injunctions or the diminished availability of huge trees, the incentive to operate in the region is lessened. All of these factors cause environmental groups to conclude that the spotted owl is a scapegoat for an economic situation that had long been in the making.

Still another component of the controversy over the appropriate proportion of wilderness areas concerns the charge made by Congress to the Forest Service to act as caretaker of the nation's timber lands. This mandate requires that much of the forests be managed to prevent natural threats from jeopardizing timber yields (see Figure 16–3). For example, when mature trees develop a disease such as mistletoe, the infection usually spreads to the younger trees

Figure 16–3 The exposed roots and rotting trunks of blown-down trees in an unmanaged forest will continue to decay, clutter the forest floor, inhibit new growth, and attract insects. On the plus side, the decaying wood will replenish the forest soil with valuable nutrients.

below. In time, the original forests can die out and be replaced by a different—and often less desirable—species. An example of this would be a Douglas fir forest being replaced by a hemlock forest. In this instance, the natural evolution of a forest is not in the best interest of America's need for structural timber.

Clearly the responsibilities of two separate federal agencies, the U.S. Forest Service and the Fish and Wildlife Service, can be in conflict with one another. Also involved in the dispute are the values of Americans at large: the cost of forest products such as lumber and paper goods matched against the sense of obligation to preserve a diminishing and irreplaceable ecosystem within our borders. To date, politicians have sought compromises as exemplified by the Timber Summit sponsored by the Clinton administration in early 1993.

After the conference, the Clinton administration issued a plan permitting the logging of 1.2 billion board feet of old-growth forest each year. Because the allowable cut was only 25 percent of what the timber industry wanted, the region's loggers were not pleased; however, because the plan allowed significant cutting of old-growth forest, the environmentalists felt betrayed (see Figure 16–4).

New Forestry

The Clinton plan also included a provision that timber cuts in National Forests be as ecologically sensitive as practical. Such a technique—called **new forestry**—has been in practice in part of Oregon's Willamette National Forest since 1990. Instead of clearcutting with the intent of creating a new forest that is homogenous in specie and age, new forestry leaves behind clusters of live trees, as well as a few dead

'So, How Do You Like My Compromise?'

Figure 16–4 The Clinton plan's compromising approach to the cutting of old-growth forest in the Pacific Northwest satisfied neither the forest industry nor the environmental community because of the complexity of the issue. © *1993 Engelhardt in the* St. Louis Post-Dispatch. *Reprinted with permission.*

trees, called *snags,* and plenty of wood debris. The goal of new forestry is to harvest timber without profoundly altering the ecology of the area. The snags can become ideal homes for woodpeckers and other birds that are cavity nesters, while the logs left on the ground provide a habitat for small animals to burrow under, as well as food for insects and fungi. Also, as the debris decomposes, nutrients will be returned to the forest soil. As the result of these practices, timber harvests do not have to permanently alter a forest's composition as do the more traditional regeneration cuts such as clearcutting, shelterwood, and seed-tree cuts. After a new forestry harvest has occurred, the forest should appear as it was before the cut, with the same mix of species for both trees and wildlife within 70 years.[3]

The biggest obstacle facing large-scale adoption of new forestry techniques is the question of whether they are capable of yielding an adequate volume of affordable timber to meet the nation's requirements. If not, large sections of National Forest land will probably receive some form of clearcutting and the forest product industry and environmental groups will surely be at odds for decades to come.

America's Growing Concern with Pollution

In 1962, thousands of Americans started to become aware of the potential danger of tampering with the environment. In that year, Rachel Carson's book *Silent Spring* was published. In it, Carson made the case that the widespread use of pesticides could threaten birds, fish, and even humans by poisoning the food chain. As the result of her arguments, pesticide use became restricted and—more significantly—environmental awareness increased among Americans in general as other books, magazine articles, and documentaries pointed out how a decision involving some sector of the environment cannot be considered as an isolated act. As Americans become increasingly aware of the interconnections between various species of plants and animals, more intense governmental efforts were channeled to assess the state of the environment, identify sources of any deterioration, and impose regulations on these sources. Eight years after the publication of *Silent Spring,* the **Environmental Protection Agency (EPA)** was created to consolidate the 15 separate federal programs involved in some way with pollution. The new agency was charged with conducting research, creating water and air protection standards, and enforcing those standards.

Few federal agencies have generated more controversy than the EPA. Both industries and municipalities have bristled at being forced to comply with EPA water and air emission standards, while states complain that they are given mandates and time tables, but inadequate funding to meet the requirements. On the other hand, environmental groups argue that the EPA is slow to move and is too tolerant of major polluters.

Still others point to the $541 billion in private and public funds spent on the environment between 1972 and 1990 and question the appropriateness of such an investment.[4] Despite its inability to satisfy many environmental groups, local governments, or industries, the EPA has had, in conjunction with Congressional legislation, a profound effect on American life—and the paper, ink, and printing industries are no exception.

Water as a Resource in Papermaking

Over 338 billion gallons of water are used daily in America. Of this tremendous volume, 36 billion gallons (nearly 11 percent) are used by industry—and the pulp and paper industries are among the leaders for water consumption (see Figure 16–5). For example, 300 million gallons of water are required nationally to produce a single day's supply of newsprint and over 25 million gallons of water can be used daily in a single integrated paper mill.

After its use, wastewater is discharged into a river or lake—either directly or indirectly through a municipal treatment plant. If untreated, a mill's wastewater is likely to threaten the water course in several ways. First, it will carry clays and other *suspended solids*. Second, lignin and other *organic materials* will be present as a result of the pulping operation. Third, the use of *elemental chlorine* in the bleaching operation probably will generate some toxic compounds.

The Clean Water Act of 1972

Prior to the 1960s many municipalities and manufacturing facilities discharged untreated wastewater into nearby waterways. This type of discharge is called **point-source pollution** because it can be traced back to a single location. In contrast to point-source pollution is **nonpoint-source pollution,** which includes the runoff of fertilization and pesticides from agricultural and urban areas.

The passage of the Clean Water Act by Congress in 1972 reflected a major commitment by the federal government to target point-source pollution. The goal of the Clean Water Act was "to restore and maintain the chemical, physical, and biological integrity of this nation's waters" (see Table 16–2). The first efforts toward this goal were directed at point-source pollution—specifically, municipalities and industries that were dumping either untreated or inadequately treated waters into the nation's lakes and rivers. During the next nine years, $30 billion was spent on either constructing new or improving existing municipal sewage plants. Between 1972 and 1988, the number of people served by effective treatment facilities increased from 85 million to 144 million.

Primary and Secondary Treatment

Both municipal sewage treatment plants and industrial facilities that discharge wastewater usually must treat their effluent in two separate stages. **Primary treatment** removes clay and other solid particles from wastewater. After passing through screens that remove large solids, the water is sent to wide, shallow enclosed basins, called *clarifiers,* that function like a huge overflowing bathtub (see Figure 16–6). The water enters the basin at the center and, with minimal agitation, eventually makes its way outward until it spills over the wall. However, on the slow journey across the basin, the suspended solids in the water settle out and accumulate as a

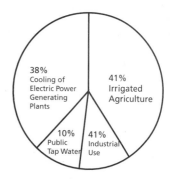

Figure 16–5 Amazingly, nearly four times as much of America's water usage is required for cooling electric generators as is needed by the remainder of the nation's industries. *Source:* 1993 Information Please Environmental Almanac.

1948	**Water Pollution Control Act**
	Provided funds to state and local governments to reduce water pollution.
1965	**Water Quality Act of 1965**
	Empowered a federal agency to develop standards for water quality.
1970	**Establishment of the Environmental Protection Agency**
1972	**Water Pollution Control Amendments**
	Regulated discharge of toxic pollutants; budgeted $5 billion for building and improving municipal sewage plants during the next fiscal year; established programs to develop new technologies to reduce pollution; required permits for discharging pollutants into water ways. Charged EPA with publishing a list of toxic pollutants and establishing effluent standards for industrial discharges.
1977	**Major revision of CWA of 1972**
	Expanded the range of pollutants to be EPA-regulated.
1981	**CWA further revised**
	Tightened criteria for wastewater treatment facility construction grants.
1987	**Water Quality Act of 1987**
	Replaced grants for wastewater treatment facilities with revolving loan programs to states; required states to identify "toxic hot spots" that would receive tighter controls.

Table 16–2 The major federal legislation concerned with the nation's water quality is listed along with significant provisions.

sludge on the bottom. Therefore, the water that leaves primary treatment does so without the solid particles with which it entered.

The sludge that is collected from primary treatment is dewatered (dried) and, in this state, represents the majority of the 12 million tons of solid waste produced annually by America's pulp and paper mills. The sludge is disposed of in three ways. Some mills *apply it to agricultural fields* as an enhancement to the soil (see Figure 16–7). However, in 1991 the Environmental Protection Agency proposed regulations on land application of sludge produced at mills that use elemental chlorine in the bleaching of pulp. Their regulations grow from a concern that the resultant sludge contains haz-

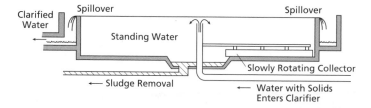

Figure 16–6 A clarifier allows water to spill over a retaining wall after previously suspended solids have precipitated to the bottom of the tank where they are collected. *Courtesy of Howard Paper Mills.*

ardous concentrations of dioxin and, as a result, not all mills will have the option of disposing of their sludge on fields.

A second option is to *burn* the dewatered sludge in the mill's boilers, thereby making a waste-to-energy conversion. The third option, and also the most traditional, is to haul the sludge to *landfills*. If the sludge contains dioxin, it must go to specially-lined landfills that will prevent the pollution of underground water.

After undissolved solids have been removed, wastewater is sent on to **secondary treatment** that deals with *organic compounds* that did not settle out in the clarifiers because they were dissolved in the water. Dissolved organic compounds include lignin and other carbohydrates that were separated from the cellulose fibers during pulping. Because lignin and carbohydrates are part of nature, they are not hazardous materials; however, their excessive presence in waterways threatens aquatic life because they *consume the water's oxygen as they decompose.*

As organic matter is consumed by microorganisms or oxidation, the process consumes oxygen that is dissolved in the water. When excessive organic substances enter a body of

Figure 16–7 Secondary treatment can occur by spraying wastewater onto fields where any organic compounds will biodegrade. *Courtesy of Howard Paper Mills.*

water, they can deplete the oxygen supply to the degree that certain species of animal life cannot be sustained; when a body of water is declared "dead," the reason is usually due to inadequate oxygen. The amount of oxygen required for the decomposition of an organic pollutant is termed its *biochemical oxygen demand* **(BOD).** Because high levels of organic matter in water bring with them a high demand for oxygen, BOD is a useful measurement of the amount of organic pollution in a mill's effluent.

To lower the BOD of a mill's wastewater, oxygen is introduced to the water in a process known as **aeration.** Generally, aeration occurs in ponds or lagoons that hold water that has already received primary treatment in clarifiers. While in these lagoons, the water is churned with motor-driven impellers that function like giant egg beaters as they mix air into the water (see Figure 16–8). Aeration can also occur through a fountainlike spraying of water into the air. As organic materials are oxidized in the lagoon, the depleted oxygen is replaced through this aeration. In addition, keeping the oxygen levels of the water at a high level accelerates the efficiency of the cleaning process. Water from pulp mills usually receives secondary treatment for between two and ten days, based on the BOD level of the mill's effluent and the efficiency of the aeration process.

The Clean Water Act of 1972 was further amended in 1977 and 1981 to fine-tune its effectiveness at reducing point-source pollution. Between 1972 and 1982, improvements in municipal wastewater treatment plants had reduced the BOD levels of their effluent by nearly one-half;

Figure 16–8 In this aerial view of a mill's water treatment facility, on the right are two large clarifiers and at left is the secondary treatment area where the water is churned by several motors that accelerate aeration.

industrial site discharges showed an even greater improvement with a reduction of over 70 percent. By 1990, 80 percent of America's surveyed river miles were found safe for fishing and 75 percent were found suitable for swimming (see Table 16–3). However, it was also becoming clear that two sources of water contamination were not being adequately addressed; one was nonpoint-source pollution and the other was toxic pollution.

The Water Quality Act of 1987

Nonpoint-source pollution includes the runoff of fertilizers, pesticides, animal wastes, and silt. According to the EPA, nonpoint-source pollution represents an estimated 50 to 70 percent of America's remaining water quality problems. By its very nature, nonpoint-source pollution is extremely expensive to control as compared with "end-of-pipe" effluents, which are easier to locate.

The other problem—*toxic pollution*—remained a problem because primary and secondary treatment does not remove dissolved substances from wastewater and some toxic compounds are water soluble. In response, Congress passed the Water Quality Act of 1987 to further strengthen the Clean Water Act by initiating controls on nonpoint-source pollution and tightening the controls on toxic pollutants, defined as substances that are poisonous to plants or animals.

Toxic Water Pollution Reduction

Because toxic pollutants can remain in river bottoms and other parts of the environment for long periods of time, as well as accumulate within living organisms, even very small levels of contamination can be significant. However, the task of monitoring the presence of the 50,000-plus chemicals in use in America is not financially feasible.

State	River Miles Tested for Fishing	River Miles Fishable	River Miles Percent Fishable	Tested for Swimming	River Miles Swimmable	Percent Swimmable
Alabama	11,857	8,703	73.3	11,857	8,703	73.3
Arizona	5,296	662	12.5	5,296	1,810	34.2
California	11,448	9,069	79.2	10,463	7,947	75.9
Colorado	28,770	28,105	97.7	31,377	9,062	28.9
Delaware	643	498	77.4	643	98	15.2
Illinois	13,123	11,476	87.4	4,525	1,144	25.3
Indiana	4,944	2,986	59.8	2,304	138	5.9
Iowa	7,155	90	1.3	7,155	158	2.2
Maine	31,672	31,282	98.8	31,672	31,506	99.5
Maryland	17,000	15,618	91.9	17,000	16,998	99.9
Michigan	36,350	35,632	98.0	36,350	36,086	99.3
Minnesota	6,079	2,292	37.7	5,021	2,045	40.7
Nebraska	7,330	5,131	70.0	12,011	1,035	86.2
New Jersey	1,719	1,315	76.5	592	91	15.4
New Mexico	3,117	2,851	91.5	3,117	3,117	100.0
New York	70,000	69,300	99.0	70,000	69,200	98.9
North Carolina	37,293	23,820	63.0	37,293	23,820	63.9
Oregon	27,739	26,197	94.4	27,739	26,773	96.5
Utah	11,779	1,303	11.1	4,320	120	2.8
Virginia	10,809	8,862	82.0	10,809	8,862	82.0
Washington	5,141	2,873	55.9	4,928	3,441	69.8
Wisconsin	13,302	8,272	62.2	13,192	8,235	62.4

Table 16–3 The suitability of fishing and swimming in sampled portions of rivers for 22 states is shown above. These states represent different regions and levels of urbanization; they also include states like Oregon, Washington, Maine, and Alabama where logging and pulping are major industries. Compare neighboring states like Illinois and Indiana or Arizona and New Mexico. Also compare eastern and western states. *Source: U.S. Environmental Protection Agency.*

Included on the list of toxic substances found in some rivers and lakes are arsenic, DDT (dichlorodiphenal-trichloroethane, formerly used as an insecticide), PCBs (polychorinated biphenyls, formerly used in numerous applications including pesticides), Dieldrin (a former insecticide), chlordane (formerly used to control termites), heavy metals (such as mercury, lead, and cadmium), and dioxin. Whereas most compounds break down fairly soon after disposal, many of the toxins just listed are extremely stable. For example, DDT was banned in 1972, PCBs in 1976, and Dieldrin in 1988, but they can still be found in America's waterways. To make matters worse, PCBs do not easily dissolve in water, so, rather than being flushed out of an animal's system, they tend to *bioaccumulate*, or build up in animal tissue, and then become further concentrated as they get passed up the food chain. As evidence of this pattern, PCB concentration in the eggs of eagles near the Great Lakes is reported to be over 20 million times greater than their concentration in the nearby lake water.[5] Clearly, we are constantly learning about technology's impact on the environment.

Technologically, two patterns are continuing to unfold. First, research is *finding toxicity* in substances previously considered benign. Second, improvements in the procedures and equipment used to detect water-borne toxins are allowing nearly infinitesimal traces to be found. For example, *dioxin* was not detected in the wastewater of pulp mills until, in the mid-1980s, its detection in parts per trillion became possible. Because of dioxin's ability to bioaccumulate, by 1992 the EPA was suggesting a limit on dioxin levels in waterways of .013 parts per quadrillion.[6]

The capability of finding such minute traces of dioxin and other chlorinated organic compounds, coupled with the lack of knowledge of their effect on humans at such small levels, has caused the EPA to closely monitor kraft paper mills. As discussed in Chapter 3, the pulp and paper industry is a major user of chlorine. In fact, 15 percent of the 12 million tons of chlorine produced annually in America is used to bleach pulp.[7] Nonetheless, because dioxin and other chlorinated organic compounds are formed when elemental chlorine (Cl_2) combines with lignin during the bleaching of wood pulp, many mills have invested heavily to replace elemental chlorine with chlorine dioxide or another bleaching agent. Still others have approached the problem by using *oxygen delignification* or *lignin extraction,* while still others use a combination of these operations. Pulp that was bleached without the use of chlorine is termed *elemental chlorine-free* (**ECF**) pulp and it usually has 75 percent lower levels of chlorinated organic compounds than does pulp bleached traditionally with elemental chlorine. Still, there is some pressure from a few food-related customers and several environmental groups to produce *totally chlorine-free* (**TCF**) paper. As of 1994, very few American mills were producing TCF, but some mills view the elimination of chlorine bleaching as the only means of meeting anticipated EPA-imposed standards of the twenty-first century.

Meeting the standards of the 1987 Water Quality Act has been expensive for the pulp and paper industry. In 1992 and 1993 alone, Boise Cascade installed (in three separate mills) a new chlorine dioxide plant ($27 million), new wastewater treatment equipment ($20 million), and a dioxin reduction program ($35 million). From 1992 to 1994, Champion rebuilt a pulp mill to bring it into environmental compliance ($310 million) and replaced elemental chlorine bleaching with chlorine dioxide and oxygen delignification ($40 million). Although this is only a partial list of mills investing in water pollution control, it provides some insight into the enormous capital being allocated to comply with federal and state standards.

After expenditures of hundreds of millions of dollars, the U.S. pulp industry reports a 90 percent reduction in its dioxin discharges from 1988 to 1993 and claims that the dioxin discharge for all U.S. mills in 1992 was only four ounces.[8] However, the standards that the EPA envisions for the future would place many presently-conforming pulp mills in violation. For pulp mills to qualify under the proposed standards, they would likely need to replace elemental chlorine as a bleaching agent with chlorine dioxide, hydrogen peroxide, or ozone. To add still more impetus to the antichlorine movement, lawsuits have been filed by downstream residents of mills that discharge dioxin traces in their wastewater. As has already been noted, the refitting of a pulp mill to eliminate elemental chlorine can cost $40 million, a capital expenditure that many mills cannot afford.

The EPA's justification for these measures centers on eliminating the chlorinated organic compounds that accumulate in fish which are caught and eaten by humans. The agency predicts that their proposed standards would allow the lifting of 25 fishing advisories currently attributed to dioxin and ultimately prevent 2 to 30 cases of cancer each year. The paper industry questions the claim that the dioxin levels presently in America's waterways are responsible for even one incident of cancer. As was pointed out in Chapter 3, laboratory research on the link between dioxin level and cancer includes no data on humans, with the result being an ongoing controversy about how much dioxin is harmful to people.

As of 1994, the proposed EPA dioxin standards have yet to become law and the pulp and paper industry is arguing the case with Congressional legislators that the standards are unreasonable and unnecessary. Even if the industry is successful in getting the standards relaxed, the trend will still be to eliminate bleaching pulp with elemental chlorine and replace it with a process that is more environmentally friendly and less likely to lead to litigation at some future time.

Air Quality Legislation

Air pollution has been a concern of societies since the industrial revolution and legislation addressing it was not far behind, as evidenced by United Kingdom emissions control laws dating as far back as 1863. Air pollution is any form (gaseous or particulate matter) of contaminant in the atmosphere and it can result from natural causes such as forest fires and volcanoes, as well as automobile and industrial exhausts. Traditionally, concerns over air pollution have revolved around its effect on human health; however, discoveries of effects such as acid rain have led to a growing awareness of the damage air pollution can cause to the environment.

In 1963, Congress passed the **Clean Air Act,** which allocated funds for research and support for the establishment of state agencies to control air pollution. Congressional legislation passed in 1965 and 1967 placed additional requirements on the states and also increased the role of the federal government in improving the nation's air quality. In 1970, Congress passed the Clean Air Act of 1970, which established the federal government as the major player in the effort to improve air quality by mandating the newly-created Environmental Protection Agency to establish emission standards for key pollutants often called **criteria pollutants:** carbon monoxide, nitrogen oxide, ozone, particulate matter, sulfur dioxide, and lead. States were encouraged to develop their own plans for implementing these standards, but the plans had to be EPA-approved or the EPA would regulate those states' air emissions. Amendments passed in 1977 increased penalties for violations and expanded the scope of the Clean Air Act from the improvement of already-polluted air to the safeguarding of clean air, such as is found over wilderness areas.

Industry's Response

As a result of this legislation, many industries that were in violation of the new standards installed systems to bring their air emissions into compliance. For pulp and paper mills, the primary expenditures were aimed at removing particulate matter from smoke stack emissions (see Figure 16–9). **Particulate pollutants,** such as smoke particles and flying ash, result from combustion. Pulp and paper mills burn materials in boilers to generate energy and steam to power their operations. In an effort to adequately reduce particulate matter in their air emissions, mills can install

Figure 16–9 America's paper and related industries have spent over $7 billion since 1972 to clean harmful particles from their releases to the air. *Courtesy of Hammermill Papers— division of International Paper.*

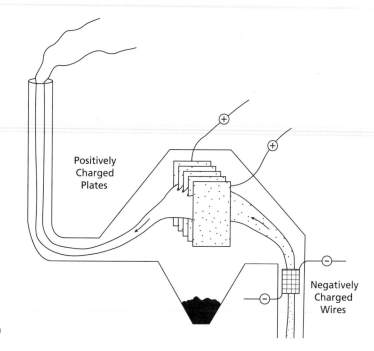

Positively Charged Plates

Negatively Charged Wires

Figure 16–10 An electrostatic precipitator charges solid particles and then attracts them to oppositely charged plates, thereby pulling them from the airstream.

electrostatic precipitators or scrubbers. *Electrostatic precipitators* function by passing waste gases between wires carrying high-voltage current that impart a negative charge to the solid particles. As the newly-charged particles then pass between positively-charged plates, they are attracted to and collected by the plates (see Figure 16–10). Electrostatic precipitators can attain efficiency levels of 99 percent. *Scrubbers* remove particulate by sending waste gases through a sheet

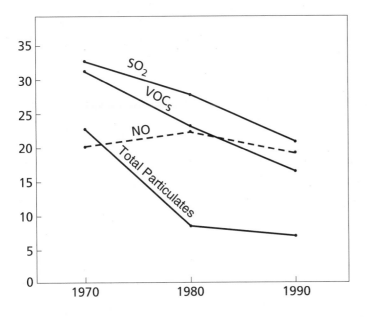

Figure 16–11 Significant progress has been made in reducing the levels of sulfur dioxide (SO₂), volatile organic compounds (VOCs), and total particulate matter in America's air. Nitrogen oxide (NO) emissions have proven to be a more formidable problem. The chart reflects millions of tons emitted annually. *Source: U.S. Environmental Protection Agency.*

of water droplets that capture the solid particles and remove them from the gas stream. Scrubbers can also be designed to remove undesirable chemicals from a gas stream that is sent through a body of liquid that absorbs the chemicals. Paper mills often use their black liquor as the absorbent agent. Because electrostatic precipitators and scrubbers can be highly efficient, America has witnessed a significant reduction in the levels of some major air pollutants since 1970 (see Figure 16–11).

The 1990 Clean Air Act

In 1990, the Clean Air Act was further amended to tighten controls on toxic emissions (see Table 16–4), including compounds that mix with moisture in the air to form acids which fall to earth as *acid rain*. Sulfur dioxide (SO₂) is a primary culprit in the formation of acid rain. According to the U.S. National Research Council, about one-half of the sulfur dioxide in the atmosphere is attributable to natural sources such as volcanoes; however, the other half of the volume is man-made. Electric utility plants that burn high-sulfur coal are clearly the largest industrial source, but the new EPA restrictions on sulfur dioxide impact the paper industry, as well. Because sulfite chemical pulping uses sulfur-containing compounds, fumes and waste gases from most kraft mills contain some sulfur dioxide, as well as other sulfur compounds that comprise a mill's **total reduced sulfur** or

Paper, Ink, and the Environment   353

1. Chemicals	886.6	10. Furniture & fixtures	45.3
2. Primary metals	215.1	11. Instruments	41.6
3. Paper	207.9	12. Textiles	34.9
4. Transportation equpt.	192.0	13. Stone, clay, glass	25.7
5. Rubber & plastics	132.0	14. Lumber & wood	25.0
6. Fabricated metals	110.2	15. Misc. manufacturing	21.7
7. Electrical & electronics	89.7	16. Food	15.7
8. Petroleum & coal	75.5	17. Leather	7.5
9. Machinery	46.2	18. Apparel	2.1

Table 16–4 The emission of toxic chemicals to the air for various American manufacturing industries in 1987 is shown in millions of pounds. Electric utility plants are not included. The Clean Air Act of 1990 was written to significantly reduce these emissions. *Source:* Statistical Record of the Environment *from the U.S. Environmental Protection Agency.*

TRS. Included in the TRS category is *hydrogen sulfide,* the air emission most responsible for the odor associated with paper mills. Although TRS compounds are generated in the digester, they can vent into the air at numerous points between the digester and the paper machine itself.

Industry's Response

To comply with the 1990 Clean Air Act's new standards, many mills will need to install more efficient systems to *capture venting gases* from process points such as the digester, blow tank, brown stock washers, black liquor evaporater, recovery furnace, and the lime kiln. After capture, the gases are *collected* and usually *burned* in a boiler or routed through a *scrubber.* Either process is highly effective in the reduction of TRS emissions, but the cost for installing such a comprehensive system can exceed $50 million. In 1992, a Mississippi mill spent $60 million on a fluegas collection and scrubber system and other modifications in order to comply with new federal and state emission requirements. In 1993, a New Hampshire mill installed scrubbers and an electrostatic precipitator, as well as a rebuilt boiler and improved a black liquor recovery furnace at a cost of over $73 million in order to reduce TRS emissions by 90 percent. However, as was mentioned in the discussion of the most recent version of the Clean Water Act, some mills will not be able to afford the necessary investment for compliance.

Solid Ink Waste Disposal

The pulping and papermaking processes generate not only liquid and gaseous wastes, but solid wastes, as well. This solid material includes the ash collected during incineration as well as the sludges produced during primary and secondary water treatment and the sludge created by de-inking waste paper to make recycled paper. In a year's time, America's pulp and paper mills generate over 12 million tons of solid waste that must be disposed of in some fashion.[9] Although some of this waste is incinerated, the vast majority is buried in landfills.

The burial of solid wastes in landfills has become a much more expensive function with the advent of stringent regulations that grew from irresponsible dumping of hazardous wastes during the post-World War II era. Fearing another disaster like the Love Canal incident of the 1970s, communities have increasingly adopted a "not in my back yard" (NIMBY) position on landfills.

Even in communities where a suitable site can be found, state and federal regulations on the construction, operation, and maintenance of landfills have made them too expensive to build. As the result of these two forces, the number of operating landfills in America has been steadily dropping, as evidenced by a 27 percent decline from 8,000 in 1988 to 5,812 by the end of 1991.[10] This trend of landfill shortages is projected to continue into the foreseeable future (refer back to Figure 5–2), thereby driving up tipping charges exponentially. In Philadelphia, for example, the $20 cost to bury a ton of trash in 1980 increased five times to $100 by 1987.[11]

In response to the reduced access to municipal facilities, most paper mills have built their own landfills. Despite being private, these landfills, like the public facilities, are regulated by state agencies and the EPA according to provisions of the Resource Conservation and Recovery Act (RCRA) of 1976.

Landfill Design

A properly-designed and operated landfill is able to secure its contents from escaping and potentially contaminating ground water in the area. Several design components are necessary to achieve this end. First, the bottom and sides of the landfill must have an impermeable 3–5-foot-thick layer of recompacted clay. In some instances, this layer of clay must be reinforced with a thick plastic liner (see Figure 16–12).

Although very little moisture is present in the materials that go into a landfill, rainfall during the period of operation contributes water to the mix and can *leach* (mix with the contents and become contaminated), which creates a **leach-**

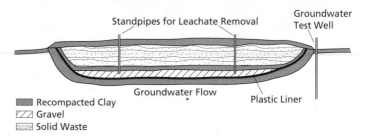

Figure 16–12 (Top) Solid waste from the paper mill that owns this landfill is disposed according to EPA regulations. The white tubes are standpipes. (Right) This cross section of a secure landfill reveals the layers of nonpermeable material that surround and contain the solid waste. Any leachate that penetrates the first clay layer will accumulate in the gravel to be pumped out through the standpipes.

Standpipes for Leachate Removal

Groundwater Test Well

Groundwater Flow

Plastic Liner

Recompacted Clay
Gravel
Solid Waste

ate (the contaminated water) that is directed to one or more low spots in the basin of the landfill from where it can be pumped out through built-in *standpipes*. If the landfill content could produce methane gas through decomposition, the pipes are installed to allow gases to escape, instead of building up and creating the possibility of explosion.

To monitor the effectiveness of the landfill's containment system, *test wells* are installed nearby so that groundwater can periodically be tested for contamination. After the facility becomes full, it is covered with the same materials used to line the bottom and sides and then usually topped off with top soil or sand for aesthetic purposes.

These provisions are required of landfills containing either **hazardous** or **nonhazardous wastes,** but facilities with hazardous contents have more vigorous standards, such as the number and thickness of the required liners as well as the number of test wells. At this time, the solid waste of the vast majority of pulp and paper mills is considered nonhazardous, but the pattern is clearly one of increasingly stringent regulations. Under existing conditions the cost to the pulping industry for solid waste disposal is growing at a steep rate (see Figure 16–13). It is estimated that the cost to build and maintain a landfill exceeds $110 million, with over half of the expense allocated to taxes, insurance, and related matters.[12]

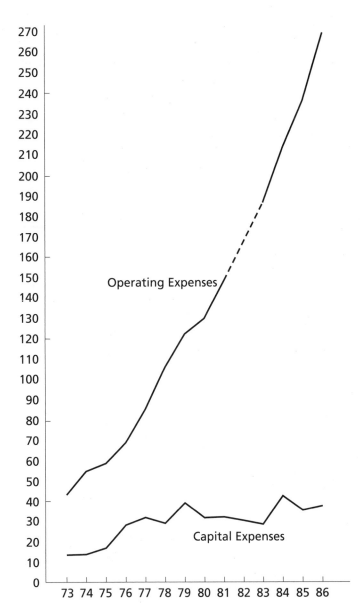

Figure 16–13 Expenditures by the pulp industry for solid waste disposal from 1973 to 1986 are shown in millions of dollars. Although equipment costs have clearly risen over the 14-year period, their growth rate is dwarfed by that of operating expenses. The broken line represents a period of no data. *Source: Council on Environmental Quality.*

Reducing the Volume of Solid Waste

In an effort to reduce their volume of solid waste, many mills are looking at evolving technologies. For example, mills in Wisconsin and Michigan have been experimenting with a Wisconsin facility that can convert 100 tons of wet mill sludge into 34 tons of pellets that can then be mixed with coal to fire boilers at electric power plants.[13] As the cost

Paper, Ink, and the Environment   357

	Water	Air	Solid Waste	Total
1973	$279.1	$225.6	$55.3	$560.0
1974	$345.2	$352.0	$68.6	$765.8
1975	$451.5	$423.9	$73.8	$949.2
1976	$517.7	$303.9	$94.6	$916.2
1977	$570.7	$267.6	$118.0	$956.3
1978	$546.6	$282.3	$134.3	$963.2
1979	$581.1	$383.6	$159.9	$1,124.6
1980	$547.9	$393.6	$160.1	$1,101.6
1981	$556.4	$379.5	$179.1	$1,115.0
1982	$548.9	$396.7	$163.8	$1,109.4
1983	$574.8	$348.8	$211.5	$1,135.1
1984	$634.3	$432.6	$255.3	$1,322.2
1985	$679.4	$503.9	$305.6	$1,488.9
1986	$662.6	$456.3	$307.0	$1,425.9
1988	$724.9	$605.8	$430.3	$1,761.0
1989	$947.8	$780.5	$429.1	$2,157.4
1990	$1,298.0	$812.0	$573.0	$2,683.0

Table 16–5 The total costs (capital plus operating expenses) to the paper and allied products industry (in millions of dollars) for water, air, and solid waste environmental control have steadily risen since 1973. *Source: U.S. Department of Commerce, Bureau of the Census.*

for landfilling continues to rise, research like this will continue to seek cost effective alternatives.

The Economic Impact of Environmental Controls

America's paper and related industries spent nearly $1.5 billion on environmental control between 1973 and 1990 (see Table 16–5). Included in these figures are both *capital* expenditures for the purchase and installation of pollution control equipment and *operating* expenses that include taxes, insurance, utilities, salaries, and raw material costs. As the data on Figure 16–13 reveal, environmental costs are increasing steadily and were projected to account for 11 percent of American mills' capital budgets for the 1993–95 biennium (see Table 16–6). This escalating trend for environmental allocations by the paper industry is expected to continue, as the Environmental Protection Agency has proposed to combine air and water regulations into stricter "cluster" rules aimed at further reducing the emission of toxins from America's pulp and paper mills as well as other major industries (see Table 16–7). According to the EPA, the proposed standards take "an ecosystem-wide approach to improve and

	Water		Air		Solid Waste	
New England	$0	0%	$93,300	33%	$1,400	1%
Mid-Atlantic	$47,509	6%	$32,619	8%	$19,895	11%
East North Central	$69,917	8%	$10,456	3%	$16,210	9%
West North Central	$45,450	6%	$40,890	10%	$25,860	14%
South Atlantic	$454,516	55%	$145,214	36%	$23,067	12%
East South Central	$51,550	6%	$22,893	6%	$58,000	31%
West South Central	$50,200	6%	$18,200	5%	$17,100	9%
Mountain Pacific	$105,780	13%	$39,280	10%	$25,290	14%
Totals	$824,922		$399,852		$186,822	

Table 16–6 Over $1.4 billion in capital expenditures has been allocated by American pulp and paper mills for environmental control during the 1993–95 biennium (figures exclude operating costs). Environmental control represents 11 percent of the total capital budget. Cost figures shown represent thousands. *Source:* Pulp & Paper.

protect public health and the environment by combining air and water requirements in the same regulation."[14]

The cluster rules are particularly threatening to mills that use any form of chlorine (except chlorine dioxide) in pulp bleaching and are financially unable to convert to another process. The EPA acknowledges that the new standards would result in 11 to 13 pulp mill closings, between 2,800 and 10,700 lost jobs, and a price increase on uncoated chemically pulped paper grades. The paper industry maintains that the number of closures is more likely to approach 30 and represent the elimination of 275,000 jobs (19,000 at the mill level).[15]

The American Forest and Paper Association supports the cluster rule concept, but believes that the air and water emissions levels desired by the EPA are impractical and unnecessarily stringent. The association believes that the EPA's proposals are based largely on unverified assumptions and projections.[16] In a statement released in 1993, the AFPA projected the industry-wide cost of compliance to be $3.8 billion and thereby "threaten the U.S. pulp and paper industry's economic viability and the jobs of tens of thousands of workers around the country. . ." and a 1994-released study placed the compliance cost at $11 billion.[17]

As EPA standards become more stringent, the cost for compliance becomes greater, as does the percentage of total capital that is allocated to environmental control. American mills' environmental control allocations of their capital budgets were 11.1 percent in 1990, 13.8 percent in 1991, and 14.1 percent in 1992. In addition to the need to divert finances away from expenditures that could increase produc-

State	Surface Air Emissions	Underground Water Discharges	Well Injection	Land Releases	Total Off-site Transfers	Transfers and Releases
Chemicals	695,358	133,479	658,662	98,518	251,193	2,107,156
Metals	351,509	12,501	20,052	315,591	364,125	1,081,447
Paper	243,936	37,676	0.07	7,368	18,441	359,510
Transportation	193,522	235	0.32	1,949	39,537	244,941
Plastics	192,541	463	15	200	22,455	224,534
Petroleum	71,299	4,987	37,851	3,114	9,249	134,214
Electrical	81,867	416	19	2,736	35,119	132,656
Food	26,886	5,404	36	8,689	9,116	92,393
Machinery	52,777	209	0.568	139	13,564	69,427
Furniture	60,051	4	0.07	7,368	4,328	64,796
Printing	52,035	1	0.35	4.7	4,383	56,734
Textiles	33,967	556	0.04	37	3,120	45,663
Stone/clay	21,869	175	7,555	2,283	12,001	45,651
Lumber	37,649	209	0.08	127	7,326	45,414

Table 16–7 Toxic chemicals released by American industries in 1990. The figures represent thousands of pounds. Note that the paper industry ranks third in surface air and underground water discharges. The printing industry's air emissions are largely volatile organic compounds that evaporate when ink dries. *Source: U.S. Environmental Protection Agency.*

tivity or enhance their products, many persons in the paper industry are frustrated by the changing regulations imposed by federal and state governments. Some observers have likened the process to trying to hit a moving target or running in a race with no finish line. For some mills, millions of dollars are spent to come into compliance with current regulations, only to find that newer standards place them back into noncompliance.

As a case in point, a southwestern Washington pulp mill that was in compliance with EPA regulations announced in early 1994 that it would close by July of the same year, citing the $46 million projected cost to meet cluster regulations scheduled to go into effect in 1998.[18]

A second source of frustration for mill executives is the possibility that a mill can update its processes to meet federal regulations, only to have its state's regulating agency impose even tougher standards at a later date.

The actual impact of the proposed cluster rules on the pulp and paper industry will likely not be seen for several years, but the trend of increasingly more stringent environmental regulations is expected to continue.

VOC Levels for Printers

One of the goals of the Clean Air Act of 1977 was to restrict the industrial emission of **volatile organic compounds (VOCs)** into the air. **Volatility** is the transformation of a liquid into the gaseous state—like the evaporation of water. Organic compounds are carbon-containing matter derived from organisms; oil and petroleum products are examples of organic compounds. From an environmental viewpoint, VOCs are hydrocarbons and a form of air pollution. Common hydrocarbons that pollute the air—propane, toluene, and ethane—usually come from automobile emissions. Many printing inks, especially some liquid inks used in flexographic and gravure printing, are highly volatile and rely on that property for quick drying. As a result of their volatility, theses inks give off VOCs to the air during printing. Other inks, such as most paste inks used in lithography, are not very volatile and release very low amounts of hydrocarbons to the air.

During the 1970s and 1980s, the EPA defined VOCs as compounds that had a *vapor pressure* of 0.1 mm of mercury at room temperature (see Figure 16–14), which exempted nearly all paste inks. The Clean Air Act of 1990 omitted the vapor pressure consideration for inks and applied a single test method to all inks. Referred to as EPA Test Method 24, the procedure weighs an amount of paste ink, places it in an 110° C (230° F) oven for an hour, and then weighs it again. Any weight difference is considered to represent a VOC loss.

Many printers and ink makers challenge the relevance of Test Method 24 because an hour of 230° heat creates a volatility level in paste inks that far exceeds normal printing conditions where 80 percent of VOCs remain in the sheet. Inks that are judged to emit excessive VOC amounts may require that air pollution equipment be installed to prevent printing facilities from violating the newest Clean Air Act. The National Association of Printing Ink Manufacturers (NAPIM) argues that VOCs would be better identified by establishing a threshold vapor pressure under standard temperatures and conditions; any ink with a vapor pressure exceeding the limit would then qualify as a VOC.[19]

Under the existing regulations, ink makers have used several techniques to develop inks with low VOC contents. In the early 1970s, as much as 55 percent of heat-set web offset inks were VOCs, but today they are available with 30 percent VOC content. Water-based inks have been developed to replace petroleum-based inks in packaging and publication

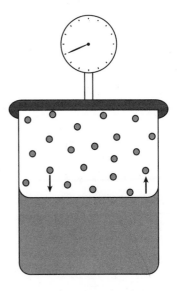

Figure 16–14 When a liquid is placed in a closed container, evaporation will occur until an equilibrium is reached between the two opposing processes of evaporation and condensation. At this point, the vapor above the liquid will exert a constant pressure, called vapor pressure, that can be measured as shown.

printing. In other inks, vegetable oils can replace petroleum oils and reduce the VOC content from the 20–35 percent range to 5–8 percent.[20]

Heavy Metals in Inks

As was discussed earlier in the chapter, dioxins and some other organochlorine compounds are considered especially hazardous to animals because they can bioaccumulate in tissue and build up their toxicity. In a similar fashion, certain naturally-occurring metals tend to collect in organs such as the liver, thereby becoming a health hazard. These metals have high atomic weights and are accordingly referred to as **heavy metals.**

The exact number of heavy metals varies somewhat from one regulatory agency or group to another, but a starting place is the acronym **CAMALS,** which includes cadmium, arsenic, mercury, antimony, lead, and selenium. Although barium is not included in the six CAMALS elements, it is usually considered a heavy metal, as well. The linkage between heavy metals and printing revolves around the presence of heavy metals in certain traditional ink pigments. Barium Red Lake C, lead chromate, and cadmium sulfide are examples of pigments used to achieve desired ink colors. In addition to instances in which the metal is a significant part of the pigment material, small traces of heavy metals also can find their way into inks as contaminants.

Heavy Metal Legislation

The presence of heavy metals in printing inks has created concern among lawmakers for two reasons. First, it was observed in the 1950s and 1960s that children who ate flaking paint or chewed on painted furniture became poisoned from the lead in the paint. Second, there is a growing concern over the potential contamination of groundwater by heavy metals buried in landfills. As a result of these concerns, both federal and state agencies have moved to place limits on the heavy metal content of toys, toy packaging, gift wrapping paper, and packaging in general.

The Consumer Product Safety Commission (CPSC) has limited the acid-soluble lead content of toy packaging to 600 parts per million, which is roughly comparable to 3 feet per mile. The gift wrap industry has applied the same standards to all of its decorative packaging.

Concern over heavy metals escaping from landfills and contaminating ground water has prompted other legislation. The Resource Conservation and Recovery Act **(RCRA)** restricts the content of waste materials to EPA-established levels of eight metals. Four of these eight metals had traditionally been found in printing ink pigments: lead, barium, cadmium, and chromium. By including barium, the RCRA restrictions forced ink makers to find alternative sources for many yellowish-red pigments. Unfortunately, some of the replacement pigments usually cost more than the traditional barium-salt colorants and are more inclined to have their hue influenced by different light sources (more subject to metamerism).[21]

At the state level, the Coalition of Northeastern Governors **(CONEG)** has spearheaded legislation that requires **content certification** for inks that have had certain heavy metals intentionally added to them. The CONEG metals include lead, mercury, hexavelent chromium, and cadmium. CONEG legislation also limits the content of these four metals to decreasing levels (600 to 100 ppm) over a three-year time span. As of 1993, roughly one-half of the states had either adopted or had pending CONEG legislation. Interestingly enough, the biggest impact of CONEG legislation on the printing and ink industries is not on ink chemistry because most heavy metals already had been adequately reduced due to previous legislation and voluntary efforts. Instead, the major problem stems from the issue of pigment and ink makers being required to provide a content analysis of their products. As a result of this mandate, consumer product manufacturers and printers are turning to their ink makers who, in like manner, are turning to their pigment suppliers for verification of heavy metal content.

Although content verification seems like a reasonable request, the fact is that the complexity of printing inks makes reliable analysis of data difficult. The presence of a range of colorants, polymers, and other additives, along with the metal being tested, creates the *matrix effect,* which basically means that the accuracy of metal-level detection is influenced by the other ink components.[22] For example, the analyses of two different inks with the same amount of barium could indicate different barium levels due to the other ingredients being dissimilar.

A second concern of the ink industry is that the list of restricted metals will someday swell to include metals like aluminum, cobalt, copper, lithium, manganese, molybdenum, silver, titanium, and zinc. Such a move would impact the printing industry in several ways. If copper becomes restricted, ink makers will be denied the use of phthalocyanine blue which creates cyan—the blue of four-color (or process) printing. At present, there is no alternative source for this pigment available. If aluminum, copper, and zinc were banned or greatly limited, the printing of metallic inks would likely cease.[23] Also, cobalt and manganese are added to some inks to accelerate drying time.

Many people within the ink industry question the concern over the heavy metal content of packaging inks because they constitute less than 1 percent of the total lead found in municipal waste. Nonetheless, where consumer goods manufacturers are required to verify the metal content of their packaging, the printer, ink maker, and pigment suppliers will need to supply the necessary data. At present, the ink industry is still able to provide the full range of colors and stay within the restrictions of RCRA and CONEG legislation. However, if limits are later reduced below 100 ppm, then certain blues, reds, and yellows may not be attainable.

Waste Ink Disposal

Although the identity of the first printer is unknown, it is probably safe to say that the person had to figure out what to do with leftover ink. Through the centuries, disposing of waste ink has been a chore shared by all printers. For printers in today's world the problem of waste ink disposal is more formidable for two reasons: (1) there is usually more of it; and (2) the Resource Conservation and Recovery Act of 1976 (RCRA) and its Solid Waste Amendments of 1984 contain regulations that make the process more complex.

RCRA is federal legislation that controls the handling of hazardous waste. Under RCRA authorization the EPA has identified four properties that constitute a hazardous waste: ignitable, corrosive, reactive, and toxic. According to these criteria, any substances used in printing facilities can qualify as a hazardous material. A substance is considered **ignitable** if it has a **flash point** (will ignite if exposed to an open flame) of 140° F or lower. Several blanket washes and other solvents used by printers qualify as ignitable. **Corrosive** materials burn the skin or dissolve metal; acids and alkaline cleaning fluids qualify as corrosive. **Reactive** substances are unstable or react violently with water, air, or other materials; included in this category are cyanide plating compounds, oxidizers, and bleaches. **Toxic** materials include inks with excessive levels of heavy metals, such as barium and chromium.[24] Because the ink industry has moved to greatly restrict the heavy metal content of inks, the most likely reason for a waste ink to be considered hazardous is for it to be contaminated with ignitable solvents such as blanket and roller washes. Even a small amount of an ignitable solvent in an amount of waste ink will make the entire mixture hazardous.

RCRA Regulations for Hazardous Wastes

If a printer becomes a generator of hazardous waste, RCRA regulations come into play and will dictate how the waste will be stored, labeled, transported, and disposed of. Cradle-to-grave record keeping is also mandated. Historically, many printers have looked to their ink suppliers to remove waste ink for them. Although few ink suppliers wanted to perform this function, they often did it as a service to their customers. With the advent of RCRA, ink suppliers are increasingly unwilling to haul away waste ink because of the possibility that it may be contaminated with a hazardous substance, thereby placing them in violation of RCRA regulations.

With the return-the-ink-to-its-supplier option becoming less viable, printers are forced to find alternatives. One option is to hire an EPA-certified hauler to transport waste ink from the premises. If some of the ink qualifies as hazardous, then the RCRA regulations go into effect. First, the printer is considered the generator and, as such, must get an RCRA permit and remains responsible for the material even after it leaves the premises. If a load of hazardous waste is dumped illegally by the contracted hauler, the printer is still liable. Hazardous waste also requires a cradle-to-grave paper trail. A printer who ships hazardous waste without taking appropriate measures is subject to a $25,000 fine per day per violation. An ink supplier who knowingly transports a hazardous waste without an RCRA permit can be fined $50,000 per day per violation.[25]

Waste Ink Disposal Options

Once an ink—hazardous or nonhazardous—leaves a printing facility, it can be dealt with in one of a few ways. First, it can still be landfilled, but RCRA regulations on landfill operations have caused costs to skyrocket. For hazardous waste ink, the cost can exceed $600 for a 55-gallon drum. As an alternative, many gravure and flexographic inks have a BTU value that is adequate for their incineration. If the ink qualifies as hazardous, there are over 25 cement kilns in the country that burn these inks as a supplement fuel source, along with paint thinner, industrial cleaning solvents, and old tires. The 2,450° F and above temperatures inside the kiln destroy 99.9 percent of the organic wastes. The metals component of the ink becomes chemically bonded with the lime, clay, and other materials, thereby becoming part of the cement.

Instead of sending a nonhazardous waste ink off-site, some printers have invested in systems that recycle waste ink into usable ink. After installing an ink recycling system, one printing plant that had been spending $2,000 a month on disposal costs was able to recover over 1,600 pounds of ink in a week.[26] Still other printers are sending their waste ink to recycling facilities that reformulate it to produce a black ink that meets the individual printer's specifications. Heat-set, nonheat-set, and forms inks that contain no hazardous materials are compatible with this process. Printers are able to reduce the volume of waste ink that they must pay to have buried or burned and also are able to buy back their recycled ink at bargain prices.[27]

Waste Ink Reduction

With the high costs of waste ink disposal, prudent printers are paying more attention to simply *reducing the volume of excess ink that becomes waste ink.* According to the RCRA definitions, **excess ink** has never been removed from its shipping container. In contrast, **waste ink** was removed from its shipping container, but did not become part of a salable product; examples include ink removed from the fountain and press rollers, as well as ink mixed on a slab and remaining in basically empty containers.

Printers wishing to minimize the transformation of excess ink into waste ink are careful to keep the two from becoming mixed. Also, printers can reduce their use of solvents and switch to blanket washes with higher flash points or other properties that make them less hazardous.

Reading an MSDS

In 1985, federal law required that employees at production facilities be aware of chemicals being used or stored on the premises, the potential hazards of working with those chemicals, and the appropriate responses to spills, fires, or potentially dangerous exposure. To make this information available, **material safety data sheets (MSDSs)** for each chemical product are required to be kept on-hand by the user of the product. Ink, varnish, blanket wash, tack reducer, stripper's opaquing fluid, and rubber cement are all examples of materials that require an MSDS be maintained. In fact, a printer could be required to keep a file of over 200 MSDS forms. For an example, examine the MSDS shown in Figure 16–15 which describes the chemical composition of a

MATERIAL SAFETY DATA SHEET
FOR PRINTING INK AND RELATED MATERIALS

Fictitious Ink Company
333 Nebulous Parkway
Sausalito, CA XXXXX

DATE OF PREP. 6/12/95
PREPARED BY M. Cook

HAZARD RATINGS		
Minimal............. 0	HEALTH	1
Slight................ 1	FLAMMABILITY	1
Moderate 2		
Serious.............. 3	REACTIVITY	0
Severe............... 4		

SECTION I	PRODUCT IDENTIFICATION

Product class: Litho ink (sheetfed)

Trade name: Ficto Maxo

MANUFACTURER'S CODE ID: FI78-549

SECTION II	INGREDIENTS

Ingredient	CAS Number	% Wt.	OSHA PEL	ACGIH TLV	OSHA Hazard	CARCIN.
Synthetic resins	Various	30-50	N/A	N/A	No	No
Organic pigments	Various	15-30	3.5 mg/M^3	3.5 mg/M^3	No	No
Petrol.hydrocarbons	64742-46-7	15-30	5mg/M^3	N/A	No	No
Polymeric wax blend	N/A	5-10	15mg/M^3	10mg/M^3	No	No
Cobalt tallate	61789-52-4	0-1	0.1mg/M^3	N/A	No	No
Manganese tallate	61788-58-7	0-1	5mg/M^3	N/A	No	No

All ingredients listed on TSCA inventory.
Cobalt and Manganese listed on SARA 313 list.
Volatile organic compounds (VOC): 1.60-1.90 lbs./gallon
190-220 gms./liter
Method 24

SECTION III	PHYSICAL DATA

BOILING RANGE: 500-590°F	VAPOR DENSITY: HEAVIER ☒ (vs. air) LIGHTER ☐	LIQUID DENSITY: HEAVIER ☐ (vs. water) LIGHTER ☒	TYPE OF ODOR: Mild odor
APPEARANCE: COLORED PASTE	EVAPORATION RATE FASTER ☐ (vs. butyl acetate) SLOWER ☒	PERCENT VOLATILE WT. 20-25% ASTM D-2369	

SECTION IV	FIRE AND EXPLOSION DATA

FLAMMABILITY CLASSIFICATION: OSHA IIIB DOT NONE	FLASH POINT: 250° F METHOD USED: Closed cup	LEL N/A

EXTINGUISHING MEDIA:
☐ FOAM ☐ "ALCOHOL" FOAM ☒ CO_2 ☒ DRY CHEMICAL ☐ WATER FOG ☐ OTHER

UNUSUAL FIRE AND EXPLOSION HAZARDS None	SPECIAL FIRE FIGHTING PROCEDURES None

Figure 16–15 A typical first page of an MSDS identifies the hazard potential of a material. Subsequent pages are less technical, usually consisting of procedures to follow in storing and handling the material, as well as how to respond to a spill or fire.

lithographic ink as provided by the ink's manufacturer. The information identifies potential hazards for employees who work with or near the ink and also describes the most appropriate means of fighting an ink-related fire. In preparing this sheet, the ink manufacturer, in turn, probably relied heavily on information contained in the MSDS that accompanied the raw materials purchased from chemical manufacturers of other suppliers.

Unfortunately, the information contained on an MSDS is written by chemists and/or lawyers and holds little meaning for people unfamiliar with either scientific terms such as vapor pressure, flash point, and reactivity or bureaucratic jargon such as OSHA PEL and SARA Title III Section 313. Also, there is no uniform format for an MSDS; federal law **(OSHA)** stipulates which information is to be contained, but allows the supplier to use its own presentation. Efforts are underway to generate a consensus standard for the industry.

The following is an explanation of some of the key terms and abbreviations on the first page of a typical MSDS as represented on Figure 16–15. The potential value that an MSDS can offer a printer in worker protection, accident prevention, and pollution control should become evident as these terms are explained.

CAS number (Chemical Abstracts Service Registration Number)—the number assigned to a chemical that allows it to be located in *Chemical Abstracts,* a research publication produced by the American Chemical Society. The CAS number is a research tool that allows an individual to electronically search the abstracts and locate the titles of articles that discuss a particular chemical.

Percentage weight—the percentage of the compound that is comprised by that chemical. If half the weight of a compound is ethyl alcohol, then ethyl alcohol's "% weight" is 50.

OSHA PEL—the Permissible Exposure Limit as established by the Occupational Safety and Health Administration. On this sheet, the maximum exposure levels are expressed as milligrams per cubic meter of air. OSHA is the United States Department of Labor's agency that regulates and enforces industrial safety and health standards.

ACGIM TLV—Threshold Limit Values as created by the American Conference of Governmental Industrial Hygienists. The highest airborne concentration of a material that is not a likely risk to a healthy worker during normal conditions. Sometimes the heading is merely "TLV."

OSHA Hazard— a chemical material that meets one or more of seven criteria established by OSHA. These seven standards revolve around the material being reactive, flammable, toxic, or caustic.

Carcinogen—a material that either causes cancer in humans, or on the basis of animal research, is judged to have that capability.

TSCA Inventory—Toxic Substances Control Act Inventory is an EPA-mandated list of industrial chemicals.

SARA—Superfund Amendments and Reauthorization Act. A section of this act requires facilities with stored hazardous materials to inform local officials of the types, amounts, and locations of these materials to assist in responding to a spill, fire, or explosion.

VOC—Volatile organic compounds are liquid carbon compounds that evaporate very rapidly at room temperature. VOCs are in most inks to accelerate in the drying rate.

Boiling range—First, boiling *point* refers to the temperature at which a liquid begins to evaporate rapidly. The boiling *range* is given for a mixture of chemicals with different boiling points. A liquid with a low boiling point will evaporate at a fast rate.

Vapor density—A numerical ratio that compares the molecular weight of a gas to the molecular weight of air. A vapor density of 1.20 indicates a gas that is 20 percent heavier than an equal volume of air. Such a gas would likely concentrate in low areas.

Liquid density (ASTM)—American Society for Testing and Materials is an organization that establishes standards for material testing and use. Liquid density is a comparison between the density of a liquid and water. Liquids lighter than water will float—useful information in the event of a fire.

Evaporation rate—a number that compares a particular material's rate of evaporation with that of butyl acetate, which has been assigned an evaporation rate of 1.0. Ordinarily, three categories exist: Fast, ER higher than 3.0 (an example is acetone at 5.6); Medium, ER between 0.8 and 3.0 (an example is ethyl alcohol at 1.4); Slow, ER lower than 0.8 (an example is water at 0.3). However, MSDS sheets merely indicate whether the material's ER is higher or lower than butyl acetate.

Percent Volatile Weight—the percentage of the weight of a liquid or solid that will be lost through evaporation.

HMIS hazard ratings—the Hazardous Materials Identification System, developed by the National Paint and Coatings Association, rates compounds for the potential hazard that they offer relative to health, flammability, and reactivity. As shown on the MSDS form, there are five ratings to assess the degree of potential hazard. The ratings are presented and defined below as written in the HMIS Implementation Manual 1.

A material's *health* hazard risk is the potential to adversely affect the body of a human or animal. The five health ratings are as follows:

0—No significant health risk.

1—Irritation or minor reversible injury possible.

2—Temporary or minor injury may occur.

3—Major injury is likely unless prompt action is taken and medical treatment is given.

4—Life-threatening, major, or permanent damage may result from single or repeated exposures.

A material's *flammability* is its tendency to ignite and burn rapidly. The five flammability ratings are as follows:

0—Materials that are normally stable and will not burn unless heated.

1—Materials that must be preheated before ignition will occur; flammable liquids in this category have flash points at or above 200° F (NFPA Class III B).

2—Material that must be moderately heated before ignition will occur, including flammable liquids with flash points at or above 100° F and below 200° F (NFPA Class II and Class III A).

3—Materials capable of ignition under almost all normal temperature conditions, including flammable liquids with flash points below 73° F and boiling points above 100° F, as well as liquids with flash points between 73° F and 100° F (NFPA Class IB and IC).

4—Very flammable gases or very volatile flammable liquids with flash points below 73° F and boiling points below 100° F (NFPA Class IA).

A material's *reactivity* is its tendency to undergo a chemical reaction as the result of being heated, burned, or in contact with another substance. The reaction may generate (a) an increase in temperature and/or pressure or (b) chemical by-products that are toxic, corrosive, or noxious. The five reactivity ratings are as follows:

0—Materials that are normally stable, even under fire conditions, and will not react with water.
1—Materials that are normally stable, but can become unstable at high temperatures and pressures. These materials may react with water, but they will not release energy violently.
2—Materials that, in themselves, are normally unstable and readily undergo violent chemical change, but will not detonate. These materials may also react violently with water.
3—Materials that are capable of detonation or explosive action, but require a strong detonating source, or must be heated under confinement before initiation, or materials which react explosively with water.
4—Materials that are readily capable of detonation or explosive combustion at normal temperatures and pressures.

Printers often place MSDS files in looseleaf notebooks that are located in departments using chemicals. Appropriate employee training in interpreting the data is necessary before their safety potential can be realized. Environmental benefits include giving MSDS copies to local emergency officials such as the fire department and using them to evaluate the potential environmental impact of competing products; for example, the VOC levels of different inks could be compared. The information in an MSDS can make a printer aware of materials and allow the printer either to limit the inventory of those products or to substitute them with safer alternatives.

Summary

CONEG laws, RCRA, the Clean Air Act, and the Clean Water Act are all examples of environmental legislation that makes putting ink on paper a more complex and more expensive task than it was 30 years ago. The reader can probably recall once-common practices—such as using leaded gasoline and spraying crops with DDT—that later were banned when their damage to the environment was more fully recognized. It has also been the case, although more rarely, that recently-enacted regulations have been relaxed when they were found to be excessive or impractical. Admittedly, it is difficult for printers and their paper and ink suppliers to function during a time of constantly changing regulations, but these groups need to remember that environmental rules usually change as the result of our learning more about the complex effects of some on-going practice. As new ecological understanding occurs and new regulations are written that affect the paper, ink, or printing industries, it is crucial to remember that we are still in a learning mode about the ecology and, for this reason, new suspicions about and insights into what is harmful will be a constant.

Notes

1. Stephen M. Meyer, "Deadwood," *The New Republic* (August 2, 1993): 12–15.
2. Betsy Carpenter, "The Light in the Forest," *US News & World Report* (April 5, 1993): 34–40.
3. Seth Zuckerman. "New Forestry, New Hype?" *Sierra* (March/April 1992): 41–45, 67.
4. Debra S. Knopman and Richard A. Smith, "20 Years of the Clean Water Act," *Environment* (January/February 1993): 17–41.
5. David Moberg, "Sunset for chlorine?"*E Magazine* (July/August 1993): 26–31.
6. Carl Espe, "Capital Spending Plans: 1991–3," *Pulp & Paper* (January 1992): 71.
7. Moberg, p. 28.
8. Bruce Fleming, "Has EPA Gone Mad?" *Pulp & Paper* (February 1994): 138.
9. "Billion-Dollar Battle Against Pollution Continues," *Pulp & Paper* (January 1992): 72.
10. A. Hammond, ed., *The 1993 Information Please Environmental Almanac* (Boston: Houghton-Mifflin, 1992): 63.
11. W. E. Cunningham, ed., *Environmental Encyclopedia* (Detroit: Gale Research, 1994): 477.
12. Hammond, p. 63.
13. Joe Hastreiter, "Mill Sludge Landfill Volume Is Reduced by Pelletizing Process," *Pulp & Paper* (March 1993): 73.
14. "EPA's Toxins Regulations Anger Paper Industry," *Pulp & Paper* (December 1993): 21.
15. Ibid.
16. Rob Galin, "Industry Says Cluster Rules, Based on Inaccurate Data, Go Too Far," *Pulp & Paper* (April 1994): 100.

17. Ibid.
18. "Longview to Shut Down Bleach Plant," *Pulp & Paper* (February 1994): 21.
19. Theodore Lustig, "VOCs: Up in the Air," *Graphic Arts Monthly* (July 1989): 166.
20. Theodore Lustig, "Working Toward a Clean Environment," *Graphic Arts Monthly* (September 1992): 158.
21. William Rusterholz, "Regulating Metals, Pigments, and Printing Inks," *Paint & Coatings Industry* (June 1993): 64.
22. Thomas N. Donvito, Thomas S. Turan, James R. Wilson, "Heavy Metals Analysis of Inks: A Survey," *TAPPI Journal* (April 1992): 169.
23. Theodore Lustig, "Inks and Heavy Metals Don't Mix," *Graphic Arts Monthly* (August 1990): 132.
24. Lisa Cross, "Beyond Recycling," *Graphic Arts Monthly* (December 1988): 32.
25. James E. Renson, "Waste Ink and RCRA," *American Ink Maker* (January 1989): 19.
26. Lisa Cross, "Ink Waste Disposal," *Graphic Arts Monthly* (May 1989): 120.
27. Lisa Cross, "Ohio Company Tackles Ink Recycling," *Printing News/Midwest* (July 1991).

Questions for Study

1. Identify three major uses of the National Forests.
2. Give two reasons presented by environmental groups for cutting no more old-growth timber in the Pacific Northwest.
3. Explain how the Endangered Species Act is involved in the controversy over logging in old-growth forest areas.
4. Describe new forestry and explain how it may be useful in the controversy in the Pacific Northwest.
5. What potential for water pollution is presented in the pulping and papermaking processes?
6. Distinguish between the functions of primary and secondary water treatment.
7. What is BOD and why is it an important measurement of water quality?
8. Why are dioxin, DDT, PCBs and other chlorinated organic compounds potentially dangerous in even small amounts?
9. What measures are being taken by pulp and paper mills to greatly reduce toxic releases to the water?

10. Why do many people in the paper industry believe that EPA-proposed dioxin standards are inappropriate?
11. Describe how electrostatic precipitators and scrubbers work.
12. What are the sources of solid waste that is generated by the paper industry?
13. Describe RCRA provisions for landfill construction, operation, and maintenance.
14. What is meant by the term "cluster rules"? What is the paper industry's response to them?
15. Why are heavy metals considered a health risk?
16. What heavy metals are represented by the acronym CAMALS?
17. Identify the impact of CPSC and RCRA regulations on the printing industry.
18. How do CONEG laws impact ink makers and printers?
19. Describe the matrix effect of heavy metal analysis in inks.
20. What are VOCs and how do they affect the printing industry?
21. Why do many people in the ink industry question the validity of EPA Test Method 24 in assessing an ink's VOC content?
22. Itemize some RCRA provisions for the handling and disposal of an ink considered to be hazardous.
25. How can cement kilns be connected to the disposal of hazardous waste inks?
26. How can printers reduce the volume of waste ink and keep nonhazardous inks from becoming hazardous?

Key Words

wilderness	nonhazardous	multiple-use management
primary treatment	EPA	aeration
secondary treatment	new forestry	criteria pollutants
TCF	BOD	TRS
particulate pollutants	ECF	volatility
leachate	scrubber	CAMALS
VOC	hazardous	content certification
RCRA	heavy metals	flash point
corrosive	CONEG	waste ink
toxic	ignitable	reactive
MSDS	excess ink	
Clean Air Act	OSHA	
nonpoint-source pollution	point-source pollution	

Appendix A

Relative Humidity

Humidity deals with the amount of water vapor in the air. Water is always in the air to some degree. If a pot of water is enclosed, as shown in Figure A–1, water molecules will evaporate until the air in the container can hold no more; that is, it would reach its *saturation point.*

The saturation point of air is greatly influenced by its temperature. Table A–1 reveals this relationship by showing the capacity of water vapor at eight different temperatures. Observe how air's capacity for water vapor increases with its temperature. The reason for this relationship lies with the influence of heat on the movement of the air's molecules. The molecules of a gas are in constant motion that causes them to collide and change direction. The collisions are not terribly frequent, though, because the molecules are so far apart that they collectively occupy a thousand times greater volume than they actually need; the remainder of the space is empty. These empty spaces between the millions of gas molecules are where water molecules go when they evaporate.

When air is heated, its molecules' movement becomes much faster and their collisions are more intense, thereby exerting more pressure against their container. This reaction

Figure A–1 When a liquid is open in a container, some of the liquid's molecules will evaporate and mix with the gaseous molecules. If the container is closed, the air's saturation point will be reached eventually and any evaporation will be balanced by vaporous molecules returning to the liquid.

Air Temperature (F)	Vapor Capacity (in grams)
32°	4.8
41°	6.8
50°	9.3
59°	12.7
69°	17.1
77°	22.8
86°	30.0
95°	39.2

Table A–1 The capacity of 1 cubic meter of air to hold water increases as the air is warmed.

explains why a balloon expands when it is placed in the sunlight or near another heat source. The increased volume means that the same number of gaseous molecules now take up more room, so there is more empty space between them and a greater capacity for other molecules, such as water vapor. Therefore, air that is 86° F can hold 30.0 grams of water vapor, but if heated to 95° F, this capacity increases to 39.2 grams.

Air is very seldom saturated with water vapor. Whether at capacity or not, the number of water vapor molecules in a cubic meter of air is its *absolute humidity*. A more commonly used means of expressing air's moisture content is its *relative humidity*, which is a ratio of the air's moisture content to its capacity at that temperature. Table A–1 indicates that 22.8 grams is the air's capacity at 77° F. If a large balloon holding a cubic meter of air were to contain 11.4 grams of water vapor at 77°, its relative humidity would be 50 percent.

$$\text{Relative humidity} = \frac{11.4}{22.8} = .50$$

If the air in the balloon is chilled to 59° F, its gaseous molecules would lose much of their kinetic energy and would move closer together, causing the balloon to shrink. As a re-

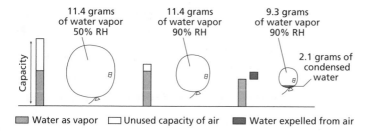

Water as vapor ☐ Unused capacity of air ▪ Water expelled from air

Figure A–2 As a container of air is chilled, its volume decreases, as does its capacity for water vapor. A balloon that contains 1 liter of air and 11.4 grams of water vapor has a 50 percent R.H. if the capacity is 22.8 grams. As its temperature falls to 59 degrees, its volume decreases and its R.H. increases as its 11.4 grams of water vapor becomes 90 percent of its new capacity, which is now 12.7 grams. As the air is further chilled, it reaches its saturation point. By the time it reaches 50 degrees, the volume is only .71 liters and the air's capacity is now only 9.3 grams of water; therefore, 2.1 grams of water have condensed out and are at the bottom of the balloon. Observe that further chilling will decrease the balloon's volume and expel even more liquid. The R.H. never exceeds 100 percent.

sult, less space would exist between these gaseous molecules and the capacity for water vapor would be reduced to 12.7 grams. Then the relative humidity would become 90 percent.

$$\frac{11.4}{12.7} = .90 \ (90\% \ \text{relative humidity})$$

The same amount of water vapor is being housed in a reduced volume (see Figure A–2).

If the same balloon is chilled further, the air's capacity for moisture will continue to drop; when it hits 11.4 grams, the air in the balloon will become saturated and have a relative humidity of 100 percent. Further chilling will begin to squeeze out vapor that can no longer be contained. At 50°F, the capacity is only 9.3 grams, so 2.1 grams would have already been expelled from the air ($11.4 - 9.3 = 2.1$). After their release, these molecules accumulate and condense into drops of liquid, a phenomenon commonly observed throughout nature. When moist air that is high in the atmosphere is chilled sufficiently, the result is rain. When moist air close to the ground is chilled sufficiently, the result is fog. The chilling of air and lowering of its capacity to hold moisture also explains why a glass of ice water will "sweat," or collect water on its outside surface. When the air immediately around the cold glass becomes sufficiently chilled, the

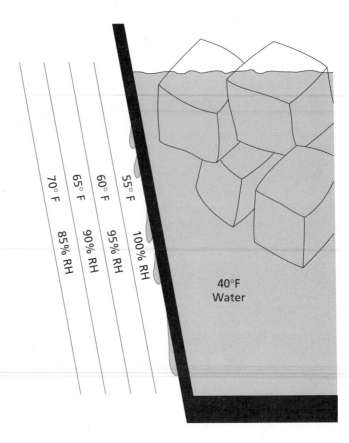

Figure A–3 The "sweating" of a cold surface is due to the chilling of surrounding air and a resultant lowering of its capacity to hold water as vapor. When this capacity is reached, water vapor begins to be expelled and it then condenses onto the cold surface.

expelled water vapor forms droplets on the glass. Obviously, a glass will sweat more quickly on a humid day when the air is already close to its saturation point (see Figure A–3).

Relative humidity is important to printers because a skid of paper brought into a pressroom from a cold warehouse will be several degrees colder than the air in the pressroom. If the skid is unwrapped and exposed to the air while still cold, it will (like the cold glass) chill the air around it, perhaps to its saturation point. Then the excess water vapor will be expelled from the air and be absorbed by the sides of the paper, thereby producing wavy edges that are likely to make the sheets unprintable.

Appendix B

Ink Wetting and Surface Tension

The concept of *surface tension* explains a number of peculiar observations that are commonly made regarding liquids: a water strider can literally walk on water, a steel needle can float, a glass of water can be filled to slightly beyond its capacity, and a drop of water always tries to form a spherical shape (see Figure B–1).

The key to all of these phenomena lies with the molecules of a liquid having an attraction for one another that requires a significant force to break. The spherical shape of rain drops results from these molecules trying to get as close as they can and, when in the shape of a sphere, they create the smallest total surface area possible.

For a water bug or a needle to break through a liquid's surface, it must overcome the attraction that the liquid's molecules have for one another. It may be helpful to think of a child stepping into one of these commonly-seen play areas that contains thousands of fist-sized plastic balls. As the child steps in, his foot pushes down between the balls until it contacts the floor of the room. However, if the attraction of each ball to its neighboring balls were somehow sufficient to exceed the force of the child's weight, then the child would actually be able to walk across the top of

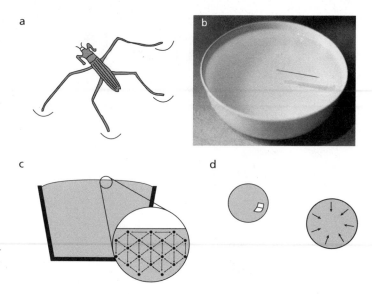

Figure B–1 Just as a water strider can walk on water (a), so can a steel needle float on the surface (b). The phenomenon of surface tension is also responsible for a container of water being able to be filled above the rim (c) and a falling drop of water's forming up in a sphere (d). In all these situations, the water molecules' attraction for one another minimizes the total area that they occupy and resists their separating.

the balls. In the case of the water strider and the needle, the weight is not sufficient to break the liquid's surface tension.

The importance of surface tension in printing concerns a liquid's ability to become more free-flowing and penetrating. When photographic film is washed, water spots will form unless a *wetting agent* (a detergent) is used to reduce the water's surface tension and help it to spread out across the film's surface (see Figure 14–9). Any chemical that alters the surface characteristics of a liquid is a *surfactant* (a word made from surface-active agent). Surfactants are used in dampening solutions to encourage total plate coverage. Ink vehicles require the addition of surfactants to enable them to more readily penetrate the nooks and crannies of pigment agglomerates. Also, liquid-in-liquid emulsification is more readily achieved when the liquids have had their surface tension lowered.

Appendix C

Solutions, Suspensions, and Emulsions

Inks and numerous other materials used in printing, such as film developer and blanket wash, are the result of combining two or more components to form a mixture. When certain components are combined, they produce a homogeneous mixture; that is to say, the mixture is uniform throughout because the two or more components have lost their individual identities and created something new. For example, if alcohol is mixed in water, a *homogeneous* mixture results, but if some mud is mixed in water, a *heterogeneous* mixture will be made. Although the particles of mud were stirred and distributed through the water, they never ceased to be tiny particles of mud and, if left to stand, will separate from the water by settling to the bottom of the container.

Homogeneous mixtures, called *solutions,* are produced when a solvent dissolves another substance, such as water dissolving salt to produce a saline solution. Also, a dye is made by dissolving a pigment with a solvent such as an acid. When two substances are combined to form a heterogeneous mixture, there is no solution; instead the result is a dispersion called a *colloid.* Smoke (solid-in-a-gas), mist (liquid-in-a-gas), and milk (liquid-in-a-liquid) are all colloids. Milk, which consists of fat particles in water, and all

Figure C–1 Emulsifying agents can prevent solute particles from coagulating by introducing molecules that surround the solute molecules. Detergents work in this fashion by surrounding oily dirt particles.

other liquid-in-liquid colloids are called *emulsions*. Some emulsions, such as oil-in-water, are temporary and, as soon as they are left undisturbed, their components will separate. Salad dressings that must be shaken to remix the separated liquids are also good examples of temporary emulsions because the tiny droplets of liquid A that have been dispersed throughout liquid B do not stay in suspension.

Not all emulsions are temporary, though. In homogenized milk, for example, the milk and cream do not separate out, so the emulsion is permanent. To produce a permanent emulsion, an *emulsifying agent* must be introduced to reduce the *surface tension* of the dispersing liquid (solvent) and surround each droplet of the dispersed liquid (solute). With all solute particles surrounded by molecules of the emulsifying agent, they cannot *coagulate* (group together) and terminate the uniform dispersion (see Figure C–1).

Printing inks usually are composed of both solutions and emulsions. The vehicle will probably contain a resin that was dissolved by, and is therefore in solution with, a solvent. At the same time, the tiny wax particles in the ink do not dissolve but rather emulsify through the vehicle. During lithographic printing, dampening solution comes into contact with the ink. Because these two materials are so different chemically they do not form a solution, but some of the dampening solution droplets do disperse through the ink. Controlled water-in-ink emulsification is a normal occurrence in lithography that actually improves the ink's performance. Excessive emulsification, however, produces a "waterlogged" ink that must be removed from the press.

Incidentally, not everything that is called an emulsion within the printing industry is really a true emulsion. Technically speaking, only a liquid-in-a-liquid colloid qualifies as a true emulsion. Nonetheless, the solid-in-a-liquid colloid that makes the light-sensitive coating on photographic film is referred to as an emulsion, as well. To apply a photographic emulsion to a film base, silver halide particles are dispersed through a liquid gel, which is then applied to a clear film and dried. The function of the gel is merely to uniformly distribute the silver halide particles across the film and then hold them in place. A solid-in-a-liquid colloid is actually known as a *sol,* but platemakers will continue to describe their exposures as being "emulsion-to-emulsion."

Appendix D

Volatility and Ink Evaporation

All liquids are capable of evaporation, but some liquids evaporate more readily than others. For example, water begins to boil at 212° F, but ethyl alcohol boils at 172° F, which is 40° "sooner." For this reason, alcohol is more *volatile* than water and, as a result, it dries faster.

Comparatively high volatility is needed in inks that dry through evaporation such as heat-set gravure inks. However, inks that are too volatile can dry on the press before they even get to the substrate. An ink's drying rate can be partly controlled by using solvents with the appropriate volatility.

Appendix E

Polymerization, Crosslinking, and Plasticizers

A very basic understanding of the units that comprise compounds is needed before discussion of *polymerization* is possible. An *atom*, the smallest particle that can exist as a stable entity, is the building block of chemistry. When two or more atoms combine, they form a *molecule*. For example, when a sodium atom links up with a chlorine atom, the result is a molecule of sodium chloride (NaCl), which is common table salt. By the same token, when an oxygen atom joins with two hydrogen atoms, the result is water (H_2O). Certain molecules are capable of combining with other molecules just like themselves to form long molecular chains. Such molecules are called *monomers* and the long chains that are formed are called *polymers* (from the Greek words for "many parts").

A good example of polymerization involves ethylene, which is a colorless gas with the formula C_2H_4. When acted on by a catalyst such as heat, UV light, or X-rays, ethylene molecules lose the internal stability that allows them to remain individual entities and they begin to link up, as shown in Figure E–1.

There are several examples of well-known or at least commonly-found polymers. When gaseous *ethylene* mole-

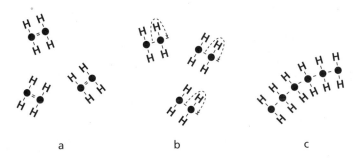

a   b   c

Figure E–1 Ethylene molecules are comprised of two carbon and four hydrogen atoms (C_2H_4). The double bond between the carbon atoms keeps the molecules stable (a). The presence of certain conditions breaks up the double bond and the potential for linking shifts to the end of each ethylene molecule (b). With this newly-found linking potential, the monomers begin to connect like railroad cars to form long polymer chains; in this instance, the new compound is polyethylene, used to make squeeze bottles and several other products on which screen printers often print.

cules become part of long chains, they lose their mobility, start to firm up and become the common plastic material known as *polyethylene*, the material used to coat milk cartons, insulate electrical wires, and form plastic squeeze bottles. In printing, it is used to package bread and other products, thus it is a common substrate in gravure plants. Other polymers are produced in a similar fashion. *Styrene* monomers combine to produce *polystyrene*, which is also used in packaging and *vinyl chloride* is the source for *polyvinylchoride* (PVC), which is used to make tubing (plumbing components), credit cards, and dozens of other products which are often printed.

The primary value of polymerization to the printer, however, is its role in the drying of ink. Ink vehicles that contain linseed, soya, or other drying or semidrying oil will polymerize when exposed to the air's oxygen and an appropriate catalyst such as a cobalt or manganese drier. Like the polymers descibed in the previous paragraph, when the vehicle's individual molecules lose their mobility, they form long chains and the entire ink film begins to lose its collective mobility and stiffen. As the chains get longer, the film becomes increasingly rigid and sets up.

For an ink film's molecules to become so immobile that the film will resist an aggressive rubbing, the process of crosslinking is effective. *Crosslinking* occurs when some of the shorter polymer chains attach themselves to pairs of long chains. When the giant polymers are linked, they can

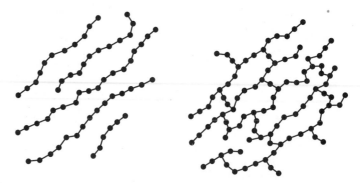

Figure E–2 After long-chained polymers have formed, the substance begins to firm up, but the long chains can still slide past each other, thereby allowing some degree of softness. When shorter polymer chains crosslink, they connect the long chains, brace the molecular structure, reduce mobility, and harden the compound.

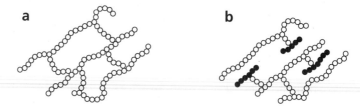

Figure E–3 Plasticizers place themselves between long polymer chains and prevent the rigidity that would result from a high level of interconnection.

no longer slide independently by one another and the resulting network produces a very firm film (see Figure E–2).

When instant polymerization is required, oxygen and a catalyst are not fast enough. In these situations, infrared or electron beam radiation serves as the catalyst for polymerization.

When an ink is applied to a flexible substrate such as a toothpaste tube or bread wrapper, the ink film must remain flexible enough to avoid flaking off when the substrate is bent or otherwise deformed. For an ink to retain enough flexibility to pass the "crinkle test," its formula must contain one or more *plasticizers*, compounds with molecules that make their way between the long polymer chains during the ink-drying stage and interfere with the crosslinking process (see Figure E–3). The optimum amount of plasticizer provides adequate hardness, while preventing excessive brittleness.

Appendix F

Specific Gravity

An old trick question goes "Which is heavier—a pound of lead or a pound of feathers?" The answer, of course, is that their weights are equal because each is one pound. The tricky part of the question lies with the fact that feathers are not very dense; that is, their weight-to-volume ratio is much lower than lead's.

The ratio between *mass* (weight) and *volume* is the *density* of a substance, expressed in the formula:

$$\text{Density} = \frac{\text{Mass (grams)}}{\text{Volume (cubic centimeters)}}$$

A substance with a high density is comparatively heavy. For example, lead has a density of 11.4, while pine wood has a density of .50. Examination of Figure F–1 reveals that a given volume of lead would weigh much more (over 22 times more) than the same volume of wood. Also, 1 pound of lead would have only $\frac{1}{22}$ of the same volume of a pound of wood.

The relative densities of substances are measured by *specific gravity*, which is the ratio between the density of a given substance and the density of pure water. The specific gravity

Figure F–1 Because of a big difference in their densities, a cubic foot of lead weighs over 22 times more than the same volume of wood.

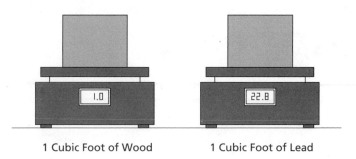

1 Cubic Foot of Wood   1 Cubic Foot of Lead

Figure F–2 A high specific gravity means a high density and a small volume per unit of weight. A pound of iron takes up less space than a pound of water.

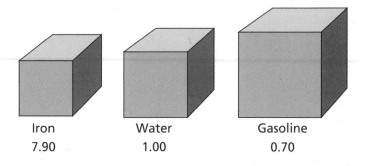

Iron   Water   Gasoline
7.90   1.00   0.70

Figure F–3 To understand the influence that an ink's specific gravity has on its mileage, think of an ink film, such as this halftone dot, in three dimensions.

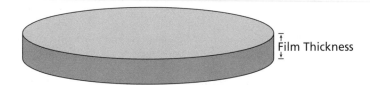

Film Thickness

(s.g.) of water is 1.0, so anything with a greater density has an s.g. of more than 1.0 and anything with less density has an s.g. of less than 1.0. For example, iron has an s.g. of 7.9, while gasoline has an s.g. of .70 (see Figure F–2).

The significance of an ink's specific gravity is its linkage of weight and volume. Although ink is usually purchased by the pound, its use is really actual volume-related in its use. Generally speaking, printers think of an ink film in terms of square inches—that is, in two dimensions. In reality, an ink film has a thickness and variations in it will profoundly affect the amount of ink used in a job. To better understand ink application in three dimensions, imagine a halftone dot as a short cylinder, as shown in Figure F–3. Because of differences in the covering ability of certain pigments, some ink colors require a higher pigment content than do others. Because pigment materials are comparatively dense, inks with high pigment content have higher specific gravities than do inks that cover adequately with less pigment material. For example, the specific gravity of a black ink could be as low as 1.2, while the specific gravity of a yellow ink could be as high as 2.2 with the result that a pound of the black ink would be nearly 2 times larger than a pound of the yellow ink and, therefore, cover a great deal more of a substrate.

Appendix G

Edible Paper?

In the movie *Night Shift,* Michael Keaton's manic character carries a tape recorder with him to capture the ideas that bombard him. The concepts that he announces into the recorder are funny in their absurdity and one of them is "edible paper." Although it sounds as goofy as his other ideas, the concept of paper as a source of food could potentially help solve two growing problems: waste paper and world hunger.

Paper is actually a carbohydrate—a compound made up of carbon, hydrogen, and oxygen. The starches found in corn and potatoes are also carbohydrates. In fact, the molecular formula for starch (amylose) and cellulose is the same. However, a subtle difference in their molecular structures makes one suitable for human consumption while the other is not. Note in Figure G–1 that, in both compounds, glucose units ($C_6H_{12}O_6$) are linked together by an oxygen atom. The subtle difference in question is the angle of the bonds around the connecting oxygen atoms; with amylose it is 120° and with cellulose it is a full 180°.

Enzymes in the human digestive system have no trouble breaking down the oxygen bonds of the amylose chains and reducing these chains to units of glucose and other starches.

Figure G–1 Only the angle of molecular bonding distinguishes the edible amylose from the inedible cellulose.

However, humans lack the enzymes needed to perform the same function with cellulose. Grazing animals—like cows, sheep, and horses—do have the appropriate enzymes that allow them to consume grasses and other sources of cellulose fiber.

Research is ongoing to find a way to either allow the human body to break down cellulose into starch or economically perform the function in a lab. If either development occurs, Keaton's character will be vindicated and real progress can be made in reducing world hunger.

Appendix H

Paper and Ink Development

1764 English patent granted for paper coating.

1774 Discovery of chlorine in Sweden.

1780 Hollander beaters widely replace stamping mills in American papermills.

1797 European paper first dyed to yellow and brown hues.

1798 Invention of paper machine by Robert (France).

1800 Invention of rosin sizing by Illig (Germany).

1801 English patent for paper machine awarded to John Gamble, who worked with Fourdrinier brothers to improve the machine's design.

1803 First papermill established in Canada (Quebec).

1812 Gamble's paper machine begins commercial operation (England).

1817 First American paper machine operates near Philadelphia.

1820 English patent granted to Crompton for drying cylinders to be connected to a paper machine.

1823 Gypsum first used as filler for paper.

1824 Invention of cylinder paper machine that forms cardboard (England).

1825 Dandy roll invented (England).

1826 Suction pumps applied to Fourdrinier machine to speed water removal during formation (France).

1827 First Fourdrinier machine operates in America (New York).

1830 Chlorine bleaching of pulp first performed in America.

1839 Ranson gets patent on improved drum driers (America).

1841 First groundwood paper produced in Western Hemisphere (Nova Scotia).

1842 First Christmas card produced (England).

1843 First production of safety paper to safeguard against erasure (England).

1850 Origination of the paper bag (handmade).

1854 Patent issued for chemical pulping of wood (America).

1856 First English patent granted for corrugated paper.

1860 Development of the Jordan refiner in America.

1871 Earliest use of rolled toilet paper in America.

1875 Earliest production of two-sided coated paper in America.

1876 First machine for producing paper bags.

1878 U.S. government begins the repulping of worn-out paper money.

1881 First large-scale production of coated paper in America.

1882 Production of first sulfite pulp in America.

1884 Invention of sulfate pulp.

1887 John W. Mullen creates the first device for testing paper.

1889 Annual paper production in America exceeds 1 million tons for the first time.

1899 Annual paper production in America reaches 2,167,593 tons—having doubled in the previous ten years.

1906 First paper milk bottles developed in California.

1909 First production of kraft paper in America.

1920 Paper machine speed reaches 1,000 feet per minute.

1923 Development of modern suction roll on the paper machine.

1933 Conical refiner.

1940 Annual American paper production exceeds 14 million tons—six times greater than during the turn of the century.

1956 Development of continuous digester.

1960 Twin-wire vertical paper former.

Glossary

-A-

abrasion resistance—a paper's resistance to being worn away by the rubbing action of an eraser, as measured by weight loss.

absolute humidity—the actual amount of water vapor in a volume of air; can be expressed as grams per cubic meter or ounces per cubic foot. See also *relative humidity*.

absorbency—the ability of a material to take in moisture that it contacts.

absorption—the act of a solid substance taking up a liquid.

acetone—a fast-drying solvent used in gravure inks.

acid-free paper—paper containing acid or acid-producing chemicals that would cause it to deteriorate prematurely; usually papers having a 7 or 8 pH value.

acid pulping—the sulfite method of chemically pulping wood chips with a cooking liquor containing sulfur dioxide.

acid rain—rain that has a pH below 7; it results from sulfur compounds being placed in the upper atmosphere, by either natural causes (such as volcanoes) or air pollution.

acid size—rosin sizing. See also *rosin size*.

acrylic resins—binding agents resulting from the polymerization of acrylic or methacrylic acid; a thermoplastic material used in many common products such as Plexiglas©.

additive—a substance added

to a compound such as ink or paper coating.

additive color—a system whereby the three primary colors of light (red, blue, and green) can be combined in varying degrees to produce a wide range of colors.

adhesion—the bond of a material (like ink) to a solid surface.

adhesive coated paper (also called *adhesive paper*)—paper coated on one side with one of several types of adhesives such as water-activated, solvent-activated, heat-seeking, or pressure sensitive.

adsorption—concentration of gas or liquid molecules adhering to a surface; adsorption can remove the drier from an ink vehicle and impair drying.

advance growth—young trees present before a clearcut is performed.

aeration—the introduction of air into a liquid; usually performed to increase oxygen content.

AFPA—American Forest and Paper Association, a New York-based group that generates weekly statistical reports and forecasts on the production and consumption of paper, paperboard, and other forest products. Formerly was the American Paper Institute.

after-tack—tack that develops after an ink has appeared to dry.

agglomerate—a cluster of pigment particles in an ink.

air-knife coating—coating method employing a blast of air to remove excess coating from the sheet.

airmail paper—a lightweight and opaque writing paper with a basic size of 17 × 22 inches and a substance weight range from 5 to 9 pounds.

air pollution—gases, liquid droplets, or solid particles in the air that threaten one or more aspects of the environment.

air shear burst—the type of web break caused by air entrapped during the winding of the roll.

alcohols—a family of organic solvents used in flexographic and gravure inks.

alkaline—containing an alkali; having a pH greater than 7.

alkaline paper—paper with a pH at or above 7.

alkaline sizing—a sizing technique that produces alkaline paper.

alkyd resins—(pronounced AL-kid) a group of synthetic resins used in formulating printing inks; they are polymers resulting from combining an alcohol with an acid.

alpha cellulose—pulped fibers with very high purity, brightness, and strength.

alpha pulp—wood pulp containing more than 90 percent of alpha cellulose.

alum (papermaker's alum)—

aluminum sulfate which is used with water to precipitate rosin onto fibers during internal sizing.

alumina hydrate—a white inorganic pigment used as an ink extender.

ambient conditions—surrounding conditions: in paper and ink applications, the temperature, pressure, and humidity of the surrounding air.

anhydrous—free of water content.

animal size—gelatinous material used to size certain papers, such as bonds and ledgers.

antioxidants—agents that retard the oxidation of drying oils and reduce skin formation of inks.

antique finish—a comparatively rough surface obtained through minimal pressure during the wet felt and calendering stages.

antiskinning agents—see *antioxidants*.

apparent density—weight per unit volume; obtained by dividing the substance weight by the caliper.

archival paper—paper that is produced to be permanent, usually made from cotton or linen fibers.

aromatic solvents—solvents used in printing ink resins.

artificial parchment—a paper resembling vegetable parchment; it possesses extremely uneven formation, a crisp feel, and high density.

asbestos—white asbestos is a fibrous crystalline magnesium silicate and can be spun or woven to create a paper that resists heat.

ash content—the percentage of inorganic material, such as filler, that remains after paper's combustion.

-B-

back-edge curl—see *tail-end hook*.

backing away from fountain—a condition in which ink with inadequate flow fails to maintain contact with the fountain roller, leading to inadequate ink supply throughout the ink train.

backtrap—the phenomenon found in multicolor lithographic printing where a just-printed image transfers from the substrate to the blanket of a subsequent printing unit; phrased differently, a blanket lifts the still-wet image from the paper; limited to multicolor presses.

bagasse (pronounced buh-GAS)—papermaking fiber derived from stalks of sugar cane after extraction of the sugar.

baggy roll—a roll of paper that unwinds with slash areas across its width; the roll commonly shows wrinkles, as well.

396 **ball mill**—a machine that uses moving balls to mix the components of a printing ink during its manufacture.

bank note paper—high-grade bond paper usually made with high cotton content; range of substance weight is 20 to 24 pounds.

base—a combination of one pigment or dye in a vehicle; more than one base can be used to formulate an ink.

base stock—paper that will be coated or laminated.

basic size—specified sheet dimensions from which are derived the substance weight of a given category of paper; for example, the basic sizes for business, book, cover, and newsprint are 17 × 22 inches, 25 × 38 inches, 20 × 26 inches, and 24 × 36 inches, respectively.

basic weight—see *substance weight*.

bast fibers—fibers obtained from the inner bark of a wood plant such as flax, hemp, and jute; linen is made from bast fibers found in the flax plant.

beater—tub-like machine that circulates a suspension of fibers in water with the purpose of flattening them between a rotating drum with metal bars and a bed plate; invented in Holland and commonly called a Hollander beater.

beating—the flattening and roughening of cellulose fibers prior to their formation into a sheet.

bending quality—the ability of folding cartons to be creased and scored accurately at high speed.

bible paper—thin, lightweight opaque paper used in books that require low bulk such as bibles and dictionaries, the basic size is 25 × 38 inches and substance weights range from 18 to 30 pounds; the predominant chemical wood pulp is generally augmented with cotton or linen fibers to improve strength.

binder (also called *resin*)—a substance used to improve the bonding of fibers, coatings, or inks.

binding varnish—a viscous varnish that toughens the film of dried ink.

bioaccumulate—the phenomenon of certain ingested compounds building up in the organism rather than being expelled.

biodegradable—an organic material that can be broken down to its basic elements or compounds.

biomass—the amount of a biological substance after water has been removed.

black liquor—the total quantity of living organisms of one or more species, at a particular time, usually expressed in weight per unit area; the spent cooking liquor washed from the pulp after it has been discharged from the digester.

blade coating—see *trailing blade coating*.

blank—a heavy type of paperboard that may or may not be coated; caliper ranges from 15 to 48 points and the basic size is 22 × 28 inches.

bleach—(1) a chemical that whitens pulp through oxidation or reduction. Commonly used chemicals are chlorine hypochlorite, chlorine dioxide, hydrogen peroxide, ozone, and sodium hydroxide. Chlorine in its elemental form (Cl_2) was once commonly used but is declining in popularity. (2) A method of measuring an ink's tinctorial strength by mixing it with a white base and observing the degree of color loss.

bleaching—the process of using chemicals to whiten pulp.

bleach test—a technique for comparing ink color strength by diluting two ink samples with white pigments, creating a thin film of each, and visually inspecting them.

bleed—the spreading of an ink to an unwanted area.

blister—the bubblelike bulging of moisture content being converted to steam during heat-set drying.

blocking-in-the-pile—the sticking together of printed sheets due to the ink's drying too slowly.

blotting paper—an unsized paper used where absorbency is important; the substance weights range between 35 to 50 pounds and the basic size is 24 × 36 inches.

blueprint paper—used to carry architectural or engineering drawings, usually made with cotton chemical wood pulp. Substance weights range from 12 to 30 pounds and the basic size is 17 × 22 inches.

board—see *paperboard*.

BOD (biochemical oxygen demand)—the amount of oxygen required in the biochemical breakdown of organic matter; a standard commonly applied to the contents of wastewater and landfills.

bodied oil—a drying oil that has had its viscosity increased, usually by heat.

body—an ink's viscosity.

body gum—linseed oil that has been polymerized to a gummy condition (an example is #8 varnish).

bodying agent—any material that increases an ink's viscosity.

bogus—a term sometimes applied to papers containing a high level of inferior or recycled fibers as in bogus bristol, bogus manila, and bogus tag.

bonding strength—see *surface strength*.

bond paper—a paper classification composed of chemical wood pulp and/or cotton usually used in letterhead stationery and business forms; the substance weight ranges from 13 to 24 pounds and the basic size is 17 × 22 inches.

book paper—a term applied to a large classification of printing papers exclusive of newsprint and bonds. Book papers are available in a variety of finishes and can be coated or uncoated. The substance weights range from 22 to 100 pounds and the basic size is 25 × 38 inches.

box board—paperboard used in the fabrication of cereal boxes and similar products; usually consists of three layers: top liner, filler, and bottom liner. Its composition often includes recycled fibers.

break—the complete rupture of a web of paper during manufacture or printing.

brightness—the percentage of light (at a prescribed wavelength) that is reflected, rather than absorbed; related to whiteness.

brilliance—a combination of brightness and strength.

bristol—a heavyweight printing paper, usually 6 or more points in caliper. The name comes from Bristol, England, where it was first produced.

brittleness—the tendency of a material to break when deformed by bending.

broadside—a large printed sheet that is folded down to a smaller size; distinguished from a folder because the images run across the folds.

brochure—an elaborate printed piece costing at least $1 per copy to produce.

broke—paper that is discarded at any point during its manufacture and repulped. *Wet broke* occurs during the wet press stage, while *dry broke* results when paper is damaged while being dried, calendered, or otherwise finished.

broken carton—a carton of paper that has been opened and had only some of its sheets removed. An order of paper for less than the entire contents of a carton is considered a broken carton order and will usually cost more per pound or sheet than will an order for a whole carton.

broken ream—an order for fewer than 500 sheets; the price per pound or sheet for a broken ream order is higher than if an entire ream were ordered.

brush coating—the process of using a series of oscillating flat brushes to smooth out wet coating that has been applied to the sheet.

buffer—alkaline compound added to paper to neutralize any acidity that a publication may acquire over a period of decades.

bulk—(1) the thickness of a sheet of paper relative to its weight; (2) the thickness of a number of sheets or pages. Thick sheets have *high bulk*; thin sheets have *low bulk*.

bulking book paper—a book paper with materials and procedures to maximize its bulk; filler content and calendering are kept low.

bulking number—the number of sheets of printing paper in one inch.

bulking thickness—the required caliper for a sheet in order for a certain number of pages made from that sheet to equal a specified thickness.

bulky—a sheet that has high bulk (high caliper for its weight).

bundle—50 pounds of paper (for shipping purposes).

burlap—a fabric coarsely woven from jute, flax, or hemp; often used to make bags.

burlap finish—a finish that resembles the texture of burlap, sometimes created by wrapping burlap around calender rollers.

burst—a web rupture that does not extend to either edge; often due to uneven winding or variations in caliper.

burst factor—the value derived by dividing the burst strength by the substance weight.

bursting strength—the ability of a sheet of paper to resist rupturing when a blunt object is pressed against it.

business communication

paper—paper used in stationery, business forms, and copier papers.

business forms bond—a bond specifically made to produce continuous forms on web presses.

-C-

C1S—coated one-side.

C2S—coated two-sides.

caking—see *piling*.

calcined—a material that has been transformed to an ashy dust through heat.

calcium carbonate ($CaCO_3$)—a white pigment used as a filler and a coating pigment; chalk is a natural occurring form of calcium carbonate in alkaline-sized papers. Ground calcium carbonate (GCC) is limestone that has been ground to microparticle size; precipitated calcium carbonate (PCC) is a purified form of GCC in which the particles assume exotic shapes that allow them to impart greater opacity to paper than does GCC.

calcium sulfate ($CaSO_4$)—a compound used as a filler; when found in nature, it is called gypsum.

calcium sulfite ($CaSO_3$)—a compound used as a filler.

calender—a vertical stack of cast-iron rollers through which paper passes immediately after leaving the dryer section of the paper machines. Calendering improves caliper uniformity and increases smoothness. See also *supercalender*.

calendering—to pass paper between metal rollers to increase smoothness.

caliper—the thickness of a sheet of paper, usually expressed in thousandths of an inch or points (.004-inch equals 4 points).

campaign bristol—a .010-inch caliper coated bristol named for its being commonly used for political campaign post cards. The substance weight is 120 pounds with a 22½ × 28½-inch basic size.

capacity—the maximum output of a paper machine over an extended period of time, usually expressed as tons per day.

capillary action—the pull of a liquid into a thin tube.

carbon black—a black pigment that results from burning oil or natural gas with a limited air supply.

carbonless paper—paper coated with two different coatings, one on each side. Sharp pressure applied to two or more of these sheets arranged correctly will generate a copy as the chemistry of the two coatings reset.

carbon paper—paper coated on one side with carbon inks which transfer an image with applied pressure.

cardboard—a general term for a bristol board .006-inch or thicker, usually made from recycled fiber; mistakenly used by the general public to refer to corrugated stock.

carload lot—the amount of paper required to ship paper at the carload freight rate, ranging from 36,000 to 100,000 pounds.

Carson, Rachel (1907–1964)—American marine biologist and writer; best known for writing *Silent Spring*, a book that warned of the dangers of pesticides.

carton—a shipping unit of sheets of paper usually weighing between 125 and 150 pounds. The number of sheets in a carton varies greatly with the sheet size and substance weight.

casein—protein derived from skim milk and used as a coating adhesive.

case lot—a quantity of flat paper or paperboard consisting of four cartons or 500 to 600 pounds.

cast-coated paper—paper whose wet coating is dried against a highly-polished chrome drum, high-gloss finish. The surface has superior ink receptivity to sheets with gloss achieved through supercalendering.

catalog paper—a lightweight coated or uncoated sheet used primarily in mail order catalogs and directories to reduce postage costs. The basic size is 24 × 36 inches and the range of substance weights is 9 to 28 pounds.

catalyst—a substance that speeds a chemical reaction without being permanently changed by the reaction.

causticizer—a stage in the pulping process where new sodium hydroxide (caustic soda) is created from spent sodium hydroxide.

caustic soda—sodium hydroxide (NaOH), a strong alkaline used in the sulfate method of chemical pulping, as well as in bleaching.

cell—a tiny, etched, ink-filled depression in a gravure cylinder.

cellulose fiber—the primary component of woody plants. Cellulose fibers move water throughout the plant and their hygroscopic nature makes them well-suited as the basis for paper.

centipoise—a unit measure of viscosity; one-hundredth of a poise.

centrifugal cleaners—devices that use a cyclone action to remove dirt and other contaminants from pulp.

chalking—the inadequate bonding of ink onto paper because the ink's vehicle was absorbed too rapidly by the paper.

chart paper—a sheet with good fold strength intended for charts and maps; it has a 17 × 22-inch basic size and a range of substance weights of 9 to 28 pounds.

chemical ghosting—faint images that appear on the back side of printed sheets due to the chemical interaction that drying inks can have on one another.

chemical oxygen demand (COD)—a measure of the oxygen-consuming capacity of organic and inorganic matter in water and wastewater.

chemical pulp—pulp that was produced by cooking wood chips or other plant sources with chemicals and applied heat. The three main processes are the sulfate (kraft), sulfite, and soda methods.

china clay—a naturally white mineral used as a paper coating pigment and as an ink extender.

chip—a small piece of wood used in chemical pulping.

chipboard—a low-density board with a caliper of .006-inch and thicker.

chipper—a machine that reduces wood logs or whole trees to chips with a rotating disk that contains heavy knives.

chlorination—bleaching wet pulp with a compound containing chlorine, such as chlorine dioxide.

chlorine—a poisonous gas used to bleach pulp.

chlorine dioxide (ClO_2)—a compound used to bleach chemical pulp.

clarifier—a wide, shallow basin that allows solid particles to precipitate from water; used in the primary treatment of wastewater.

clay—a natural, fine-grained material, primarily aluminum silicate, used in paper coatings and filler.

Clean Air Act—1963 federal legislation, amended in 1977 and 1990, that charges the EPA with researching and measuring air quality, setting standards, and enforcing those standards.

cleaning—elimination of competing vegetation from timber stands not taller than 4-½ feet.

Clean Water Act—*see Federal Water Pollution Control Act Amendments.*

clearcut—a tree harvesting technique that removes all trees from an area with the intent of producing a new, even-aged forest.

clear-strip cutting—a shelterwood cut in which parallel strips no wider than the trees' mean height are clearcut perpendicular to the prevailing wind.

cluster rules—pollution rules generated by the United States Environmental Protection Agency that integrate air and water pollution controls.

coating—the layer of pigment and adhesive applied to one or more sides of a sheet of paper.

coating binder—the adhesive component of a coating mixture, either a starch, protein, or synthetic compound.

coat weight—the amount of coating applied to paper, expressed as pounds of air-dried coating on a ream of 25 × 38-inch stock.

cobalt drier—a cobalt-containing material used to speed oxidation and polymerization of an ink film.

Cobb size test—a procedure of measuring the degree that paper has been sized by measuring the amount of water absorbed by the sheet during a specified period of time.

cockle—a puckered surface resulting from nonuniform drying.

cockle finish—a puckered surface that resembles handmade paper and, for this reason, is considered prestigious; the cockle finish is difficult to produce consistently and is, therefore, comparatively expensive.

cockling—the act of creating a cockled or puckered surface on a sheet of paper.

coefficient of friction—a numerical value that reflects the force required to (a) get one surface to slide across another (starting friction); or (b) maintain that movement (sliding friction). The higher the coefficient of friction, the more force is required. For example, the starting friction for steel on Teflon is 0.04; for rubber on wet concrete, 1.5; and for rubber on dry concrete, 2.0.

colloid mill—a machine used to disperse pigments throughout an ink formulation during its manufacture.

color—the property of a substance that determines the wavelengths of light that are reflected and visually perceived by an observer.

colorant—component of an ink that supplies its color.

color burn-out—an undesirable color change in a printing ink.

color gamut—in printing, the range of colors that can be achieved by a set of inks, usually the process colors of cyan, magenta, yellow, and black.

colorimeter—a test instrument that analyzes a color's hue, saturation, and lightness.

color strength—see *tinctorial strength*.

commodity grade papers— general term used for low-end grades of bond and book paper, usually #5 grades.

communication papers—a broad term that includes fine papers such as business, book, and cover sheets as well as newsprint because of their frequent use in the communication process. Formerly, the term was applied to paper to be perforated and used with telegraph machines.

complementary colors—colors that are on opposite sides of the color wheel from one another.

composition—the quantitative representation (usually by percent) of a tree species within a forest.

compressibility—the degree of thickness reduction that results from applied pressure.

CONEG—Coalition of Northeastern Governors.

cones—(1) nerve cells found in the rear wall of the eye that are sensitive to light and are responsible for the perception of color; (2) the seed cluster of coniferous trees.

coniferous—trees that produce cones, are usually evergreens, and are also called softwoods.

content certification—a document that lists the chemical composition of a material.

continuous pulping—the pulping of chips in a continuous digester instead of in a batch digester; the chips progress through a large tube and are cooked as they move along.

contrast ratio—the ratio between the amount of light striking a sheet of paper and the amount of light reflected with a black surface under the sheet, a test for measuring *opacity*.

conversion—the process of transforming paper into envelopes, bags, or other containers.

conversion coating—the process of coating paper after it has been removed from the paper machine.

coppice—a method of tree generation in which new growth originates from remaining stumps or roots.

cord—a volume of pulpwood measuring 8 feet long, 4 feet

wide, and 4 feet high; contains 128 cubic feet.

core—a tube around which paper is wound.

core waste—paper nearest the core that is not used for printing because of the severity of its curl.

corrugated board—a board usually comprised of a fluted (wavy) medium faced by one or more flat sheets; among the general public, this is incorrectly called cardboard.

corrugated medium— paperboard used to produce the fluted (wavy) component of corrugated board.

cotton—a plant with seed hair cellulose that produces paper with excellent strength and permanence.

cotton content—the percentage of cotton fiber in a sheet of paper, usually in amounts of 25, 50, 75, and 100 percent.

cotton linters—cotton fibers too short to be useful in textile manufacturing and sometimes used in papermaking.

couch roll (pronounced COOCH)—the large roller at the end of a Fourdrinier machine's wire; the point of departure for the paper from the wire.

cover paper—a grade of medium-weight, chemically-pulped paper that can be either coated or uncoated. Cover sheets are commonly manufactured to match book or text sheets. The basic size is 20 × 26 inches and the range of substance weights is 40 to 130 pounds.

CPSC (Consumer Product Safety Commission)—a federal agency created in 1973 to issue and enforce safety standards and performance requirements for consumer products.

cracking—the breaking of the coating layer when a sheet is folded.

cradle-to-grave—an expression describing the process of monitoring a material from its creation, through its applications and uses, and to its final disposal.

crash finish—paper with a textured surface created by contact with a roller covered with a coarse material such as burlap.

creped (pronounced KRAYPT)—the crinkly effect imparted to wet paper by crowding it with a doctor blade as it moves forward.

criteria pollutants—six air pollutants that have had national air quality standards established for them by the EPA according to criteria such as health effects, damage to buildings, and visibility. The seven criteria pollutants are sulfur dioxide, carbon monoxide, nitrogen dioxide, ozone, hydrocarbons, lead, and total suspended particles.

crop tree—a tree determined to be valuable enough to be left to mature within the forest.

crosslinking—the linking of two or more parallel long polymer chains by either a single molecule or short-chain polymer; the bracing effect of crosslinking prevents the long polymer chains from gliding past one another, thereby solidifying the material.

crystallization—term used by printers to describe the repelling by one ink film of a second ink film printed over it.

curing—conversion of a wet ink film or coating to a solid film through polymerization as opposed to mere vehicle evaporation, which is called drying.

curing inks—inks that cure instead of dry.

custom order—an order for the manufacture of a material with specifications created by the customer.

cut—to use solvents to thin an ink or varnish.

cut size—bonds and writing papers cut into 8½ × 11-inch and 8½ × 14-inch sheet sizes.

cutter dust—small coating particles loosened from the sheet during the cutting operation.

cyan—a greenish-blue ink that is one of the four process colors used in printing.

cylinder machine—a papermaking machine in which one or more mesh-covered cylinders replace the wire of a Fourdrinier machine; the flow of furnish through the revolving cylinder causes the fibers to form against the mesh and ride it onto a felt. Several cylinders in a machine can be used to form paperboard too heavy to be made on a Fourdrinier machine.

-D-

dandy roll—a mesh-covered cylinder that revolves above the fibers matting on the wire of a Fourdrinier machine; the purpose may be to burst any air bubbles or carry the metal designs if the paper being made is being watermarked.

deciduous—those tree species that lose their leaves each year; also known as hardwood trees.

deckle—in hand papermaking, the removable, rectangular wooden frame that runs above the perimeter of the mold and contains the pulp over the wire mesh.

deckle edge—originally, the outside edges of a sheet of handmade paper that touched the deckle as the sheet was being formed; machine-made paper is artificially given a deckle edge by various means to simulate the look of handmade paper.

deed paper—paper made from cotton to increase permanence and durability.

defoamer (also called *antifoam*

agent)—a chemical agent that inhibits foam formation and breaks bubbles that form.

degree of contamination—a measurement of the ratio between a desired material and foreign matter.

de-inking—the removal of nearly all of the ink from reclaimed paper.

delamination—separation of layers of a multiply sheet of paper.

delignification—the process of removing lignin from wood fibers prior to bleaching.

densimeter—a machine that measures a sheet's porosity.

densitometer—an instrument that measures a reflected or transmitted light.

density—weight per volume.

diffusion—the spreading of a substance into its surroundings.

digester—the container in which usually wood chips are reduced to fibrous pulp with chemicals, high pressure, and high temperature.

diluent—a liquid that mixes with a heavier liquid and reduces its viscosity.

dimensional stability—the degree that paper will change in its length and width due to fluctuations in moisture content or applied strength.

diploma parchment—paper made to resemble animal parchment; made from cotton fiber and sized with animal glue; basic size is 17 × 22 inches and the substance weight range is 48 to 88 pounds.

direct dyes (also called *substantive dyes*)—a class of light-fast aniline dyes with a high affinity for cellulose fiber.

dished—sheets of paper that are concave; that is, the corners are higher than the center; usually caused by exposure to dry air.

disk refiner—a device consisting of two revolving disks between which fibers are forced, flattened, and roughened.

dispersion—a uniform mixing of solid particles and a liquid medium.

dispersion agents (see also *wetting agent*)—materials that improve dispersion of pigments through a liquid medium.

doctor blade—a blade that presses against the surface of paper during its manufacture to lightly scrape off excess pulp or sizing material.

document parchment—a vegetable parchment resembling animal parchment; used for diplomas and other documents. The basic size is 17 × 22 inches and the substance weight range is 48 to 88 pounds.

dot slurring—the elongated smearing of halftone dots at their trailing edge.

double coated—a sheet that has been coated twice, usually by different methods, on the same side.

doubling—the printing of an image twice, with the second image faint and out of register.

doughnut hickey—appearance of an unintended tiny white circle within the image area of a printed sheet.

drawdown—to lay down a film of ink by pulling across a sheet of paper a smooth edge blade that is behind an amount of ink; the purpose is to compare an ink's *masstone* with its *undertone*.

drier—an ink additive that speeds drying.

drum driers—a series of heated metal cylinders that the paper web contacts after leaving the wet press section of a paper machine.

dry back—color change of an ink due to drying.

drying—the solidification of a liquid film due largely to evaporation.

drying oil—oils that harden upon oxidation.

drying rate—a measurement of the time required for ink to lose its liquid components due to evaporation.

drying time—the length of time between the printing of an ink onto a surface and its forming a tack-free surface.

dry pigment (also called *dry color*)—pigment material in a dry or powder form.

ductor roller—the roller of an ink or fountain train that makes intermittent contact with the fountain roller.

dull coated—a coated sheet with comparatively coarse coating pigments and light supercalendering to reduce gloss to less than 55 percent.

duplex—a sheet consisting of two plies or sheets, usually of two different colors.

duplicating paper—a hard-surfaced paper designed for use in spirit duplicators. The basic size is 17 × 22 inches and the substance weight range is 16 to 20 pounds.

duplicator—a small lithographic press that prints by either the lithography, spirit, or mimeograph process.

durability—the ability of a sheet to retain its original properties after prolonged use.

dusting—the accumulation of slitter dust particles onto the nonimage areas of an offset blanket.

dyes—coloring materials that are soluble in a vehicle or solvent.

-E-

ecology—the study of the relationships among organisms and their environment.

edge tear—a short tear that originates at the edge of the web and can extend into a web break.

E.F.—English finish.

effluent—polluted matter discharged into the environment from a single source.

efflux cup—a device that measures the viscosity of flexographic or gravure inks by measuring the amount of time required for an amount of ink to flow through a hole in the bottom of a metal cup.

eggshell finish—a paper's finish that resembles an eggshell and, therefore, is rather rough; created during the wet press stage with special felts before the sheet is dried.

electrostatic precipitators—devices that remove solid particles from gases by electrically charging the particles and then pulling them out of the gas with screens that have the opposite electrical charge.

elemental chlorine—chlorine that has not combined with other elements to form a compound.

elemental chlorine-free (ELF)—in papermaking, the process of bleaching pulp without the use of elemental chlorine.

Elmendorf test—a method of determining a paper's tear strength.

embossed finish—a raised or depressed pattern imparted to paper by running it between an engraved metal roller and a soft roller.

emulsification—(in lithographic printing) the distribution of fountain solution in the ink.

emulsion—a mixture of two liquids that are insoluble with one another; an example is a shaken container of water and oil.

enamel coated—paper with a high gloss.

endangered species—a plant or animal considered to be in danger of extinction.

Endangered Species Act—1973 federal legislation, amended in 1988, that charges the Fish and Wildlife Service with conserving the ecosystems of species whose existence has been determined to be endangered.

end sheet—the heavy, strong pages that connect a hardbound book's cover to its contents or text.

end-use—the intended and usually final application of a product.

English finish—paper that is uncoated, yet smooth through heavy calendering; the basic size is 25 × 38 inches and the substance weight is 45 pounds.

engraving—an intaglio printing process in which ink is carried in grooves physically carved into the plate.

envelope paper—paper intended to be made into envelopes; must be opaque and perform well with die cutting, folding, and gumming.

envelope stock—see *envelope paper.*

EPA (Environmental Protection Agency)—an independent federal agency established in 1970 to research the causes and effects of pollution as well as establish and enforce adherence to pollution control standards.

epoxy resins—synthetic thermosetting resins that contain highly-reactive epoxy groups that are quick to crosslink and cure rapidly.

equilibrium relative humidity—the relative humidity in which a given pile of paper will neither gain nor lose moisture.

equivalent weight—the weight of a ream of paper in a given size as calculated from its substance weight.

erasability—the ability of paper to withstand the erasure of a typed or written image on its surface.

erasable bond—a very heavily surface-sized bond that erases with minimum fiber disturbance.

esparto—a grass found in North Africa and Spain that is used in papermaking.

etch—to create an image by using acid to remove areas from the plate material.

evaporation—moving from a liquid to a gaseous or vapor state, as when an ink's solvent leaves the ink film.

even-aged—a forest that has a very small range of tree ages.

expansivity—the degree that paper changes its dimensions due to a change in relative humidity.

extender—a white pigment or varnish that can affect an ink's color strength and/or working properties, but not its hue.

exudation—the migration of solid materials to the surface of an ink film.

-F-

fade-ometer—a machine that uses an intense light to accelerate and measure ink fading.

fading—the loss of an ink's hue or color strength due to light or heat exposure.

Federal Water Pollution Control Act Amendments—1972 federal legislation, amended in 1977 and 1987, authorizing the EPA to identify toxic chemicals in lakes and streams and to set standards regulating the discharge of these chemicals.

felt—continuous belt on which the wet web of paper is carried after leaving the wire at the couch roll; while on the felt, water is pressed out of the paper.

felt finish—a texture created on paper by using a felt with a nonstandard weave.

felt side—the top side of the paper when it was being formed on the Fourdrinier wire.

fiber—in papermaking, tiny thread-like unit of vegetable growth.

fibrillation—the process of generating fibrils during the beating stage.

fibrils (also called *fibrillae* and *fibrilla*)—threadlike elements that are loosened from a cellulose fiber's walls during beating.

fill—maximum trimmed width of paper that can be formed on a paper machine.

filler—mineral particles added to pulp, generally to improve opacity in the finished sheet.

filler clay—paper that has received a coating in which pigment content is less than 50 percent.

fill-in—the undesirable condition in flexographic printing of ink flowing into the open areas of small type.

film coated—paper that has been coated with a sizing that contains small amounts of coating pigments.

filtration—the separation of solid particles from a liquid by straining the liquid through a material with pores small enough to stop the solid matter.

fineness-of-grind—a measure of the size of an ink's pigment particles.

fine papers—broad category for papers used for printing and writing other than *coarse* papers; see *communication papers.*

fines—cellulose fibers and fiber fragments that are too short to make strong paper.

405

fine screening—a stage in the production of recycled paper that consists of passing the pulp through a wire screen to remove foreign material and fiber clumps.

finish—the surface contour of a sheet of paper including smoothness, glass, and overall appearance.

finishing—any of several operations that can be performed on paper after it leaves the paper machine such as slitting, embossing, and packaging.

first-down color—the first color of a multicolor job to be printed.

flag—small strip of paper protruding from a roll or skid of paper to mark a place.

flat rolls—damaged rolls of paper that have been dropped or stored in a manner to create a flat area on their circumference.

flax—the bast fiber of the flax plant, used to impart strength to cigarette and other thin papers.

flexography—a printing process that is relief in nature and uses rubber plates that contact the substrate.

flocculation—in papermaking, a clustering of paper fibers in a sheet, undesirable except in parchment; in ink, the clustering of pigment particles due to inadequate dispersion; the grouping together of separate particles to form clumps.

flotation de-inking—the pumping of air into a tub of recycled pulp so that ink particles will cling to the rising bubbles and be carried off.

flow—the tendency to level out or run as a liquid, the opposite of viscosity.

fluorescence—the transformation of invisible ultraviolet light into visible light by certain pigments.

flushed pigment (also called *flushed color*)—a color base prepared by flushing.

flushing—to transform pigment-in-water dispersions into pigment-in-oil dispersions by displacing water with oil.

flying—see *misting*.

foils—stationary plastic bars that support the moving wire of a Fourdrinier machine and help pull water through the wire.

fold strength (also called *fold resistance* and *folding endurance*)—the number of folds that a paper can undergo before breaking.

folio—a sheet of 17×22-inch paper.

forest genetics—the science of tree reproduction and its effect on growth rate and other characteristics.

Forest Service, U.S.—the agency of the U.S. Department of Agriculture that oversees the protection and management of National Forests, Rangelands, and Grasslands.

formation—the degree that a sheet's fibers are uniformly distributed.

form roller—the roller of an ink train that transfers ink to the plate.

forms bond—lightweight paper made from chemical pulp and designed for continuous form printing such as green bar computer paper.

fountain—the ink or fountain solution.

fountain roller—the roller of an ink train that revolves inside the ink or fountain reservoir.

fountain solution—in lithography, a mixture of water, acid, buffer, and a gum to prevent ink from transferring to a plate's nonimage areas.

Fourdrinier—a paper machine with a horizontal moving screen.

free sheet—paper that contains no groundwood fiber.

fugitive colors—inks made from nonpermanent pigments and dyes.

furnish—the pulp-in-water combination that forms paper on the paper machine.

fuzz—tiny fibers that project from a paper's surface.

-G-

galvanized—ink films that vary in gloss, density, or color.

gampi—a Japanese shrub with

a bast fiber used in papermaking.

GATF—Graphic Arts Technical Foundation, a Pittsburgh-based center for printing-related research and education.

GCC (ground calcium carbonate)—see *calcium carbonate*.

ghosting—unintended images that appear in printed areas of a press sheet; there are two types—mechanical ghosting and chemical ghosting.

glassine—semitransparent and smooth paper made from chemical pulp that received an excessive amount of beating.

gloss—a shiny or lustrous appearance, the result of specular reflection exceeding diffuse reflection from a surface.

gloss ghosting—see *chemical ghosting*.

gloss ink—an ink that dries with a sheen due to minimal penetration into the substrate.

gloss meter—an instrument that measures specular reflection from a surface.

grade—(1) a quality ranking for papers; (2) a term sometimes used to refer to paper's category.

grain—the predominant alignment of the paper fibers.

grain direction—the predominant direction of fiber alignment.

grain long—a sheet in which the grain direction is parallel to the long dimension of the sheet.

grain short—opposite of grain long.

grammage—grams per square meter, metric expression of substance weight.

gravure—the intaglio printing process that carries the image area in tiny cells etched into the image carrier's surface.

grindometer (also called *fineness of grind gauge*)—a device used to measure the size of an ink's pigment particles.

groundwood free—paper made from chemically pulped wood.

groundwood pulp—pulp created by the action of a grindstone revolving against logs. Groundwood pulping provides a high yield of comparatively low quality pulp because it contains materials that greatly reduce paper's permanence.

growing stock—a forest's trees as expressed in either number or volume.

guild—an association of medieval artisans of the same trade.

gum—in ink formulation, any natural resinous binder; in lithographic printing, a water soluble colloid (commonly gum arabic) used to treat plates.

gummed paper—a paper coated on one side with a water-activated adhesive.

gum rosin—rosin extracted exclusively from the gum of living trees.

-H-

halo effect—the accumulation of ink at the edges of printed images such as halftone dots and type.

hammer mill—see *stamping mill*.

handling stiffness—a measure of a sheet's stiffness in practical situations.

handmade paper—sheets of paper made individually as a mold is manually lifted from a pulp slurry.

hand sheet—a sheet formed individually, usually as part of an evaluation procedure such as testing pulp brightness.

hard beating—extended beating of pulp.

hard pulp—pulp that received limited cooking and still contains significant levels of noncellulose materials.

hard-sized—paper that has received a large amount of sizing.

hard stock—pulp made from rags, jutes, or rope.

hardwood—deciduous trees such as oaks, maples, and poplars.

harvest cutting—the removal of desirable mature trees.

hazardous—dangerous, perilous; a compound that has the potential of becoming dangerous to living organisms.

hazardous waste—according to the EPA, waste with one or more of the following characteristics: ignitible (flash point of 140° F or less), corrosive (dissolves metal or burns the skin), reactive (reacts violently with water, air, or other materials), toxic (harmful to plants or animals).

head-box—large chamber that uniformly disperses dilute pulp onto the wire of a paper machine.

header—round, treated board that covers and protects both ends of a roll of paper.

heatseal paper—a label paper that carries an adhesive coating that is activated by heat.

heat-set inks—inks that dry as applied heat evaporates their solvents.

heavy bodied inks—inks with high viscosity or stiffness.

heavy metals—metals with comparatively high densities.

heavy weights—the heaviest weights of paper available in a given grade or category.

hemp—a plant that grows in central Asia and contains a bast fiber called jute; used to produce burlap and papers requiring high strength levels.

hickies—see *doughnut hickey.*

high-bulk papers—antique finish sheets with a 25 × 38-inch basic size and a 45 pound substance weight that bulks from 344 to 440 papers to an inch.

high finish—paper surface with a smooth coating.

high-solids inks—inks that have a comparatively high percentage of pigment material relative to its liquid vehicle.

hue—the color that results from combining two different colors; e.g. blue-green is a hue of combining green and blue; hues are measured by wavelength.

hydration—in papermaking, causing cellulose fibers to retain water and produce stronger paper; achieved through a proper amount of beating.

hydrocarbons—organic compounds comprised of only hydrogen and carbon.

hydrogen peroxide—(H_2O_2) used to bleach wood pulp through oxidation.

hydrometer—an instrument that measures specific gravity.

hydrophilic—the property of readily accepting moisture or being wetted.

hydrophobic—the tendency to receive oils or lithographic ink, but repel water.

hygroexpansivity—the degree that an increase in moisture content causes a sheet to expand or contract. High hygroexpansivity means poor dimensional stability.

hygrometer—an instrument that measures relative humidity.

hygroscopic—the property of readily absorbing moisture.

hysteresis—the phenomenon that a property's value is affected by whether it was arrived at from a higher or lower level.

-I-

imitation parchment—formerly, paper formed on the Fourdrinier machine with inconsistent fiber distribution to resemble vegetable parchment; now the term refers to a type of wrapping paper.

impregnated paper—sheets that were film-coated during manufacture.

improvement cut—removal of less valuable trees from a forest.

indelible ink—an ink used on cloth and resistant to laundering.

index bristol—a heavy paper commonly used for index cards. The basic size is 25½ × 30½-inches and the substance weight range is 90–220 pounds.

infrared—invisible light waves that can be used to cure certain inks.

infrared ink—inks that cure (harden) when exposed to sufficient amounts of infrared light.

inhibitor—a compound that slows or terminates a chemical reaction, sometimes regarded as negative catalysts.

inking system—the part of a printing press that carries ink from a reservoir to the substrate.

inkometer—an instrument that measures an ink's tack.

ink receptivity—the ability of a substrate to accept ink.

inorganic pigments—ink pigments that contain metallic compounds such as chrome orange and iron blue.

integrated mill—a paper mill that performs both the pulping and sheet-forming functions.

intermediate cut—the removal of some trees for any one of several reasons such as thinning or improvement of the forest.

internal bond—a sheet's resistance to being pulled apart by an external force pulling against its surface.

internal sizing—the process of applying material to pulp fibers to reduce their absorbency after they are later formed into a sheet.

International Standards Organization (ISO)— an organization that coordinates and establishes standards throughout the world.

intolerant—trees unable to survive or thrive under specific conditions, such as drought, shade, wind, and flooding.

iodine number—a number that identifies the drying potential of vegetable oils,

with fast drying oils having higher numbers.

iron oxides—a group of iron/oxygen compounds used as ink pigments; some are appropriate in magnetic inks.

ISO 9000—standards for manufacturing operations that ensure quality control through process controls; these standards were originally published by the Geneva, Switzerland-based International Organization for Standardization in 1987 as part of the movement toward a European common market, but they have evolved into an important credential for companies wishing to sell goods to European buyers.

ISO paper sizes—paper sizes that have a 1:1.414 ratio between width and length.

ivory finish—the result of calendering paper that has been rubbed with beeswax.

-J-

Japan paper (also called *Japan art paper*)—a strong paper with a mottled, irregular formed appearance; it has a 25 × 38-inch basic size and a substance weight range of 50–150 pounds.

jobber—a business that purchases paper from mills and sells it to printers, sometimes after performing a converting (conversion) operation.

job lot—paper that enters the secondary market from overproduction or fails to meet one or more specifications.

Jordan—a type of pulp refiner that consists of a conical casing and a rotating similarly-shaped plug that flattens fibers.

jumbo roll—paper rolls exceeding 24 inches in diameter and 500 pounds in weight.

junior carton—a carton that contains 8 to 10 wrapped reams of 8-½ × 11-inch or 8-½ × 14-inch paper.

jute fiber—used to produce papers with high strength such as tag and wrapping grades; commonly acquired from old burlap and gunny sacks.

-K-

kamyr digester—a vertical continuous digester that converts wood chips into chemical pulp.

kaoline (KAY-o-lin)—white clay mass which is ground to fine particle size and used as filler and coating pigment in papermaking; also known as kaolinite, and china clay.

kappa number—measures the degree to which pulp is free of lignin.

K.B. value (Kauri-butanol value)—a numeric scale from 20 to 105 that measures the solvent power

409

of hydrocarbon oils and solvents.

kenaf (pronounced keh-NUF)—an annual plant with long fibers used in papermaking.

ketones—a class of organic solvents commonly used in gravure inks.

kiln— a furnace or oven for burning, drying, or baking.

knots—clumps of incompletely digested fibers in paper.

kraft paper—paper, used primarily for wrapping and bags, in its unbleached form. The basic size is 24 × 36 inches and the range of common substance weights is 25–60 pounds.

kraft pulp—pulp produced by the sulfate process.

-L-

lacquer—a transparent resin coating that can protect paper from water and grease; lacquer is mixed with a solvent which evaporates from the substrate leaving the dried lacquer film.

lagoon—a shallow pond useful in purifying wastewater.

laid finish—a simulation of handmade paper formed on a mold; a combination of watermarking and embossing usually creates the effect.

lake—an insoluble organic pigment made by staining solid organic particles with a soluble organic dye.

laminate—to bond two or more layers of material.

laminated paper—a sheet that consists of two separate sheets that have been fused.

lampblack—a carbon black pigment that produces a dull black ink.

latex—a combination of water and a natural or synthetic rubber derivative that is used as both a coating adhesive and pigment.

latex-treated paper—paper with latex-impregnated fibers that provide increased durability, flexibility, and fold strength.

leach—the action of soluble chemicals being washed from a substance by an invading (percolating) liquid.

leachate—the contaminated liquid leaving a substance; the product of leaching.

lead driers—printing ink driers that contain lead and various acids.

ledger—medium-weight papers used in record keeping; characterized by a hard surface and good erasability. The basic size is 17 × 22 inches and the range of substance weights is 24–36 pounds.

length—the property that allows an ink to be pulled before snapping; inversely related to viscosity.

letterpress—a printing process in which the ink is carried upon raised letters or other images.

letterpress paper—paper that has been manufactured with properties appropriate to being used in letterpress printing.

levelness—degree to which a sheet's caliper is uniform; dependent upon an even dispersion of fibers across the wire of the paper machine.

level paper—a supercalendered sheet made from chemical pulp and usually gummed on one side.

light reflectance—the degree that visible light is reflected from a surface as opposed to being absorbed.

lightweight papers—sheets with substance weights below the normal minimum weights for a given grade.

lignin—the natural bonding agent found in wood; also, the main cause of poor permanence in groundwood paper.

lime kiln—a furnace that burns lime sludge ($CaCO_3$) created during chemical pulping to produce calcium oxide (CaO) and carbon dioxide (CO_2).

linen and **linen fiber**—bast fibers found in the flax plant that add strength to a pulp mixture.

linen finish—an embossed texture that simulates the appearance of linen fabric.

linen paper—if labeled as such, paper containing at least some linen fibers; otherwise, paper that has been embossed with a linen finish.

linerboard—the flat-facing component of corrugated board.

linoleaters—compounds derived from linseed acids and used as ink driers.

linseed oil—oil obtained from flaxseed and used in lithographic varnish.

lint—poorly-bonded surface fibers that are lifted by ink during printing.

linters—see *cotton linters*.

lithographic inks—inks prepared for the lithographic printing process; also called *offset inks*.

lithographic varnish—lithographic ink vehicle derived from linseed oil.

lithography—a printing process based on the chemical law that fatty, oily inks do not mix with water; usually referred to as *offset*.

livening—the irreversible gelation of ink.

loading—adding clay and other filler to pulp.

long grain—see *grain long*.

Love Canal—an abandoned residential area near Buffalo, NY that was built over a previous dumping ground for hazardous industrial waste. A series of illnesses among the residents was traced to the pollution.

low finish—see *antique finish*.

lump—see *knots*.

luster—see *gloss*.

-M-

M—Roman numeral for 1,000.

machine broke—wastepaper created by a web break on the paper machine.

machine direction—the direction of a sheet of paper that is parallel to the direction the paper was moving on the paper machine.

magenta—pinkish-red ink used as one of four colors in process printing.

magnetic inks—inks with pigments that can be magnetized after printing and used with MICR (magnetic ink character recognition) equipment.

making order (M.O.)—a custom order for paper made from the customer's specifications.

maleic resin—a synthetic polymer binder derived from rosin and maleic acid.

manganese—a metal nearly always found and used in oxide form; an ingredient used in oxidation-drying inks to speed drying.

manifold paper—a lightweight bond paper commonly used for air mail correspondence or in conjunction with carbon paper. The basic size is 17 × 22 inches and the substance weight range is 7–9 pounds.

manila stock—any paper or paperboard made with manila (hemp) fiber.

marble paper—a decorated paper sometimes used as end sheets in hardcover books.

market pulp—pulp that is sold on the open market rather than used internally in an integrated mill.

mass—in recycling, the volume or concentration of the material to be recycled.

masstone—the reflected color of a bulk ink instead of an ink film.

matrix effect—the phenomenon of metal detection being impaired by the presence of certain other metals in the material.

matte finish—coated papers with a dull finish; minimal supercalendering produces a gloss level of 20 or less.

M.C.—machine coated.

mean annual increment—annual cubic feet of growth per acre of forest.

mechanical ghosting—a faint image created in an area of high ink coverage caused by an inadequate ink supply on a portion of the plate.

mechanical wood pulp—wood pulp produced by a mechanical process such as groundwood or thermomechanical.

merchant's brand—see *private brand*.

metallic inks—inks with aluminum and bronze powders in varnish that appear to be silver or gold.

metamerism—the phenomenon of a color's appearing differently under different light sources.

meter—as a verb, to control and/or record the rate of flow or distribution of a material; as a noun, a device that measures flow.

411

M.F.—machine finish.

M.G.—machine glazed.

M.I.C.R.—magnetic ink character recognition.

microencapsulation—the packing of liquid particles into tiny capsules that release their contents under pressure; it is the technology behind carbonless paper.

microparticle retention system—a chemical treatment of the pulp to reduce the amount of filler particles that falls through the wire of the paper machine.

microwave drying—using microwaves to dry inks that contain polar materials.

mileage—the number of square inches that can be covered by 1 pound of an ink on an given substrate.

milking—an accumulation of coating particles on the nonimage areas of a lithographic blanket because of coating that is inadequately water-resistant.

mill—.001-inch.

mill blanks—paperboards that are between 3 and 10 plies and consist of a mechanically-pulped center between two white liners; basic size is 22 × 28 inches.

milling—to grind or pulverize a solid material into finer particles.

mimeograph paper—a bond paper intended to be used on a mimeograph duplicator and characterized by high bulk to ensure stiffness and absorbency. Basic size is 17 × 22 inches and the substance weight range is 16–24 pounds.

mineral oils—high molecular weight hydrocarbon oils derived from petroleum.

misting—the spraying of tiny ink particles during high-speed printing.

mixing—to blend two or more substances into a uniform mass or compound.

mixing white—a white ink used to make tints.

moisture content—the percentage of a paper's total weight that is moisture.

moisture-set inks—inks that dry or set primarily when their solvent absorbs moisture and causes the binder to precipitate.

moisture welts—ridges parallel to the paper's grain direction that results from uneven moisture content.

monochromatic colors—two or more colors that differ only in tone (lightness/darkness).

monomer—the basic unit of a polymer.

mottle—the nonuniform appearance of an area of solid ink coverage.

M papers—imperfect, but not seriously flawed, paper.

MSDS (Material Safety Data Sheet)—a document that contains information about the potential hazards of a material.

Mullen test—measures a sheet's burst strength.

multiple-use management—the concept of managing a resource, such as the National Forests, so that several benefits can be achieved.

multistage bleaching—the bleaching of pulp in two or more stages, designed to bleach without destroying the fibers' strength.

M wt—the weight of 1,000 sheets of paper.

-N-

NAPIM—National Association of Printing Ink Manufacturers.

napthas—low hydrocarbon petroleum-derived solvents.

NCR paper—carbonless papers that transfer images through applied pressure.

neutral pH—paper with a pH value between 6.0 and 8.0.

new forestry—an approach to tree harvesting that attempts to minimize ecological impact.

news inks—basic inks formulated to dry by absorption on newsprint stock.

newsprint—the stock that newspapers are printed on, consisting mostly of mechanical pulp with some chemical pulp; the least expensive printing substrate; basic size is 24 × 36 inches and the substance weight range is 28.5–35 pounds.

nip—the point or line of contact between two rollers or cylinders.

nitrocellulose—a film-forming substance commonly used in gravure and flexographic inks.

noiseless paper—any paper that is treated to minimize its rattle.

nominal weight—the weight per ream or M used for billing purposes, not necessarily the actual weight.

nondrying oils—oils that do not dry after exposure to air.

nonhazardous—materials that do not pose a threat to living organisms.

nonpoint-source pollution—pollution that does not emanate from a single place; an example is the run-off of fertilizer from agricultural fields.

nonscratch inks—inks that resist abrasion when dry.

novel paper—newsprint with maximum bulk commonly used in paperback novels and coloring books.

N paper—a classification of paper that is wrinkled or dirty.

NPTA—National Paper Trade Association, an association of paper merchants.

nurse crop—a crop of trees or other plants that protect another crop from frost or wind.

-O-

OCR inks (optical character recognition)—inks with low reflectance pigments that are easily read by optical scanners.

OCR paper—paper made to work well with the optical character recognition process.

offcut—sheets in nonstandard sizes that result from paper left over on the roll after standard sizes are cut.

offset curl—the curling of paper that results from excessive moisture gain on one side of the sheet due to exposure to dampening solution on a lithographic press.

offset paper—paper made with internal sizing to resist the moisture contact inherent with lithographic printing. The basic size is 25 × 38 inches and the substance weight range is 45–70 pounds; also called *litho paper*.

offset printing—see *lithography*.

onionskin paper—a lightweight sheet used for making typewritten carbon copies; usually made with a cotton and chemical wood pulp mixture. The basic size is 17 × 22 inches and the substance weight range is 7–10 pounds.

on-machine coating—coating applied to paper while it is still on the paper machine or in line with it.

ONP—old newspapers.

opacimeter—an instrument that measures paper's opacity.

opacity—the property of a medium that reduces the transmission of light through a sheet of paper, ink film, or other film.

opaque—the property of totally preventing the transmission of light.

opaque paper—usually refers to paper that contains high levels of filler to enhance opacity for a given weight or caliper.

optical properties—physical properties of a material that can be visually recognized; examples are color, opacity, and brightness.

orange peel—a type of mottle found with gravure and flexography.

organic—material originating from living organisms.

organic pigments—coal-tar-derived pigments.

OSHA—Occupational Safety and Health Administration, an agency within the U.S. Department of Labor that was established in 1970 to develop and enforce job safety and health regulations.

overprint varnish—a clear varnish applied over a printed image.

overrun—the amount of paper produced in excess of the amount ordered.

oversize—paper made larger than ordered and intended for subsequent trimming.

overstory—trees in a forest that form the uppermost canopy.

overweight—paper made with

413

a higher substance weight than was intended.

oxidation—(1) the chemical process of combining with oxygen; (2) the action of oxygen combining with an ink's vehicle to produce a surface film.

oxygen bleaching—using oxygen in an alkaline medium to reduce lignin prior to subsequent bleaching.

ozone—an unstable molecule that consists of three oxygen atoms (O_3); used to bleach pulp.

-P-

pages per inch (p.p.i.)—a bulking term, the number of pages in a 1-inch pile of pages.

Pantone Matching System (formerly known as PMS)—a system of cataloging hundreds of colors that can be achieved from combining two or more inks from a list of 14 (12 colors plus Pantone Black and Pantone Transparent White).

paperboard—heavier and stiffer than paper, generally sheets that are .012-inch think.

papeterie (pronounced PAP-uh-tri)—distinctive paper used for greeting cards and stationery.

papier mache—a combination of wood pulp and an adhesive used for molding.

papyrus—a plant that grows in the Nile region and was pressed to matted sheets by ancient Egyptians to form a writing medium.

parchment—a writing medium made from the skin of a sheep or goat; see *artificial parchment* and *vegetable parchment.*

parchment bond—an imitation of parchment made with cotton and wood pulp for 17 × 22 inches and a substance weight range of 20–40 pounds.

particulate pollutants—air pollution in the form of solid or liquid particles that are small enough to remain in suspension for long periods of time; examples include smoke, dust, fumes, and mists.

PCC (precipitated calcium carbonate)—see *calcium carbonate.*

pebble mill—a revolving cylindrical device that mills ink through the cascading of pebblelike media within it; a rough comparison can be made to a rock polisher.

penetration—the degree to which a liquid is absorbed in a substrate.

perfecting press—a rotary printing press that applies ink to both sides of the substrate during a single pass.

perilla oil—a vegetable drying oil.

permanence—the ability of paper to retain its properties over centuries instead of decades.

permanent inks—inks that resist a change in their properties despite lengthy exposure to light and other external forces.

pH (potential hydrogen)—a 1–14 logarithmic scale that measures acidity and alkalinity with low numbers being acidic, high numbers being alkaline, and 7 being neutral.

phenolic resin—a synthetic resin that imparts gloss and hardness to an ink formula.

photoinitiator—the component in ultraviolet-cured inks that responds to UV rays by generating radicals (atoms with an unpaired electron) that prompt the linking of the ink's molecules into polymers.

photopolymer—a polymer that is affected by light.

pick—the result of picking; fiber pick is the lifting of clumps of fiber from a press sheet's surface and coating pick is the lifting of chunks of the coating layer.

picking—the rupturing of a substrate's surface due to the pull of high-tack ink.

pigment—insoluble, finely-ground solid particles held in suspension by a vehicle; their function in ink formulas is to supply the desired color.

pigment piling—the accumulation of tiny mineral particles from the substrate onto the nonimage portion of the blanket during offset printing; see *caking*.

pigment-to-vehicle ratio—the mathematical ratio between these two components of an ink; a high ratio reflects a high pigment level.

PIMA—Paper Industry Management Association.

pinholing—small holes that appear in a printed area because the ink failed to form a continuous film.

pioneer—a plant capable of invading bare spots and growing.

pitch—an asphaltlike material that makes an ink longer (increases flow).

plasticizers—additives that increase the flexibility of a dried film.

plastic viscosity—the flow property of printing inks.

plastisol—particles suspended in an organic liquid instead of in a solvent.

point—one-thousandth of an inch.

point-source pollution—pollution that can be traced to a single source, such as a smoke stack or effluent pipe.

points per pound—a measure of burst strength.

poise—a measurement of viscosity; see *centipoise*.

polar solvents—solvents that contain oxygen.

pollution—undesirable matter that contaminates a gaseous, liquid, or solid body.

polyamide resins—in ink applications, synthetic polymer binders derived from acids and amines; nylon is an example of a polyamide resin.

polymer—a macromolecule composed of five or more identical units (monomers). Pure cellulose, for example, consists of 3,500 monomers.

polymerization—the union of two or more of the same kind of molecules to form a larger molecule with different properties.

pop test—see *Mullen test*.

porosity—having connected pores through which liquids can pass; measured by the amount of air that can pass through a medium.

postconsumer waste—recovered materials that have fulfilled their intended final use.

pounds per point—used to measure the ratio between a sheet's substance weight and its caliper.

powdering—the accumulation of visible particles from the paper on the lithographic blank's nonimage areas; also called *dusting*.

practical maximum capacity—the tonnage of paper that could be produced by a mill with no interruptions.

preconsumer waste—paper that was reclaimed before it reached its intended final use; e.g. mill broke, trim waste.

preparatory cutting—removing trees with the purpose of opening the canopy and improving seed production before a shelterwood cut.

preservatives—chemicals added to pulp to prevent or inhibit the presence of organisms that could impair permanence.

press cake—a pigment-in-water dispersion (with a 30 to 80 percent water content) that is used to produce a flushed pigment.

pressure-sensitive papers—paper that is coated on one or both sides with an adhesive that is sensitive to pressure.

primary colors—those colors from which all others can be made; for light, the primary colors are red, blue, and green; for inks, the primary colors are cyan, magenta, and yellow.

primary treatment—removing sediments from wastewater by allowing them to precipitate.

printability—the subjective ability to transfer an image with faithful reproduction.

private brand—a brand of paper that is offered by a merchant rather than the mill that manufactured it.

process color—the printing of an infinite number of hues by combining dots of four inks—cyan, magenta, yellow, and black; sometimes called *four-color* and *full-color*.

proof press—a specially-designed press that is used to print a single copy for proofing purposes.

proprietary alcohol—denatured ethyl alcohol.

PSIA—Paper Stock Institute of America.

pseudoplasticity—the property of a material that causes it to lose viscosity when stirred and regain viscosity when left to stand.

psychometer—a type of hygrometer using the wet-and-dry bulb method of measuring relative humidity.

pucker—the bumpy effect that results on the surface of a sheet that has been dried unevenly; see also *cockle finish*.

pulp—a slurry of cellulose fiber in water that is prepared for papermaking.

pulpboard—see *paperboard*.

pulper—a machine that converts sheets of pulp or wastepaper into a slurry through agitation.

pulping—reducing wood, rags, or wastepaper into a pulp.

pulp substitute—reclaimed unprinted wastepaper that requires minimum treatment before making new paper.

pulpwood—wood logs used in the production of wood pulp and, subsequently, paper.

pyroxylin paper—paper with a sulfite base and coated with a pyroxylin lacquer that imparts a pearly, iridescent appearance.

-Q-

quadrat—a small sample area on which ecological observations are made.

quick-set ink—a term applied to any ink with vehicles that include a heavy oil and a lighter solvent; when printed, the solvent penetrates the paper surface and leaves behind the oil which sets up immediately.

quire—25 sheets or 1/20 of a ream.

quired—the packing of sheets folded in half, rather than flat.

-R-

radical—an atom or molecule that contains an unpaired electron. Such substances are highly reactive or eager to bond with other substances. The creation of radicals is a key part of the ink curing process.

rag book—see *bible paper*.

rag content—the percentage of a sheet that is made of cotton fibers, usually 25, 50, 75, or 100 percent.

railroad board—a heavy, colored paperboard, the basic size is 22 × 28 inches and the substance weight range is 130 to 300 pounds.

rattle—the noise generated by shaking a piece of paper, sometimes associated with high-quality bond paper.

rayon—a synthetic textile fiber made from solidifying cellulose solutions into fibers.

rayon rejects—rayon or cellophane fibers unsuitable for their intended use that are instead used in making soft, absorbent paper.

RCRA (Resource Conservation and Recovery Act)—1976 federal legislation that regulates the handling and disposal of waste materials.

ream—ordinarily, 500 sheets of printing paper; however, in certain situations a ream is 480 or 1,000 sheets.

ream weight—weight of a ream of paper.

ream wrapped—a package that contains a ream of paper.

recycled fiber—fibers obtained from processing wastepaper.

recycled paper—paper made exclusively or in part with recycled fiber.

reducers—varnishes, solvents, oils, or waxes that lower ink tack.

reel—the roll of paper that is created at the dry end of the paper machine.

reel curl—the curling of paper created by being tightly wound near the core of a roll.

reel samples—samples removed from the reel for testing.

refiner—a machine that rubs and flattens to impart desired characteristics to a sheet.

refiner groundwood—papers that are primarily composed

of fibers that were pulped in a refiner.

reflectance—the ratio between the amount of light reflected by the sample in question and the amount of light reflected by a standard reflector.

reflectivity—a sample's reflectance.

refraction—the bending of light as it passes through media with different densities.

regeneration— the renewal of a tree crop.

regeneration cut—any removal of trees that assists regeneration.

register bond—see *business forms bond.*

relative humidity—the ratio between the amount of moisture in the air and the moisture capacity of the air at a given temperature and pressure.

release—freeing a tree or group of trees from competition by removing overtopping or surrounding growth.

repulping—the process of reducing paper to individual fibers with the intent of forming new paper.

resiliency—the ability of paper to recover from being physically distorted.

resin—any organic substance that is insoluble in water, but soluble in organic solvents; usually used as adhesives.

retarders—compounds that slow the drying of ink by slowing the rate of either evaporation or oxidation.

retention—the percentage of filler or other material that does not fall through the mesh of the Fourdrinier machine during the papermaking process.

rewinder—the machine that moves paper from one reel to another, usually for the purpose of slitting, embossing, or otherwise finishing the paper.

ridge—the lumpy ring around the circumference of a roll of paper caused by an excessive caliper in that portion of the sheet.

rheology—the study of the flow of matter and how it is affected by external forces.

rice paper—a flat material formed from the spongy central portion of a small tree grown in Taiwan. The material is neither paper nor made from rice.

right side of paper—the top or felt side; also, the side where the watermark is right-reading.

rigidity—resistance to bending or flexing.

ripple finish—a type of embossed finish that is composed of small dimples.

rods—nerve cells found in the rear wall of the eye that can function in low light levels; they cannot distinguish colors.

roll—a continuous sheet of paper wound around a shaft.

roll coating—a method of coating paper and paperboard by applying a metered amount of coating onto the sheet without a subsequent removal of any excess.

R.O.P colors—run of press colors; ink colors considered standard for newspaper printing.

ropes—raised bands or welts that run around a roll of paper; a rope-like pattern of diagonal lines is also visible.

rosin—what remains after turpentine is boiled from gum; see also *gum rosin.*

rosin size—a suspension of rosin that is added to pulp to improve the fibers' resistance to water.

rotary cutter—a machine that transforms a roll of paper into sheets.

rotogravure paper—paper manufactured to possess characteristics appropriate to gravure printing.

runnability—the ability of paper to move through a printing or finishing operation without jamming, rupturing, or otherwise causing problems.

rupture—to break or come apart.

-S-

safety factor—an amount of ink beyond what should be needed that is added to the original amount to ensure

417

not running out of ink during the run.

safety paper—paper that carries a printed design or hidden message or both to discourage mechanical or chemical means of altering a typed or written image; commonly used for checks, certificates of title, and lottery tickets.

salvage cutting—to remove one or more trees that are dead or damaged before their timber becomes worthless.

sanitation cutting—the removal of dead or infected trees to prevent the spread of pests or disease through the forest.

SAPI—Sales Association of the Paper Industry.

satin finish—a smooth surface on paper.

S/C—supercalendered.

S and S/C—sized and supercalendered.

Schopper's tester—an instrument that measures a sheet's fold strength.

scoring—the creasing of paper to improve its folding ability.

screening—to pass either wood chips or pulp over or through screens to remove components that are the wrong size.

scrubber—a device that removes solid particles from waste gases by spraying water through the gas.

scumming—the result of the inability of the nonimage

area of a lithographic plate to resist ink.

seasoning—allowing an entire pile or roll of paper to adjust to the temperature and humidity of any new environment.

secondary color—any color that results from combining equal amounts of any two primary colors.

secondary fiber—wastepaper.

secondary treatment—treating wastewater by adding oxygen or otherwise assisting bacteria to consume organic waste material.

seconds—paper that fails to meet one or more established standards.

seed tree cut—a tree harvesting technique that removes nearly all the trees in a stand except for a few of a desired species that are left for reseeding.

semichemical pulp—pulp that results from both mechanical and chemical processes.

semidrying oils—oils that harden upon oxidation, but do so comparatively slowly; examples include soya, tung, and linseed oil.

setoff—the transfer of printed but not yet dry ink to the sheet above it while in the delivery pile.

setting of ink—the "drying" stage that occurs after ink transfer, but before actual drying, during which the substrate can be handled without smearing.

shade—the result of adding black to a color.

shake—the sideways oscillation of a Fourdrinier's wire to enhance uniform sheet formation and fiber alignment.

shear—the relative movement of adjacent portions of a liquid or semisolid during flow; ink moves in layers and the action between two layers moving at different rates is called shear.

sheen—gloss or luster.

sheeted—paper that has been cut from a continuous roll.

sheeter—a device that converts a web of paper into sheets.

sheet-fed press—a press designed to print on individual sheets instead of on a continuous web.

shelf life—the length of time that an ink can resist oxidation or an internal chemical action that alters its properties.

Shell cup—a brand of efflux cup.

shellac—a natural resin that is alcohol soluble and used in flexographic ink.

shelterwood cut—a tree harvesting technique that removes all trees except enough to provide a protective canopy for young growth.

shives—a bundle of undercooked fibers that must be removed from the pulp.

short grain paper—a sheet with the grain running in the short dimension.

shortness—the property of ink that resists flow.

shot mill—a device that pumps ink through a large cylinder that contains small moving balls with the purpose of dispersing the ink's ingredients.

show-through—the passage of light through a sheet that allows images to be visible from the other side.

shrinkage—any decrease in the dimensions of paper, usually through the loss of water.

side run—a roll in a nonstandard width that is left over after a log has been slit into several standard widths; also called *butt rolls.*

Silent Spring—a 1962 book by Rachel Carson that traced the dangers of pesticides through the food chain. The book alerted millions of Americans to the concept of ecology.

silviculture—the science of forest management that includes tending, harvesting, and regeneration.

size—any material used in the internal or surface sizing of paper.

sizing—the process of making paper less absorbent of moisture.

skid—a wooden pallet or the amount of paper in flat sheets carried by a pallet.

skim coat—see *film coated.*

slack-sized—a lightly-sized paper.

slice—the horizontal rectangular orifice through which pulp leaves the headbox and enters the wire of the paper machine.

slime—a collection of contaminants in pulp such as microbiological growths.

slimeicide—a pulp additive that controls microbiological growths in pulp.

slime spots—small areas on finished paper caused by fungal or bacterial growth in the pulp.

slip compound—any ink additive that lowers the dried ink film's coefficient of friction.

slit—to slice a roll of paper into two or more smaller rolls.

sludge—in pollution control, a slurry that results from solids settling out from suspension in the water during primary treatment.

slur—distortion of a printed image due to slippage between the plate and blanket, plate and substrate, or blanket and substrate.

slurry—watery suspension of pigments and additives used in coating paper.

smoothness—the property of a surface that is without irregularities.

soda—sodium carbonate.

soda ash—a commercial anhydrous sodium carbonate.

soda process—a chemical method of reducing wood chips to pulp by using sodium hydroxide and heat.

soda pulp—a chemical pulp produced by cooking wood chips in the soda process.

sodium bisulfate—($NaHSO_3$) also called *sodium acid sulfite;* used in chemical pulping and to remove excessive chlorine from bleached pulp.

sodium chlorite—used to bleach pulp.

sodium hydrosulfite—($Na_2S_2O_4$) also called *sodium dithionite;* used to bleach mechanical pulp.

sodium hydroxide—($NaOH$) also called *caustic soda* and *soda lye;* a powerful alkaline used in the soda and kraft processes.

sodium hypochlorite—($NaOCl$) used in bleaching pulp.

sodium peroxide—(Na_2O_2) used to bleach mechanical pulp.

sodium sulfate—(Na_2SO_3) used in sulfate and kraft pulping to maintain a uniform level of sodium hydroxide.

sodium thiosulfate—($Na_2S_2O_3$) used to neutralize excessive chlorine during bleaching.

soft paper—a sheet with little or no surface sizing.

soft roll—a roll of paper that was wound too loosely.

softening point—temperature that is sufficient to alter the shape of a plastic material.

softwood—wood from coniferous trees (e.g. pine, spruce, fir).

solid—an adjective that

signifies that the paper product it describes is made of the same material throughout, such as solid bristol.

solvent—a liquid that will dissolve a solid.

solvent release—the influence that a binder has on a solvent's rate of evaporation.

soya bean oil—a vegetable semidrying oil used in producing ink vehicles and derived from soybeans.

soybean protein—a soybean derivative used as a coating adhesive and a sizing material.

specialty papers—paper manufactured or converted with specific properties, usually not suited to other purposes.

specific gravity—the ratio between the weight of a substance and the weight of an equal volume of water.

spectrophotometer—an instrument that identifies the color of paper by measuring its wavelength.

specular reflection—the consistent reflection of light from a surface at the same angle at which it struck the surface; specular reflection results in gloss.

splice—the joining of two lengths of paper to make one continuous roll.

splice tag—a piece of paper protruding from a roll to identify the location of a splice.

splitting—the tearing apart of

plies of paper, often by high tack inks during printing.

stability—see *permanence.*

stamping mill—mechanical device that consists of large hammers that are raised and then dropped on pulp to flatten the fibers.

stand—a group of trees of any age that bear enough similarity to be distinguished from adjacent trees.

stand density—a measure of the degree of crowding of trees within a forest.

standpipe—a vertical pipe that allows a liquid to be introduced or extracted from beneath the earth's surface.

starch—a vegetable derivative used in surface sizing and as an adhesive.

starred roll—inconsistently wound roll of paper resulting in a star pattern on the ends.

statistical process control (SPC)—the use of statistics to control the manufacturing process and thereby avoid generating products that are outside of specifications.

Stefan's equation—a formula that demonstrates an ink's tack to be related to its viscosity, film thickness, and press speed. The tack, defined as the Force required to split a thin ink film, equals the product of the Viscosity, the press Speed, and the Area of the film divided by the cube of the film Thickness. Expressed as a formula :

$$F = \frac{VSA}{T^3}$$

stiffening agent—an ink additive such as a heavy gum or varnish that increases the ink's vicosity.

stiffness—the property of paper that resists bending.

stock—(1) pulp that has been refined and is ready for the paper machine; (2) finished paper to be printed.

stock order—direct sale to a customer.

strike through—the observable penetration of an ink's vehicle through the sheet.

stringiness—the property of an ink that allows it to be drawn into threads.

stripping—the failure of an ink to uniformly distribute itself across an ink train's metal rollers.

structural curl—permanently curled paper resulting from a flaw in its manufacture.

stub roll—the remainder of a roll of paper after most of its paper has been removed.

sublimation—the phenomenon of a substance moving directly from a solid to a gaseous state without passing through the liquid state; also the reverse action.

substance weight (also called *basis weight*)—the weight of 500 sheets of a paper cut in its basic size.

substrate—any material that receives printing.

suction box—a vacuum under the wire of a Fourdrinier

machine that accelerates the removal of water from the fibers.

sulfate process—an alkaline chemical pulping process; also known as the *kraft process*.

sulfite bond—bond paper made from wood fibers instead of cotton fibers; the term is a carryover from before sulfite pulping was largely replaced by sulfate pulping for making bond paper.

sulfite process—an acidic chemical pulping process.

supercalender—a stack of alternating steel and soft-covered rollers that are used to polish paper after it has been coated.

supercalendered finish—the smooth surface that results from being sent through the supercalender; the degree of smoothness can be controlled to produce a range from dull-coated to gloss-coated.

surface sizing—to seal the surfaces of a sheet of paper by applying a sizing (usually starch); generally done on the paper machine.

surface strength (also called *bonding strength*)—the resistance of a sheet to rupture as a result of a pulling force, such as high-tack ink.

surfactant—a material that alters the surface properties of a liquid or solid; e.g., wetting agents, emulsifying agents, dispersing agents. Derived from *surface active agent.*

sword hygroscope—an instrument that can be inserted into a pile of paper to measure its moisture content.

synthetic paper—a sheet that resembles paper but is formed from noncellulose materials such as nylon or polyolefins.

-T-

T4S—trimmed four sides.

tablet paper—paper manufactured to be used in writing tablets.

tack—the amount of force required to split an ink film; inks with high tack do not split easily.

tag board—a dense, heavily calendered paperboard made from rope, jute, and wood fibers; the basic size is 22-½ × 28-½ inches and the substance weight range is 80–300 pounds.

tail-end hook—the curl that occurs at the back end of a press sheet when the sheet is released too late by the blanket's ink tack; also called *back-edge curl.*

TAPPI—Technical Association of the Pulp and Paper Industry.

tare weight—the weight of a container's contents arrived at by subtracting the container's weight from the total weight.

TCF (totally chlorine-free)—in papermaking, the process of bleaching pulp without the use of chlorine in any form.

tear ratio—the ratio between the tear strengths of the machine direction and the cross-machine direction of a sheet.

telescoped roll—a roll with concave or convex-shaped ends that results from slippage during winding.

tensile strength—the maximum steady force required to break a given width of paper; sometimes expressed as *pounds per inch* width of paper; of most significance in web printing.

tensiometer—an instrument that measures a liquid's surface and interfacial tensions.

tertiary treatment—the removal of phosphorous, nitrogen, and suspended solids from wastewater.

text papers—a category of high-quality sheets in a variety of textures often used in advertising and promotional pieces. The basic size is 25 × 38 inches and the range of common substance weights is 60–80 pounds.

texture—the characteristics of a sheet that relate to its appearance and feel.

thermomechanical pulping (TMP)—a method of pulping wood chips in a series of refiners after they have been presoftened by heat. The result is less damage to the fibers than occurs in mechanical pulping.

thermoplastic—solid materials that repeatedly soften in reaction to applied heat, but without changing their characteristics; materials that will soften if reheated.

thermosetting—a material that reacts to heat by polymerizing into a permanent solid; that is, it will not soften if reheated.

thickness—see *caliper.*

thinners—ink additives that reduce tack.

thinning—the removal of one or more trees in an immature stand to maintain or accelerate the overall forest growth; a type of intermediate cut.

thixotropy—the ease with which inks soften (lose viscosity) when agitated and stiffen when left to stand.

threatened species—a plant or animal that is considered likely to become endangered.

three-roll mill—a device that sends paste inks between rollers to reduce the size of the pigment particles and disperse the ink's ingredients.

through drier—an ink additive that serves as a catalyst for uniform drying throughout the depth of an ink film; often a manganese salt.

tight edges—a dishlike deformation of sheets of paper in which the outside edges have lost moisture to surrounding dry air and shrunk; also called *shrunken edges.*

timberland—forest land that is producing or capable of producing commercial timber and not legally protected from logging operations.

tinctorial strength—the ability of a pigment or dye to maintain color strength despite tinting; sometimes measured by the amount of white base required to produce a given tint.

tint—the result of adding white to a color.

tint base—the white ink used to produce tints of colors.

titanium dioxide—a bright synthetic filler pigment that imparts good opacity to a sheet.

toner—a highly concentrated pigment or dye.

tooth—a surface that is rough enough to function well as a paper for pencil drawing.

top—the felt side of a sheet of paper.

top drier—an ink additive that serves as a catalyst for a speedy hardening of an ink film's surface; usually a cobalt salt.

tough check—a multiply bristol of high strength usually made of unbleached chemical pulp; the range of calipers is 12 to 30 points.

toxic—harmful, deadly, poisonous; a substance that can cause illness or death by disrupting an organism's functions.

toxin—natural or synthetic substances that are poisonous to plants or animals.

trailing blade coating—a coating process in which a thin rigid blade is used to scrape excess coating from the paper's surface.

translucent—the ability to transmit light with a diffusion level that is high enough to prevent its being transparent.

translucent bond—paper manufactured to be the master copy for blueprints; the basic size is 17 × 22 inches and the substance weight range is 11–16 pounds.

transparent—the ability to transmit light without its diffusion, permitting objects on the other side of the material to be clearly identified.

transparent ink—inks that allow light to pass through them.

trap—the ability of a fresh ink film to receive a second ink film.

tristimulus—color readings based on the three primary colors.

TRS (total reduced sulfur)— a category of sulfur that contains compounds that can be released during the kraft pulping process and are regulated by the EPA. The compounds include hydrogen sulfide, H_2S;

methyl mercaptan, CH_3SH; dimethyl sulfide, $(CH_3)_2S$; and dimethyl disulfide $(CH_3)_2S_2$. Limits on TRS emissions are imposed to control the odor found near kraft mills.

tub sizing—applying surface sizing by sending the paper web through a tub of sizing and removing the excess; usually occurs between the first and second sets of driers.

twin-wire paper machine—a paper machine with two vertical forming wires between which the pulp is forced.

two-pot systems (also called *two-part systems*)—inks or coating with two components that are kept separate until the actual printing.

two-sidedness—a condition in which the wire side and the felt side of a sheet of paper are significantly different.

-U-

Ultra Former—a cylinder paper machine, usually used to make multiply board.

ultraviolet—light radiation in the 315–400 millimicron range.

unbleached—pulp or paper that is not white and, therefore, retains the brownish color that results from pulping.

under-runs—paper shipments that are less than the amount ordered.

understory—trees and wood species growing beneath the forest canopy.

undertone—the color of an ink when its film is thin enough to allow the transmission of light.

uneven-age stand—a forest consisting of trees with a wide range of ages.

uniformity—the lack of irregularity in fiber distribution, color, finish, etc.

UPI—United Paperworkers International.

utility papers—a residual classification comprised of several types of paper that do not belong in any of the other classifications.

UV—ultraviolet light invisible light that can be used to dry inks on high-speed multicolor presses.

-V-

vacuum filtration—use of suction to enhance the trapping of solids against a filter through which the liquid passes.

vapor permeability—the passing of a vapor through a sheet, not to be confused with *porosity*.

vapor pressure— the air pressure that is generated by the evaporation of a liquid in a closed container at the point of equilibrium between evaporation and condensation; a method of measuring an ink's volatility.

vaporproof—paper that resists vapor passage.

varnish—(1) a thin, transparent, protective coating applied to a printed sheet, usually by a printing press; (2) fluid compositions such as oils, resins, and solvents used as ink vehicles.

vat machine—see *cylinder machine*.

vegetable parchment paper— a paper made from cotton and/or chemical wood pulp that resembles animal parchment.

vehicle—the liquid portion of an ink that carries the undissolved pigments, provides its drying properties, and holds the pigment materials to the substrate.

vellum finish—a smooth finish that has a finer grain than an eggshell finish.

vellum paper—a quality, strong creme-colored paper made to resemble calfskin vellum.

velvet finish—a dull, soft surface.

virgin stock—pulp made from fibers not previously formed into paper.

viscometer—an instrument that measures a liquid's viscosity.

viscose—a cellulose derivative used in the coating or impregnation of paper; also used to produce viscose rayon and cellophane.

viscosity—the property of a liquid or semiliquid to resist flow.

423

visible spectrum—light with wavelengths (400–700 nanometers) that are visible to humans.

VOC (volatile organic compound)—organic (carbon-based) compounds that readily evaporate; organic compounds with a sea level vapor pressure of approximately 1 percent of atmospheric pressure.

volatility—the tendency of a liquid to quickly evaporate; for example, alcohol is more volatile than water.

volume regulation—a direct method of controlling and determining the amount of timber to be periodically cut by calculations based on stock volume and growth rate.

-W-

washout inks—inks applied to textiles that wash away easily.

waste—(EPA definition) any solid, liquid, or contained gaseous material that is no longer used, and is either recycled, thrown away, or stored until sufficient quantities are accumulated for treatment or disposal.

wastepaper—(1) paper that remains after the manufacture of a product; (2) paper that has served its intended final use.

water-based inks—inks with a water soluble binder.

waterleaf—unsized papers that are highly absorbent, such as towels of blotter paper.

watermark—the pattern or design built into a sheet as it is formed on the wire of a paper machine by the relief image on a dandy roll selectively displacing fibers.

water pollution—a concentration of substances that threaten living organisms or impair the aesthetics of a lake, stream, or estuary.

wavelength—the distance from one peak in a wave pattern to the next peak; the wavelength of light determines its color.

wavy edges—the warping of the outside portions of a pile of paper, caused by moisture absorption at the edges.

wax test—a test for measuring paper's surface strength that consists of placing a series of wax sticks of increasing tack against a sheet of paper.

waxes—ink additives intended to increase rub resistance and/or reduce tack.

web—(1) a roll of paper that feeds into a rotary press; (2) the ribbon of paper that winds through the press.

web break—a break in the web of paper moving through the press.

wedding bristol—a multiply, high grade bristol used for cards and announcements. The basic size is 22-½ × 28-½ inches and the substance weights are 120 pounds (2-

ply), 180 pounds (3-ply), and 240 pounds (4-ply).

welts—elongated humps running around a roll of paper.

wet broke—paper taken off a paper machine (usually during web breaks) before it has dried.

wet end—the sections of a paper machine that include the headbox, moving wire, and wet press section.

wet-end finish—a finish created by treating the paper at the wet end of the paper machine.

wet Mullen—the burst strength of paper after it has been saturated with water.

wet strength papers—papers that retain 15 percent or more of their normal strength when saturated with water.

wettability—the relative affinity of a liquid for a solid surface.

wetting—the stage of ink manufacture in which pigment particles are surrounded with the vehicle.

wetting agent—(1) a substance that improves the wettability of a solid surface by a liquid; (2) a material that reduces the surface tension of water and water solutions.

wet trapping—to successfully apply an ink film to a just-printed and still-wet ink film on a multiple-color press.

whiteness—the physical

characteristic that describes whether a white sheet of paper has a bluish or yellowish tint.

white water—milky-colored water that has been used in the papermaking process.

whole tree chipping—the reduction of an entire tree into small, thin wafers.

wild—inconsistent distribution of fibers across a sheet, imparting a mottled appearance.

wind firmness—the ability of a tree to withstand the force of the wind and resist being blown over.

wire side—the bottom side of paper when it is being formed on a Fourdrinier machine.

wood chips—small, thin wafers of wood cut from a log or an entire tree prior to being chemically or mechanically pulped.

wood fibers—cellulose fibers found in wood.

wood free—see *groundwood free.*

wood harvesting—cutting down trees with a commercial intent.

wood pulp—pulp created from wood.

working properties—used to describe ink properties that heavily affect how an ink works on the press and the substrate; examples include tack, viscosity, drying rate, and color strength.

wove—the regular finish imparted to paper by a dandy roll with nothing more than a woven wire over it.

wove dandy—a dandy roll with no watermark design.

wrap curl—see *reel curl.*

writing paper—a general term applied to the full range of papers commonly used for writing with pen and ink or pencil. The basic size is 17 × 22 inches and the substance weight range is 13–24 pounds.

wrong side—the wire side of a sheet of paper.

-X-

xerographic paper—a paper suitable for use in electrostatic (photocopying) machines; generally these are chemical wood papers with a 17 × 22-inch basic size and a substance weight range of 16–24 pounds.

xerographic dual-purpose paper—bond paper that can be used in either xerography or lithography.

-Y-

Yankee drier—a method of drying paper with one large drying drum instead of dozens of small ones.

yield—the percentage of a material that remains after processing; in forestry, the amount of timber that may be harvested over a period of time in accordance with stated management objectives.

yield value—(1) the total flow properties of a printing ink; (2) the force required to initiate flow.

-Z-

Zahn cup—a brand of efflux cup.

zein—a protein found in corn; used to coat and surface-size paper.

zinc hydrosulfite—a chemical (ZnS_2O_3) used to bleach mechanical pulp.

zinc white—zinc oxide (ZnO) used as a paper filler; also used in papers made for use in an electrofax duplication machine.

Bibliography/ for Further Reading

History of Paper

Carter, Thomas Francis. *The Invention of Printing in China and Its Spread Westward*. New York: Columbia University Press, 1925.

Heller, Jules. *Papermaking*. New York: Watson-Guptill, 1978.

Hunter, Dard. *Papermaking*. New York: Dover Publications, 1978.

Leif, Irving P. *An International Sourcebook of Paper History*. Hamden, CT: Archon Dawson, 1978.

Sutermeister, Edwin. *The Story of Papermaking*. New York: R.R. Bowker, 1962.

Weidenmuller, Ralf. *Papermaking*. San Diego: Thorfinn International Marketing Consultants, 1980.

Wilkinson, Norman B. *Papermaking in America*. Greenville, DE: The Hagley Museum, 1975.

Sources of Paper Fiber for Papermaking

Cote, Wilfred A. *Papermaking Fibers*. Syracuse: Syracuse University Press, 1980.

Smith, David M. *The Practice of Silviculture*. 8th ed. New York: John Wiley & Sons, 1986.

Zobel, Bruce and John Talbert. *Tree Improvement*. New York: John Wiley & Sons, 1984.

The Manufacture of Paper

Evans, John C.W., ed. *Trends and Developments in Papermaking*. San Francisco: Miller Freeman, 1989.

——. *Modern Paper Finishing*. San Francisco: Miller Freeman, 1988.

Patrick, Ken L., ed. *Bleaching Technology for Chemical and Mechanical Pulps*. San Francisco: Miller Freeman, 1991.

——. *Modern Mechanical Pulping in the Pulp and Paper Industry*. San Francisco: Miller Freeman, 1989.

Coated Papers

Harper, Donald T. *Paper Coatings*. Park Ridge, NJ: Noyes Data Corporation, 1976.

Patrick, Ken L., ed. *Paper Coating Trends in the Worldwide Paper Industry*. San Francisco: Miller Freeman, 1991.

Recycled Paper

Cross, Lisa. "New Age for Recycled Paper," *Graphic Arts Monthly*, October 1991, 45–48.

Ducey, Michael J. "Closing the Recycling Loop," *Graphic Arts Monthly*, August 1992, 124–5.

——. "New Recycled Definitions Stir the Market," *Graphic Arts Monthly*, May 1992, 128, 130.

Holmes, John R. *Refuse Recycling and Recovery*. New York: John Wiley and Sons, 1981.

Lowe, Kenneth, ed. *Fiber Conservation and Utilization*. San Francisco: Miller Freeman Publications, 1975.

Lustig, Theodore. "Coming Clean on De-inking," *Graphic Arts Monthly*, August 1990, 90–92, 96.

Quimby, Thomas E. *Recycling: The Alternative to Disposal*. Baltimore: Johns Hopkins University Press, 1975.

The Issue of Recycling in America and, Consequently at Champion. Stamford, CT: Champion International Corporation, 1990.

Using Paper in Printing

Bureau, William H. *What The Printer Should Know About Paper*. Pittsburgh: Graphic Arts Technical Foundation, 1982.

Classifications of Paper

Beach, Mark and Ken Russon. *Papers for Printing*. Portland, OR: Coast to Coast Books Inc., 1989.

Buying Paper

Ruggles, Phillip Kent. *Printing Estimating*. 4th ed. Albany, NY: Delmar Publishers, 1995.

The Competitive Grade Finder. Exton, PA: Grade Finders Inc., published annually.

The Paper Buyers' Encyclopedia. Exton, PA: Grade Finders, Inc., published annually.

Walden's Paper Catalog. Ramsey, NJ: Walden-Mott Corp., published semiannually.

Walden's Handbook for Paper Salespeople & Buyers of Printing Paper. Ramsey, NJ: Walden-Mott Corporation, 1991.

Ink Ingredients

Printing Ink Handbook. 4th ed. Harrison, NY: National Association of Printing Ink Manufacturers, 1992.

Ink Manufacture

Butler, P. The Origin of Printing in Europe. Chicago: University of Chicago Press, 1968.

Carter, Thomas Francis. The Invention of Printing in China. New York: Oxford University Press, 1966.

Clair, Colin. A History of Printing in Britain. New York: Oxford University Press, 1966.

De Vinne, Theodore L. The Invention of Printing. New York: Francis Hart & Co., 1876.

McMurtie, Douglas C. The Golden Book. New York: Covici Friede Publishers, 1927.

Ink Properties and Testing

Eldred, Nelson R. and Terry Scarlett. What the Printer Should Know about Ink. Pittsburgh: Graphic Arts Technical Foundation, 1990.

Ink and Color

Field, Gary G. Color and Its Reproduction. Pittsburgh: Graphic Arts Technical Foundation, 1988.

Zakia, Richard D. and Hollis N. Todd. Color Primer I and II. Dobbs Ferry, NY: Morgan & Morgan, 1986.

Ink and Paper for the Printing Processess

Adams, J. Michael, David D. Faux, and Lloyd J. Rieber. Printing Technology. 3d ed. Albany, NY: Delmar Publishers, 1988.

Cross, Lisa. "Watershed Year for Waterless," Graphic Arts Monthly, February 1993, 51–53. The first of a six-part bimonthly series on waterless printing.

Karsnitz, John R. Graphic Communication Technology. 2nd ed. Albany, NY: Delmar Publishers, 1993.

Wainberg, Peter. "Not Only Flexo Ink, but Technology," FLEXO, June 1995, 26–28.

Estimating and Ordering Ink

Ruggles, P. K. *Printing Estimating.* 4th ed. Albany, NY: Delmar IPC, 1995.

Paper, Ink, and the Environment

Bryner, G. C. *Blue Skies, Green Politics: the Clean Air Act of 1990.* Washington D.C.: CQ Press, 1993.

Copeland, C. *Water Quality: Implementing the Clean Air Act.* Congressional Research Service, The Library of Congress, 1993.

Ferguson, K., ed. *Environmental Solutions for the Pulp and Paper Industry.* San Francisco: Miller Freeman, 1991.

Luoma, J.R. *The Air around Us: an Air Pollution Primer.* Raleigh, NC: The Acid Rain Foundation.

Moyer, C.A. *Clean Air Handbook: a Practical Guide to Compliance.* New York: Clark Boardman Callaghan, 1992.

Sittig, M. *Pulp and Paper Manufacture: Energy Conservation and Pollution Prevention.* Park Ridge, NJ: Noyes Data, 1977.

Index